INTERNATIONAL TRADE IN SUSTAINABLE ELECTRICITY

Cross-border trade in electricity is rapidly expanding as a result of technical innovations, economic and geopolitical developments and the ongoing decarbonisation of the electricity sector in response to climate change. The expansion of electricity networks and the integration of increasing shares of electricity produced from renewable energy (RE) sources into the grid have made long-distance electricity flows both feasible and desirable. Drawing on the work of experts in trade and energy law and policy, and offering novel, multidisciplinary perspectives on the rapidly evolving landscape shaping international trade in electricity, this book examines the most important challenges – technical, economic, legal and policy-related – posed by long-distance and sustainable electricity trade. The book explores the regulatory implications of the policy instruments aimed at supporting RE electricity and considers how best to promote greater overall coherence in international electricity governance.

THOMAS COTTIER is Professor Emeritus of European and International Economic Law at the University of Bern, a senior research fellow at the World Trade Institute and adjunct professor at the University of Ottawa, Faculty of Law. He was educated at the University of Bern, University of Michigan Law School, and has been a visiting fellow at the University of Cambridge, UK. He was previously Deputy-Director General of the Swiss Intellectual Property Office and has served as a member or chair of several GATT and WTO panels.

ILARIA ESPA is a senior research fellow at the World Trade Institute (WTI), University of Bern, Academic Coordinator of the WTI Doctoral Programme and a lecturer at the University of Milan. A former Marie Curie fellow, she received her PhD in International Law and Economics from Bocconi University with a dissertation on *Export Restrictions on Critical Minerals and Metals: Testing the Adequacy of WTO Disciplines* (Cambridge University Press, 2015).

INTERNATIONAL TRADE IN SUSTAINABLE ELECTRICITY

Regulatory Challenges in International Economic Law

Edited by

THOMAS COTTIER

World Trade Institute, University of Bern

ILARIA ESPA

World Trade Institute, University of Bern

CAMBRIDGE
UNIVERSITY PRESS

CAMBRIDGE
UNIVERSITY PRESS

University Printing House, Cambridge CB2 8BS, United Kingdom

One Liberty Plaza, 20th Floor, New York, NY 10006, USA

477 Williamstown Road, Port Melbourne, VIC 3207, Australia

4843/24, 2nd Floor, Ansari Road, Daryaganj, Delhi – 110002, India

79 Anson Road, #06–04/06, Singapore 079906

Cambridge University Press is part of the University of Cambridge.

It furthers the University's mission by disseminating knowledge in the pursuit of education, learning, and research at the highest international levels of excellence.

www.cambridge.org
Information on this title: www.cambridge.org/9781107163348
DOI: 10.1017/9781316681275

First published 2017

A catalogue record for this publication is available from the British Library.

Library of Congress Cataloging-in-Publication Data
Names: Cottier, Thomas, editor. | Espa, Ilaria, editor.
Title: International trade in sustainable electricity : regulatory challenges in international economic law / edited by Thomas Cottier (University of Bern), Ilaria Espa (University of Bern).
Description: Cambridge, United Kingdom ; New York, NY : Cambridge University Press, [2017] | Includes bibliographical references and index.
Identifiers: LCCN 2016046857| ISBN 9781107163348 (hardback : alk. paper) | ISBN 9781316615140 (pbk. : alk paper)
Subjects: LCSH: Electric utilities–Government policy. | Electric utilities–Law and legislation. | International economic relations. | Sustainable engineering. | Renewable energy sources. | International trade–Environmental aspects.
Classification: LCC HD9685.A2 I583 2017 | DDC 343.09/29–dc23
LC record available at https://lccn.loc.gov/2016046857

ISBN 978-1-107-16334-8 Hardback

CONTENTS

FIGURES

TABLES

CONTRIBUTORS

FREDERICK M. ABBOTT is Edward Ball Eminent Scholar Professor of International Law at Florida State University College of Law. He has served as an expert consultant for numerous international and regional organisations, governments and nongovernmental organisations, mainly in the fields of intellectual property, public health, trade and sustainable development. Professor Abbott is the author of many books and articles in these fields. Professor Abbott is on the editorial board of the *Journal of International Economic Law*. He regularly serves as a panelist for the WIPO Arbitration and Mediation Center. He has served as counsel to governments in WTO dispute settlement proceedings, and in national court proceedings. Prior to entering academia, Professor Abbott was partner in a major US law firm and specialised in international business regulation and technology-related transactions. He was based in San Francisco (1977–83) and Silicon Valley (1984–9). He is a graduate of the Yale Law School and the University of California, Berkeley.

SPYRIDON CHATZIVASILEIADIS is an assistant professor at the Technical University of Denmark. Spyros holds a PhD from ETH Zurich, Switzerland (2013) and a MS in electrical and computer engineering from the National Technical University of Athens (NTUA), Greece (2007). From 2014 to 2016 he held positions as a postdoctoral researcher at the Massachusetts Institute of Technology (MIT), Cambridge, MA, USA and at Lawrence Berkeley National Laboratory (LBNL), CA, USA. His research interests are power system optimisation and control, HVDC grids and electricity markets.

THOMAS COTTIER, former Managing Director of the World Trade Institute (WTI) and the Institute of European and International Economic Law, is Emeritus Professor of European and International Economic Law at the University of Bern and adjunct professor of law at the University of Ottawa. He is a former director of the national research programme on trade law and policy ('NCCR Trade Regulation: From Fragmentation to Coherence') located at the WTI. Professor Cottier is an associate editor of several journals. He was a visiting professor at the Graduate Institute, Geneva and currently teaches at the University of Ottawa, at the Europa Institut Saarbrücken, Germany and at Wuhan University, China. From 1997 to 2004 he was a member of the Swiss National Research Council. He served the Baker & McKenzie law firm as Of Counsel from 1998 to 2005. Professor Cottier has a long-standing involvement in General Agreement on Tariffs and Trade (GATT)/World Trade Organization (WTO) activities. He served on the Swiss negotiating team of the Uruguay Round from 1986 to 1993, first as Chief Negotiator on dispute settlement and subsidies for Switzerland and subsequently as Chief Negotiator on the Agreement on Trade-Related Aspects of International Property Rights (TRIPs). He has held several positions in the Swiss External Economic Affairs Department and was the Deputy-Director General of the Swiss Intellectual Property Office. In addition to his conceptual work in the fields of services and intellectual property and legal counselling, he has also served as a member or chair of several GATT and WTO panels. Professor Cottier has written, and publishes, on a wide range of trade and international law issues.

STEIVAN DEFILLA has been Director in the Energy Charter Secretariat since September 2010. Steivan obtained a Bachelor's degree in Economics in 1984 and a Master's degree in Statistics in 2004 from Neuchâtel University, Switzerland, and holds postgraduate degrees in Energy (Federal Polytechnical Institute Lausanne, 1986) and Public Administration (Institute for Public Administration Lausanne, 1989). Prior to his appointment at the Energy Charter Secretariat, Steivan worked for the State Secretariat for Economic Affairs (SECO) in Switzerland as Senior Energy Advisor, advising in particular on foreign energy policy and on the Swiss electricity sector reform. During that time Steivan was also Chair of the Energy Charter Group on Trade from 2000 to 2004, and was actively involved in the negotiations of the Energy Charter Trade Amendment (entered into force in 2010) and the Energy Charter Transit Protocol. Steivan has authored a book and numerous articles on energy policy.

JOËLLE DE SÉPIBUS is Senior Research Fellow of the World Trade Institute, Assistant Professor of Law at the University of Bern and Visiting Professor of Law at the College of Europe (Bruges). She teaches

European constitutional and economic law as well as international law on climate change and sustainability at the University of Bern. Her current research interests focus on European and international law, in particular in the fields of climate and energy policies, including the new regime on climate change, carbon and electricity markets. She was formerly Associate Professor of International and European Law at the University of Fribourg (Switzerland). In 2010 she worked as an expert for the EU Presidency in the run-up to the international climate negotiations in Cancun. She holds a Master's in European Law (College of Europe, Bruges) and a PhD in European Law (University of Bern). Admitted to the Bar, she has practised Swiss and international business law in law firms in Bern, Geneva and Zurich. After completing her doctoral thesis she was Visiting Research Fellow at the Robert Schuman Centre for Advanced Studies (European University Institute, Florence).

BELISA E. ELEOTERIO is a PhD candidate in International and Comparative Law at the University of São Paulo (USP), Brazil. She holds a Master's in Law from the University of São Paulo. She has been a research assistant at the Centro do Comércio Global e Investimento (CGTI) and is also a qualified Brazilian lawyer.

DAMIEN ERNST is a professor at the University of Liège, where he holds the EDF-Luminus Chair on Smart Grids. He received MS and PhD degrees in engineering from the University of Liège in 1998 and 2003, respectively. His doctoral thesis focused on reinforcement learning to solve control and decision-making problems in electricity networks. Before becoming Professor at the University of Liège, he was a research fellow of the Fonds de la Recherche Scientifique (FNRS), and before this Assistant Professor at SUPELEC, a French Grande Ecole pour Ingénieurs. He has also held Visiting Researcher positions at CMU in Pittsburgh, ETH in Zurich and MIT in Boston. Professor Ernst is well known for his research work in reinforcement learning and power system control. He has co-authored more than 200 hundred papers and two books. He has also received numerous awards for his research work. Professor Ernst is regularly consulted by the press, the industry and governments for his deep understanding of the mutations the energy sector is currently facing.

ILARIA ESPA is a senior research fellow at the World Trade Institute (WTI) and Academic Coordinator of the WTI Doctoral Programme.

She is also a lecturer at the University of Milan, where she has taught environmental and food law since 2015. Awarded a Marie Curie fellowship from the European Commission for her post-doctoral studies, Ilaria has been a member of the work package on 'Trade and Climate Change' of the NCCR Trade Regulation Programme since 2013. Within this programme, she has conducted research on the trade law implications of policies and policy instruments implemented at various levels of governance in the areas of climate, energy and commodities, as well as on the law governing the sustainable management of natural resources. Ilaria holds a PhD in international law and economics from the Department of Legal Studies of Bocconi University and was a visiting scholar at Columbia Law School in 2012. She also holds a BA in political science and an MA in international relations from Luiss University. Ilaria is the author of a monograph on *Export Restrictions on Critical Minerals and Metals: Testing the Adequacy of WTO Disciplines* (Cambridge University Press, 2015). In 2012, she received the SIDI (Italian Society of International Law) prize, awarded to a younger author for meritorious scholarship published in Revue Internationale de Droit Économique.

KAROLIS GUDAS joined the World Trade Institute (WTI) as a PhD researcher under the Swiss National Centre of Competence in Research Trade Regulation project. He has been a visiting fellow at the University of Cambridge, Cambridge Centre for Environment, Energy and Natural Resource Governance. At present Karolis is the head of the legal division at the Vilnius CHP plant (Lietuvos Energija group), a project of national significance to Lithuania. Previously he worked as Professor's Assistant at the Institute of International and European Union law at Mykolas Romeris University. He gained his main legal experience at one of the leading Baltic States dispute resolution law firms, Motieka & Audzevicius, and the European Commission Joint Research Centre Institute for Energy. He has a Master's degree in European Union Law from Mykolas Romeris University and a postgraduate diploma in European Union Law from King's College, University of London. In addition, he studied International Law at the University of Iceland and has attended Summer Academy courses at the WTI.

KATERYNA HOLZER is a post-doctoral researcher in the NCCR Trade Regulation programme at the University of Bern. Her areas of research include the interface of trade, energy and climate change, with emphasis on trade-related instruments and their World Trade Organization

(WTO) compliance, as well as public policy initiatives undertaken under Regional Trade Agreements (RTAs) and sectoral agreements. She is an author of the Working Group III Fifth Assessment Report of the Intergovernmental Panel on Climate Change (IPCC). Before joining the MILE programme at the World Trade Institute in Bern in 2005, she served as Assistant Professor and Vice Dean at the Ukrainian Academy of Foreign Trade. Lecturing in Strategic Management, she co-authored *Interactive Methods of Teaching Business Disciplines* (2004), approved by the Ministry of Education of Ukraine as a teaching handbook. As well as her PhD in Law from the University of Bern (*Carbon-related Border Adjustment and WTO Law*, 2014), she holds a PhD in Economics ('kandidat nauk') from Ukraine (*Management of Automobile Industry Enterprises of Ukraine under Protectionism*, Kiev 2003).

ANDERS PLEJDRUP HOUMØLLER is the owner and CEO of Houmoller Consulting. Mr Houmøller has more than twenty years of experience in the electricity supply industry and seven years of experience in the IT and software industry. Mr Houmøller has taught and consulted in more than thirty-five countries in Europe, Asia, Africa, the Middle East, North America and South America. Mr Houmøller spent twelve years as head of the Nordic electricity exchange Nord Pool's office in Denmark. While in this job, he introduced the Danish electricity exchange. He also participated in the establishment of the Danish gas exchange. Before joining Nord Pool, Mr Houmøller worked for the Danish power pool Elsam, and – later – for the transmission system operator in Western Denmark. Mr Houmøller holds a Master's degree in Physics, a Bachelor's degree in Mathematics, a Bachelor's degree in Computer Science and a Bachelor of Commerce.

SUFIAN JUSOH is Deputy Director and Senior Fellow at the Institute of Malaysian and International Studies, Universiti Kebangsaan Malaysia. Sufian is also External Consultant for the World Trade Institute (WTI), University of Bern and Director of WTI Consulting in Malaysia, as well as Distinguished Fellow at the Institute of Diplomacy and Foreign Relations, Ministry of Foreign Affairs, Malaysia and Barrister-at-Law (England and Wales) of Lincoln's Inn, London. Sufian was educated at the University of Wales, Cardiff and University College London, and has a Doctor Iuris (*summa cum laude*) from the University of Bern. Sufian leads a Malaysian based public–private funded research project for generating heat and electricity using specially designed biomass-powered industrial boiler and combustion system. Sufian has also been a consultant to countries, international organisations and multinational companies on international trade, free-trade agreements, WTO accession and international investment. He is now an investment policy expert at the World Bank, serving Myanmar, Laos and Timor Leste.

CALLIXTE KAMBANDA is Chief Infrastructure Specialist at the Infrastructure Consortium for Africa (ICA), which Secretariat is housed by the African Development Bank in Tunis, Tunisia. He holds an MBA in Project Management from the Maastricht School of Management and a Master's degree in Electrical Engineering from the University of Applied Sciences, Kaiserslautern, Germany. He has extensive experience working in the energy sector, particularly with African countries, Regional Economic Communities (RECs) and specialised institutions for the Energy Sector. He has led work on various ICA publications on energy, especially on independent power producers (IPPs) in Africa and on power pools. Before joining the ICA in January 2008, he was Executive Secretary of the Eastern Africa Power Pool (EAPP). He is a member of various professional associations, including the Germany Engineer Association (VDI), and until July 2006 he was the chairman of the Rwanda Electrotechnical Committee of Standardization.

CHRISTIAN KUNZE holds a PhD in Management Science and started his professional career in the energy industry in 1999. For ten years he worked in international management positions in the electricity trading environment, for example, at E.ON Supply and Trading, Hanover/ Munich; Edison Trading SpA, Milan; and Shell Energy Trading Ltd., London. In his previous position at Swissgrid AG as well as in his current position at the Norwegian Centre of Expertise (NCE) for Smart Energy Markets, he focuses on market design activities. In addition, Christian is a frequent speaker at industry events as well as a visiting lecturer in electricity-related university programmes.

CAROLINA M. LEMBO is a consultant at the Inter-American Development Bank. She has been a research fellow at the Harvard Electricity Policy Group at Harvard Kennedy School and holds a PhD in International Law from the University of São Paulo, an LLM in International Economic Law and Policy from the University of Barcelona, an MS in State Law from the University of São Paulo and a Bachelor's degree from the Mackenzie Presbyterian University in Law. Previously, she worked as

Manager of the Infrastructure Department of the Federation of Industries of the State of São Paulo, a major industry chamber in Brazil, where she managed a multidisciplinary infrastructure department with focus on domestic policy and international projects.

JÉRÔME LE PAGE is an independent energy market and regulation consultant and has worked with the European Federation of Energy Traders (EFET) since February 2012. He is in charge of coordinating EFET's public policy advocacy work on regulatory matters pertaining to the wholesale electricity market design at European level, the harmonisation of national frameworks and the integration of renewable energy sources into the wholesale market. He also coordinates policy advocacy and implementation facilitation initiatives with regard to over-the-counter (OTC) derivatives clearing and transaction reporting obligations under the European Market Infrastructure Regulation (EMIR). Prior to joining EFET, Mr Le Page was part of MSCI Inc., where he led the Energy Utilities analysis department of their Responsible Investment business. Mr Le Page holds a Master's degree in International Law and Economics from the Paris I-Sorbonne University and an LLM from the University of Cologne.

SOFYA MATTEOTTI is a qualified lawyer who has developed extensive academic and practical experience in the fields of international and European energy and trade law in Russia and Switzerland. In her current capacity as a legal counsel with Swissgrid AG (the Swiss transmission system operator), she is responsible for providing legal advice on implementation of the Swiss Energy Strategy 2050 (RES support schemes and electricity labelling). Her research interests include the status of energy sources under the international trade and investment regimes, competition in the energy sector, the concept of state sovereignty over natural resources, energy security and climate change mitigation. From 2007 to 2014 she was a research fellow at the World Trade Institute, University of Bern, Switzerland (interdisciplinary projects on Energy in WTO Law and Policy and on Trade and Climate Change). The theme of her PhD research is analysis of the oil supply management practices of the Organization of Petroleum Exporting Countries (OPEC) under the World Trade Organization rules. This research was further explored during her internship at the Legal Office of the Secretary General of OPEC in Vienna, Austria. She has published and presented on trade and climate change, energy law and trade law and policy.

MADISON MENG was an associate at VVGB Advocaten/Avocats in Brussels, where she specialised in international trade law with a particular focus on Asia. She is currently working at Sidley Austin in Hong Kong.

OLGA NARTOVA is a law department chair and an assistant professor at Dar Al-Hekma University in Jeddah, Saudi Arabia, and an adjunct professor at Loyola University College of Law, New Orleans. Olga was a research fellow at the World Trade Institute in the period 2007–2010, where she was a co-leader on trade and climate change research. Olga holds a PhD degree from the University of Bern and is a qualified lawyer in Russia, having received her law degree from Moscow State Academy of Law. She also obtained a Master of International Law and Economics degree from the World Trade Institute in Bern and holds a Master's degree in economics. Olga is a co-editor of *International Trade Regulation and the Mitigation of Climate Change: World Trade Forum* (Cambridge University Press, 2009) and has published and lectured on trade and climate change, energy law and trade law and policy.

TETYANA PAYOSOVA is a lecturer at the Institute of European and International Economic Law, University of Bern, and a legal adviser at the International Rail Transport Committee (CIT). Tetyana is a qualified lawyer in Ukraine, having graduated with the highest distinction from the National University Odessa Law Academy. She also holds a Master of Law degree (2009) from the University of Bern, Switzerland. Tetyana was a research fellow at the Institute of European and International Economic Law, University of Bern (2009–2014), where she closely cooperated with NCCR/WP 5 on Trade and Climate Change. The topic of her PhD research is analysis of framework conditions for the promotion of renewable energy, with a focus on solar energy, and international trade regulation. Tetyana was a visiting assistant professor at Florida State University in 2012, teaching a course on International Energy Law. She has also worked on several research projects with the International Centre for Trade and Sustainable Development (ICTSD) and the United Nations Conference on Trade and Development (UNCTAD). She has published and lectured on World Trade Organization (WTO) law, trade, energy, climate change and selected topics of EU law.

DAVID ROBINSON holds a doctorate in economics from the University of Oxford. He consults on competition, regulation and corporate strategy in energy, among other sectors. His current research and publications cover energy and climate policy in the US, the EU and China; Sino–EU collaboration in renewable energy; the impact of the shale gas revolution; and the process and implications of energy sector decarbonisation. He is Senior Fellow of the Oxford Institute for Energy Studies and was formerly a Member of the Faculty of Business Administration and Economics at St. Louis University (Madrid campus), where he was also on the Board of Regents. He has been the director of two leading international economic consulting firms and has run his own consulting firm since 2003. He lives in Madrid with his wife, Maria Gloria Andrade, and they have two children, Nicolás and Amalia.

LUCA RUBINI is Reader (Associate Professor) in international economic law and Deputy Director of the Institute of European Law at Birmingham Law School. He served as a *référendaire* (law clerk) with Advocate General Jacobs at the European Court of Justice. Dr Rubini has held visiting positions at the European University Institute, the Institute of International Economic Law (Georgetown), the World Trade Institute (University of Bern) and Bocconi University. He is a fellow of the Centre of European Law, King's College London, and Visiting Professor at ASERI (Catholic University, Milan). Dr Rubini has law degrees from the Catholic University, Milan (JD) and King's College London (MA, PhD). Dr Rubini's main research interest lies in subsidy law and policy, and more generally in the governance of regulation. In this regard he published the book *The Definition of Subsidy and State Aid – WTO Law and EC Law in Comparative Perspective* (2009). He is currently working on various projects on the governance of climate change subsidies in the World Trade Organization (WTO).

YULIA SELIVANOVA is an independent consultant specialising in energy and trade regulatory affairs. She advises governments, companies and intergovernmental organisations on different aspects of energy regulation and trade in energy. From 2005 to 2013 she worked in the Energy Charter Secretariat, where she was responsible for trade issues. Before joining the Energy Charter, Ms Selivanova worked for four years for an international law firm, Baker & McKenzie, in Geneva, focusing on international trade and WTO issues. She was also a consultant in the Rules Division of the WTO, where she conducted research on subsidies, anti-dumping and safeguards for the WTO dispute settlement panels. Ms Selivanova's work in Russia – in an international law firm and an international bank in Moscow – concentrated on legal support for international companies' activities in Russia. Ms Selivanova graduated from the International Law Faculty of the Moscow Institute of International Relations (MGIMO) with a specialisation in international trade law. She graduated from the Master of International Law and Economics programme in the World Trade Institute in Bern (Switzerland) and in 2006 earned a doctorate in law from the University of Bern. Ms Selivanova's doctorate thesis focused on energy dual pricing under WTO law in the context of Russia's accession to the WTO. Ms Selivanova publishes regularly on a range of trade and WTO issues, in particular trade in energy. In 2011 she edited and published the book *Regulation of Energy in International Trade Law: WTO, NAFTA and Energy Charter.*

FEREIDOON P. SIOSHANSI is President of Menlo Energy Economics, an energy consulting firm in San Francisco, and the editor and publisher of *EEnergy Informer*, a monthly newsletter with international circulation. His professional experience includes Southern California Edison Company, the Electric Power Research Institute, National Economic Research Associates and Ventyx, now part of ABB. Since 2006, he has edited eight books: *Electricity Market Reform* (2006); *Competitive Electricity Markets* (2008); *Generating Electricity in a Carbon Constrained World* (2009); *Energy Sustainability and the Environment* (2011); *Smart Grid* (2011); *Energy Efficiency* (2013); *Evolution of Global Electricity Markets* (2013); and *Distributed Generation and Its Implications for the Utility Industry* (2014). He has degrees in Engineering and Economics, including an MS and PhD in Economics from Purdue University.

EDWIN VERMULST has practised international trade and EU law and policy since 1985 and is a founding partner of VVGB Advocaten. He is a member of the Brussels Bar A-list. He graduated from the University of Utrecht in 1983 and received LLM and SJD degrees from the University of Michigan Law School in 1984 and 1986. He specialises in the representation of multinationals, governments, trade associations, exporters and importers in EU and World Trade Organization (WTO) Trade Defence Instruments (TDI) and customs proceedings, for example the recent anti-dumping and/or anti-subsidy proceedings involving various steel products, solar panels and solar glass, organic coated steel, wireless WAN modems and footwear. He is, among others, the trade counsel of the World Federation of Sporting Goods Industry (WFSGI) and the Federation of European Sporting Goods Industry (FESI). He was a WTO panellist in Mexico–HFCS from the United States and has been involved in various WTO dispute settlement proceedings as a member of the delegation, most recently in EU–FiT schemes from Italy and Greece and in EU–Footwear as counsel for the Ministry of Commerce People's Republic of China (MOFCOM). He has co-authored nine books, including landmark comparative analyses of the anti-dumping systems of Australia, Canada, the EU and the US (with Professor John Jackson, in 1989) and of rules of origin (in 1994), as well as numerous articles. The second edition of his book on *EU Anti-Dumping Law and Practice* was published in October 2010. His book on *The WTO Anti-Dumping Agreement* was published in 2006. He is editor-in-chief of the *Journal of World Trade*, a member of the faculty of the World Trade Institute in Bern and of the International Economic Law and Policy (IELPO) programme in Barcelona and a member of the Advisory Board of the European Centre for International Political Economy (ECIPE). He is invariably selected as a top trade practitioner by publications such as *Chambers*, *Euromoney's Expert Guides and Best of the Best*, *Who's Who Legal*, *Legal 500* and *The Rushford Report*. *Chambers* 2016, for example, ranked him in Band 1 in the international trade/WTO area in the EU-wide group and as one of a select few Star individuals in Belgium. *Who's Who Legal Trade and Customs* has ranked him as the trade lawyer of the year for the last three consecutive years.

PREFACE

The World Trade Forum series was established in 1997 as a meeting hub for academia, policy-makers, international organisations and civil society representatives, with the aim of addressing some of the most pertinent questions facing the international trading system. In its 2014 event, the World Trade Forum focused on the regulatory challenges posed by the expansion of international trade in electricity, triggered by the climate-driven, ongoing process of decarbonisation of the electricity sector, and, how it relates to the international trade and investment regimes.

The cutting-edge innovations that are reconfiguring the landscape of electricity trade and the shift to renewable energy (RE) sources that has occurred in response to the prominence of climate change mitigation in government agendas have prompted cross-border electricity trade. As the uptake of intermittent RE electricity grows, this will require the progressive integration of national, regional and, eventually, continental grids. These trends point to the need for an informed debate in the rapidly evolving area of international trade in electricity, which holds great interest for the public and private sectors alike. The idea of a global grid, of course, reflects a long-term vision which, if it materialises at all, will do so over a long period of time. Yet it is important to work on building blocks in a national and regional context, and to prepare the tools for inter-regional interfaces in the future.

Based upon experiences in the field of trade regulation and investment, the World Trade Forum 2014 sought to contribute to this debate by analysing, from a multidisciplinary perspective, the regulatory challenges that such developments pose to trade and investment in RE electricity. This book is based upon the papers presented at the World Trade Forum held in September 2014 at the World Trade Institute, University of Bern, Switzerland. The Forum brought together leading scholars and practitioners and policy-makers from the trade, investment and energy communities in an effort to learn more about the limits and opportunities inherent to the current configuration of the trade and investment frameworks for the development of the electricity network over larger regions.

Particular thanks are owed to the authors of the chapters who agreed to present and debate the preliminary versions of their papers at the Forum and promptly went on to produce revised versions for publication. For their participation and most enriching comments during the debates, we should very much like to thank Réne Burkhard, Vanda Markovicova, Ioannis Retsoulis, Bernhard Thies, Jesse Kreier, Robert Howse, Krista Nadakavukaren Schefer and Ludivine Tamiotti.

We extend special thanks to Margrit Vetter and Christian Steiger of the World Trade Institute for support in organising the conference in Bern. We are most grateful to Susan Kaplan, the scientific editor of the World Trade Institute, for reviewing the manuscripts. Finally, our thanks go to Kim Hughes and her excellent team at Cambridge University Press for their support throughout the project and their strong commitment to the World Trade Forum Series.

ABBREVIATIONS

AB	Appellate Body
AB 32	Assembly Bill 32
AC	alternating current
ACER	Agency for the Cooperation of Energy Regulators
ADA	Anti-Dumping Agreement
AEC Blueprint	ASEAN Economic Community Blueprint
APAEC	ASEAN Plan of Action for Energy Cooperation
APG	ASEAN Power Grid
APS	alternative policy scenario
ARPP	African Regional Power Pools
ASCC Blueprint	ASEAN Socio-cultural Blueprint
ASCM	Agreement on Subsidies and Countervailing Measures
ASEAN	Association of Southeast Asian Nations
ATC	available transfer capacity
BAU	business-as-usual
BITs	bilateral investment treaties
BNDES	Brazilian Development Bank
BTA	border tax adjustment
CACM	Capacity Calculation and Congestion Management
CAPP	Central African Power Pool
CARB	California Air Resources Board
CCCME	China Chamber of Commerce for Import and Export of Machinery and Electronic Products
CCL	climate change levy
CDM/JI	Clean Development Mechanism/Joint Initiative
CEER	Council of European Energy Regulators
CHP	combined heat and power
CIER	Regional Energy Integration Commission
COMELEC	*Comité Maghrébin de l'Electricité*
CO_2	carbon dioxide
CO_2e	CO2 equivalent
COP	coefficient of performance

CRMs	capacity remuneration mechanisms
C&T	cap-and-trade
CVD	countervailing duty
dena	*Deutsche Energie-Agentur* [German Energy Agency]
DC	direct current
DER	distributed energy resources
DGAD	Directorate General of Anti-Dumping and Allied Duties (India)
DRC	Democratic Republic of the Congo
DSM	demand-side management technologies
DSOs	distribution system operators
EAPP	Eastern Africa Power Pool
EB	Electricity Balancing
ECCAS	Economic Community of Central African States
ECOWAS	Economic Commission of West African Countries
ECREEE	ECOWAS Centre for Renewable Energy and Energy Efficiency
ECT	Energy Charter Treaty
EEC Treaty	Treaty establishing the European Economic Community
EE Directive	Energy Efficiency Directive
EEGI	European Electricity Grid Initiative
EERA	European Energy Research Alliance
EERS	energy efficiency resource standards
EIA	Energy Information Administration (United States)
EIIs	European Industrial Initiatives
ELTAM	Egypt–Libya–Tunisia–Algeria–Morocco
EMCC	European Market Coupling Company
ENTSO-E	European Network of Transmission System Operators for Electricity
EPA	Environmental Protection Agency (United States)
EPEX	European Power Exchange
ERERA	ECOWAS Regional Electricity Regulatory Authority
ERGEG	European Regulators' Group for Electricity and Gas
ESCO	energy service company
ESO	European standardisation organisation
ETS	Emissions Trading System (European Union)
EU	European Union
EUAs	EU allowances
FACTS	Flexible AC Transmission Systems
FBA	flow-based allocation
FCA	Forward Capacity Allocation
FERC	Federal Energy Regulatory Commission
FET	fair and equitable treatment
FiT	feed-in tariff

FTA	free trade agreement
GATT	General Agreement on Tariffs and Trade
GATS	General Agreement on Trade in Services
GCC	Gulf Cooperation Council
GCEG	global clean electricity grid
GENI	Global Energy Network Institute
GHG	greenhouse gas
GPP	green public procurement
HDI	Human Development Index
HS	Harmonized Commodity Description and Coding System
HVDC	high-voltage, direct current
IADB	Inter-American Development Bank
IAI	Initiative for ASEAN Integration
ICSID	International Centre for Settlement of Investment Disputes
ICT	information and communications technology
IEA	International Energy Agency
IEC	International Electrotechnical Commission
IPCC	Intergovernmental Panel on Climate Change
IT	information technology
ITC	investment tax credits
LCFS	low carbon fuel standards
LED	light-emitting diode
LEC	levy exemption certificates
LNG	liquefied natural gas
MEDRING	Mediterranean Ring
MER	Regional Electricity Market
MiFID	Markets in Financial Instruments Directive
MFN	most favoured nation
MMTCDE	million metric tonnes of carbon dioxide equivalents
MoU	memorandum of understanding
MPP	Medicines Patent Pool
MTI	merchant transmission investment
Mtoe	megatonne of oil equivalent (Mtoe)
NAPs	National Action Plans
NEM	net energy metering
NEMO	Nominated Electricity Market Operators
NERC	North American Electric Reliability Council
NGO	non-governmental organisation
NIST	National Institute of Standards and Technology
NME	non-market economy
NREL	National Renewable Energy Laboratory
NSCOGI	North Seas Countries Offshore Grid Initiative

NTC	net transfer capacity
OECD	Organisation for Economic Co-operation and Development
OLADE	Latin American Energy Organization
ONS	National System Operator (Brazil)
PCIs	projects of common interest
PCR	price coupling regions
PDAM	post day ahead market
PIDA	Program for Infrastructure Development in Africa
PLEF	Pentalateral Energy Forum
PPA	power purchase agreement
PPMs	processes and production methods
PTC	production tax credits
PV	photovoltaic
RE	renewable energy
REC	Renewable Energy Certificate
RERA	Regional Electricity Regulators Association of Southern Africa
REIPPPP	Renewable Energy Independent Power Producer Procurement Program
RES	renewable energy sources
RGGI	Regional Greenhouse Gas Initiative
RMOC	Regional Market Operation Center
ROC	Renewable Obligation Certificate
RPP	regional power pools
RPS	renewable portfolio standards
SADC	Southern African Development Community
SAPP	Southern African Power Pool
SCM Agreement	Agreement on Subsidies and Countervailing Measures
SETIS	Strategic Energy Technologies Information System
SG-CG	Smart Grid Coordination Group
SGTF	Smart Grid Task Force
SIEPAC	Central American Electrical Interconnection System
SIN	Brazilian Interconnected National System
SINEA	Andean Electrical Interconnection System
SOEs	state-owned enterprises
STEM	short-term energy market
TBT	Agreement on Technical Barriers to Trade
TDIs	trade defence instruments
TEN-E	trans-European energy networks
TPA	third party access
TRIMs	Agreement on Trade-Related Investment Measures
TSO	transmission system operator
UCTE	Union for the Coordination of the Transmission of Electricity

UNASUR	Union of South American Nations
UNCITRAL	United Nations Commission on International Trade Law
UNCTAD	United Nations Conference on Trade and Development
UNEP	United Nations Environment Programme
UNFCCC	United Nations Framework Convention on Climate Change
UPDEA	Union of African Power Utilities
US	United States
USW	United Steelworkers Union
WAPP	West African Power Pool
WTO	World Trade Organization
ZNE	zero net energy

1

Introduction and Overview

THOMAS COTTIER AND ILARIA ESPA

A Renewable electricity, local production and long-distance trade

Modern societies rely extensively on the supply and consumption of electricity. A multitude of daily life and work activities depend on electricity-powered technology, machinery and infrastructure, the use of which either has transformed or is still transforming the residential, commercial, industrial and transportation sectors.[1] The use of electricity drives economic growth, facilitates international commerce and enhances the welfare of people around the world. Accordingly, global electricity consumption has steadily increased,[2] driven by the ever faster pace of technological innovation and a number of structural factors that include demographic trends, the exceptional growth performances of developing Asian and other emerging economies in recent decades and the rapid electrification of countries that are currently at earlier stages of development.[3] At the same time, approximately two-thirds of the world's electricity is today generated through fossil fuel combustion, and half of this is still produced in coal-fired power plants. Electricity is the principal source of energy-related greenhouse gas (GHG) emissions,[4] which in turn account for two-thirds of

[1] Euroelectric, *A Sector in Transformation: Electricity Industry Trends and Figures* (2015), www.eurelectric.org/media/161808/electricityindustrytrendsandfigures2015_lr-2015-030-0064-01-e.pdf (accessed 8 February 2016).

[2] International Energy Agency (IEA), *World Energy Outlook Special Report: Trade and Climate Change* (2015), www.iea.org/publications/freepublications/publication/WEO2015 SpecialReportonEnergyandClimateChange.pdf (accessed 15 December 2015), p. 27.

[3] IEA, *World Energy Outlook Report* (2015), www.worldenergyoutlook.org/media/weo website/2015/WEO2015_Chapter01.pdf (accessed 8 February 2016), pp. 37–40; IEA, *World Energy Outlook Electricity Database* (2015), www.worldenergyoutlook.org/ resources/energydevelopment/energyaccessdatabase/ (accessed 8 February 2016).

[4] IEA, *World Energy Outlook Report*, pp. 300–19.

global GHG emissions.[5] The decarbonisation of the electricity sector has accordingly figured prominently on national and supra-national climate change agendas.[6] Efforts have converged towards the promotion of electricity produced from renewable energy (RE) sources and towards allowing for differential treatment of fossil and RE electricity within the multilateral trading system.[7] The shift to renewables in power generation carries the greatest potential for GHG emissions abatement in the energy sector as a whole.[8]

Climate change and international trade was the subject of a previous World Trade Forum.[9] The present volume turns to international trade in electricity, more specifically addressing the challenges posed by the transition towards renewably produced electricity and the role of international trade in this process, supporting growth as well as decarbonisation and climate change mitigation. Until very recently, electricity itself had been traded rather limitedly compared to the large volumes of primary energy supplies, namely oil and gas, being imported into many countries for electricity production. Electricity generation has mainly occurred locally, regionally and nationally, with international trade essentially limited to trade in surplus production between neighbouring countries. Technical and political constraints hampered long-distance, inter-regional and transcontinental trade in electricity supplies. This has been changing in recent years due to advances in technology, and for a number of other reasons.

5 IEA, *World Energy Outlook Special Report 2013: Redrawing the Energy-Climate Map* (2013), www.worldenergyoutlook.org/media/weowebsite/2013/energyclimatemap/RedrawingEnergyClimateMap.pdf (accessed 8 February 2016), p. 15.

6 See, e.g., EurElectric, *Power Statistics and Trends* (2015), www.eurelectric.org/media/249736/power-statistics-and-trends-the-five-dimentions-of-the-energy-union-lr-2015-030-0641-01-e.pdf (accessed 8 February 2016); European Commission, *A Policy Framework for Climate and Energy in the Period from 2020 to 2030* (2014), http://eur-lex.europa.eu/legal-content/EN/TXT/PDF/?uri=CELEX:52014DC0015&from=EN (accessed 8 February 2016), p. 5; The White House, *President Obama's Climate Action Plan* (2015), www.whitehouse.gov/sites/default/files/docs/cap_progress_report_final_w_cover.pdf (accessed 8 February 2016), p. 6.

7 See T. Cottier, *Renewable Energy and Process and Production Methods*, E15 Initiative (Geneva: International Centre for Trade and Sustainable Development (ICTSD) and World Economic Forum, 2015), http://e15initiative.org/publications/renewable-energy-and-process-and-production-methods/ (accessed 8 February 2016).

8 IEA, *World Energy Outlook Report*, pp. 300–3.

9 T. Cottier, O. Nartova and S. Z. Bigdeli (eds.), *International Trade Regulation and the Mitigation of Climate Change: World Trade Forum* (Cambridge: Cambridge University Press, 2009).

First, local production of renewable electricity based upon hydro, solar, wind, thermal and tidal power offers an important contribution to supplies, but is unable on its own to provide the necessary stable baseload without recourse to oil, gas and coal or nuclear energy. Fluctuations caused by climatic conditions are too strong. The integration of increasing shares of RE electricity is therefore dependent upon the reinforcement of transmission interconnectors and the extension of grids over larger regions.[10] Second, the tremendous progress in electricity transmission technologies achieved in recent times has made long-distance electricity flows possible both technically and economically, facilitating cross-border electricity interconnections.[11] Modern high-voltage, direct current (HVDC) transmission lines, both on land and sea, offer the technology for long-distance transmission without substantial voltage losses. Long-distance transmission lines are able to reduce the requirement for balancing power, especially for fluctuating carbon-neutral sources such as wind and solar energy, from fully dispatchable conventional generators. They therefore contribute to increasing the reliability of supply while avoiding GHG emissions.[12]

The benefits associated with large-scale electricity transmission projects include the expansion of electricity trade areas, the promotion of competition in electricity generation (and thus the reduction of electricity prices) and the enhancement of reliability and security of supply.[13] Extending the grid over larger regions can facilitate the exploitation of comparative advantages in electricity production while carrying the potential to create stable supplies to countries located in those parts of the world that suffer from chronic shortages, by means of international trade. Furthermore, it can help to harvest RE from remote

[10] J. Sauvage and H. Bahar, 'Cross-border trade in electricity and the development of renewables-based electric power', *OECD Trade and Environment Working Papers* 2013/02 (2013), www.oecd-ilibrary.org/trade/cross-border-trade-in-electricity-and-the-development-of-renewables-based-electric-power_5k4869cdwnzr-en (accessed 8 February 2016).

[11] Massachusetts Institute of Technology Energy Initiative, *The Future of the Electric Grid: An Interdisciplinary MIT Study* (Cambridge, MA: Massachusetts Institute of Technology, 2011); T. Gönen, *Electrical Power Transmission System Engineering: Analysis and Design*, 3rd edn (Boca Raton, FL: CRC Press, 2014).

[12] S. Chatzivasileiadis, D. Ernst and G. Andersson, 'The Global Grid', *Renewable Energy*, 57 (2013), 372.

[13] International Energy Agency, *Secure and Efficient Electricity Supply during the Transition to Low Carbon Power Systems* (Paris: IEA, 2013), www.iea.org/publications/freepublications/publication/secureandefficientelectricitysupply.pdf (accessed 8 February 2016).

locations with abundant potential and very low production costs.[14] Accordingly, it could generate new sources of income, for example, from electricity produced from wind energy in the North and solar energy in the South, in regions that have faced difficulties in creating adequate economic growth.

Trends confirm these insights. The increasing global demand for electricity and the need for renewable production reflect the development and expansion of electricity networks and the intensification of cross-border electricity trade.[15] Because electricity trade relies on fixed physical infrastructure, the former is dependent on the latter, as is the case for all other network-bound industries (e.g. telecommunications and railways).[16] The progressive integration of electricity grids at the national level has paved the way for the creation of the first regional electricity systems, with some prominent examples including the European Union (EU) internal electricity market,[17] the Integrated Power System (IPS),[18]

[14] In recent years, various projects have undergone first feasibility assessments and have been launched in Europe, Asia and the Atlantic. For instance, projects such as Medgrid and OffshoreGrid will interconnect Mediterranean states with Europe, permitting the transfer of high shares of solar energy to the major load centres. Initiatives such as Gobitec in Asia and the Atlantic Wind Connection in the United States will interconnect the Asian power grids or transmit offshore wind energy to the US east coast. See Chapter 2.

[15] Energy Charter Treaty Secretariat, *Model Intergovernmental and Host Government Agreements for Cross-Border Electricity Projects*, www.energycharter.org/fileadmin/DocumentsMedia/Legal/EMAs_en.pdf (accessed 8 February 2016).

[16] P. J. Slot and A. Skudder, 'Common features of community law regulation in the network-bound sectors', *Common Market Law Review*, 38 (2001), 87.

[17] See P. Mäntysaari, *EU Electricity Trade Law: The Legal Tools of Electricity Producers in the Internal Electricity Market* (Heidelberg: Springer International Publishing, 2015); Directive 2009/72/EC of 13 July 2009 concerning common rules for the internal market in electricity and repealing Directive 2003/54/EC, http://eur-lex.europa.eu/LexUriServ/LexUriServ.do?uri=OJ:L:2009:211:0055:0093:EN:PDF (accessed 8 February 2016); and Regulation (EC) 714/2009 of 13 July 2009 on conditions for access to the network for cross-border exchanges in electricity and repealing Regulation (EC) No 1228/2003, http://eur-lex.europa.eu/legal-content/EN/ALL/?uri=CELEX:32009R0714 (accessed 8 February 2016).

[18] The Integrated Power System is an integrated transmission network based on the former Soviet Union electricity system. It currently connects the national grids of Latvia, Lithuania, Estonia, Armenia, Azerbaijan, Belarus, Georgia, Kyrgyzstan, Moldova, Tajikistan, Uzbekistan, Ukraine and Mongolia, while supplying electricity to Finland and some regions of China. See I. Tchakalov, T. Mitev and I. Hristov, 'Bulgarian power relations: the making of a Balkan power hub' in P. Högselius, A. Hommels, A. Kaijser and E. van der Vleuten (eds.), *The Making of Europe's Critical Infrastructure: Common Connections and Shared Vulnerabilities* (Basingstoke: Palgrave Macmillan, 2013), pp. 138–40.

the Southern African Power Pool (SAPP)[19] and the Central American Power System (SIEPAC).[20] Much, of course, remains to be done, and many countries in the developing world continue to suffer from insufficiently connected grids.

B Towards a Global Grid

Based on current and projected power trends and the estimated impact of technological innovations on electricity networks, experts envisage that the gradual implementation of cross-border electricity transmission projects on a large scale and over long distances will lead to the creation of regional supergrids and, eventually, a globally interconnected network.[21] The idea of a so-called global grid, of course, reflects a long-term vision which, if it materialises at all, will do so over a long period of time. It builds upon the philosophy and reality of an increasingly interconnected world, starting with the telegraph and moving through the telephone and now on to the Internet. Although technically feasible, it will require large volumes of investment in highly capital-intensive infrastructure projects, availability of fixed infrastructure and non-discriminatory access to transportation networks and distribution systems, along with mechanisms for transparent and efficient cross-border trading to facilitate the power exchange between regions. We perceive the idea of a global grid as

[19] The Southern African Power Pool is the largest and most advanced power pool in Africa and covers the networks of Democratic Republic of Congo, Tanzania, Angola, Zambia, Malawi, Namibia, Botswana, Zimbabwe, Mozambique, Swaziland, South Africa and Lesotho. See Infrastructure Consortium for Africa, *Regional Power Status in African Power Pools*, African Development Bank (2011), www.icafrica.org/fileadmin/documents/Knowledge/Energy/ICA_RegionalPowerPools_Report.pdf (accessed 8 February 2016).

[20] The Central American Electrical Interconnection System is the largest power system in Latin America and comprises the national electricity grids of Panama, Costa Rica, Honduras, Nicaragua, El Salvador and Guatemala: Proyeto Mesoamerica, *Costa Rica completa línea de transmisión eléctrica SIEPAC*, San José, Costa Rica (16 October 2014), www.proyectomesoamerica.org/joomla/index.php?option=com_content&view=article&id=731&Itemid=85 (accessed 8 February 2016) and Economic Consulting Associates, *The Potential of Regional Power Sector Integration: Central American Electric Interconnection System (SIEPAC) Transmission & Trading Case Study* (2010), www.esmap.org/sites/esmap.org/files/BN004-10_REISP-CD_Central%20American%20Electric%20Interconnection%20System-Transmisison%20&%20Trading.pdf (accessed 8 February 2016).

[21] See, in particular, Chatzivasileiadis, Ernst and Andersson, 'The Global Grid'; C. Barker, 'Practical management of variable and distributed resources in power grids' in L. E. Jones (ed.), *Renewable Energy Integration: Practical Management of Variability* (Amsterdam: Elsevier Academic Publishing, 2013), pp. 189–202.

a long-term perspective, an agenda for change. A global grid may eventually be, and should be, built upon existing national and inter-regional grid structures. As this will necessitate interfacing different systems, the legal framework, both domestically and internationally, should respond to this challenge by fostering and allowing the development of the necessary structures of multilevel governance. While it is often felt that progress in RE electricity is a matter of creating domestic policy space, it is more a question of properly allocating regulatory powers at the different levels of governance in order to produce the public good of safe and abundant supplies.[22]

In this perspective, this volume aims at identifying and discussing the regulatory challenges that the changing landscape of electricity trade (from neighbouring through contiguous to distant states) poses to the international trade and investment regimes as a way of preparing the tools for inter-regional interfaces in the future. Because international trade and investment rules were not drafted with electricity in mind,[23] the main overarching question, which is woven through the analyses in this book, is whether the trade and investment legal frameworks are sufficiently well equipped to enable, facilitate and possibly encourage long-distance electricity trade and, in particular, trade in electricity produced from RE sources.

In attempting to answer this question, this volume breaks new ground in several respects. First, the role of international trade and investment law in the field of electricity, as already indicated, has until recently been a limited one: for a long time, electricity – and the energy sector in general – has been regulated through a sovereignty-based approach, classically falling within the scope of national jurisdictions.[24] Second, despite the gradual expansion of cross-border electricity trade, the still predominant regional dimension has so far delayed the discourse on the extent to which international trade and investment rules can facilitate the

[22] T. Cottier, 'Renewable energy and WTO law: more policy space or enhanced disciplines?' *Renewable Energy Law and Policy Review*, 1 (2014), 40–52.

[23] T. Cottier, G. Malumfashi, S. Matteotti-Berkutova, O. Nartova, J. de Sépibus and S. Z. Bigdeli, 'Energy law and policy' in T. Cottier and P. Delimatsis (eds.), *The Prospects of International Trade Regulation: From Fragmentation to Coherence* (Cambridge: Cambridge University Press, 2011), pp. 214–5.

[24] J. Pauwelyn, 'Global challenges at the intersection of trade, energy and the environment: an introduction' in J. Pauwelyn (ed.), *Global Challenges at the Intersection of Trade, Energy and the Environment* (Geneva: Centre for Trade and Economic Integration, 2009), p. 3.

reinforcement of transmission interconnections and increase their security and stability along with the share of renewables fed into the grid.[25] Moreover, the focus on the EU international electricity market has left other regional experiences in the shade and contributed to putting EU law centre-stage in the debate on cross-border electricity trade.[26] Discussions in the international trade and investment domains have focused on the question of the policy space left to national governments with regard to introducing and/or terminating RE public support schemes, as well as trade remedies aimed at levelling the playing field in response to such measures.[27] Yet, the expansion of international trade in electricity and, in particular, RE electricity inevitably raises electricity-specific issues related to the physical peculiarities of electricity transmission and trade.[28]

C The Goals and Structure of This Study

Against this backdrop, the purpose of this volume is to explore the implications of the most recent technical, geopolitical and economic developments affecting international trade in electricity for the multilateral legal framework on trade and investment. The book will also provide suggestions on how to improve the existing framework in a way that ensures there is coherence throughout the various levels of governance which are going to be affected by the expansion of the electricity grid over

[25] In recent years, discussions on the importance of a multilateral legal framework for trade in energy have gradually gained prominence. See, e.g., Y. Selivanova (ed.), *Regulation of Energy in International Trade Law* (Alphen aan den Rijn: Kluwer Law International, 2011). Yet, none have focused on electricity-specific issues.

[26] So far, the most comprehensive analyses of the electricity sector, including regulatory aspects of trade and investment, have focused on the European experience. See, e.g., M. Roggenkamp, C. Redgwell, A. Rønne and I. del Guayo (eds.), *Energy Law in Europe: National, EU and International Regulation* (Oxford: Oxford University Press, 2007), and J. Bielecki and M. G. Desta (eds.), *Electricity Trade in Europe: Review of the Economic and Regulatory Challenges* (Alphen aan den Rijn: Kluwer Law International, 2004).

[27] Many challenges have recently been filed against RE public support programmes before the WTO Dispute Settlement Body and many claims have been initiated by investors under international investment treaties. For an overview see, among others, J. Salzman and M. Wu, 'The next generation of trade and environment conflicts: the rise of green energy policy', *Northwestern University Law Review*, 108 (2014), 401; K. Talus (ed.), 'Special issue on renewable energy disputes', *Oil, Gas & Energy Law*, 3 (2015).

[28] Once generated from energy sources, electricity has to be transmitted through high-voltage networks to major demand centres and distributed to the final consumers. Electricity trading is thus dependent on the interconnectivity of physical infrastructure and the availability of transmission capacity. W. Patterson, *Transforming Electricity* (London: Earthscan, 1999).

larger regions. To achieve its objectives, the book is divided into four parts, which, taken together, can provide the reader with a comprehensive understanding of the issues at stake in the field of electricity and the lessons that can be drawn from an analysis of the state of play in the current international trade and investment regimes. The four parts correspond to the following four thematic areas:

(i) the technical and economic foundations of a global grid;
(ii) regional experiences with cross-border electricity trade;
(iii) the specific interconnectivity issues posed by electricity transportation and their implications for international trade and investment law in working towards inter-regional trade in electricity;
(iv) the regulatory challenges raised in the trade and investment fields by the quest for a level playing field in the promotion of RE electricity.

Part I of the volume takes stock of the technical advances that have been made in electrical engineering and cross-border electricity trade, and identifies the current challenges posed by the expansion of inter-regional electricity grids. It also examines the economic and geopolitical determinants of international trade in electricity and addresses electricity governance issues. Chatzivasileiadis and Ernst build on the analysis of recent technical improvements in electricity network development to conceptualise the idea of a global grid as a means to efficiently integrate renewables into the grid while ensuring reliability and security of supply. They show that long-distance interconnections can be both technically feasible and economically competitive, in addition to being environmentally desirable. They identify the main market operation and regulatory challenges posed by cross-border electricity trade and underscore the need for further research to find appropriate solutions to market-coupling obstacles when interconnecting larger regions in the longer term. In the same vein, Houmøller elaborates on the possible practical solutions to the political problems raised by market coupling over vast areas, focusing on the role and tasks to be attributed to regulators. He draws on the European example to argue that a single spot exchange can ensure a reliable and cost-efficient day-ahead congestion management system, again in the spirit of facilitating the shift to renewables in the electricity sector. Robinson complements the technical stock-taking of cross-border electricity trade with an analysis of the economic and geopolitical drivers of trade in electricity. He points out that, irrespective of technical feasibility, sound economics, political stability and shared values are necessary preconditions to overcome the costs and limits of building interconnectors and trading electricity, especially when generated from RE sources. He argues that any reform process aimed at fostering RE electricity trade must be based on political stability and a shared long-term vision for the peoples of the different countries involved. Defilla maps the global intergovernmental and non-governmental organisations that play a role in the electricity industry. Based on an analysis of the most pressing regulatory challenges posed by the technological improvements occurring in the sector, he examines various options for good governance for the future global electricity grid. He contends that the Energy Charter Treaty (ECT), as a unique multilateral and legally binding framework dealing exclusively with energy issues, could play a key role in this respect, provided that renewables receive sufficient priority and that a specific instrument (a treaty or a protocol) for electricity can be developed.

Part II provides an in-depth analysis of the experiences of the most important regions of the world in promoting electricity trade integration, and RE electricity in particular. Le Page starts by describing the EU experience. He recollects the fundamental steps in the creation of the internal energy market, from the early 1990s to the adoption of the Third Energy Package in 2009. He gives an account of the set of policies pursued in parallel by the EU and its Member States to set out the path towards decarbonisation, with a particular focus on the power generation industry. In his opinion, greater coherence between the EU energy and climate frameworks is the only way to limit distortions to the internal energy market to a minimum while still ensuring a level playing field for investment in RE production and the achievement of RE targets in a cost-efficient manner. Sioshansi discusses the North American experience (United States and Canada). He notes that, to date, green electricity generation and trade has mainly been encouraged at the level of individual states, provinces, cities and communities (as well as a growing number of corporations), rather than at the federal level. Accordingly, he suggests that policy-makers focused on encouraging RE electricity trade must remain cognisant of such bottom-up developments in the region and revisit their approaches with a consideration of the challenges posed by decentralised RE electricity produced by so-called 'prosumers'. Jusoh describes the Asian experience, focusing on the Member States of the Association of Southeast Asian Nations (ASEAN). After giving an account of the state of play in RE electricity trade in ASEAN Member States, he underscores the many challenges they face in ensuring that green electricity trade in the region meets the objectives of the ASEAN

Community 2015, from rural electrification to technology development and innovation in the field, from raw material supply to trade facilitation. He urges ASEAN Member States to make a concerted effort to stimulate a new policy approach for the promotion of RE electricity which tackles issues incidental to energy transformation, such as sustainable development. Kambanda focuses on the African experience. He notes that, with the exception of the SAPP, African power pools are still in their developmental stages, and electricity trade is mostly based on bilateral agreements. He contends that regional electricity trade in Africa remains limited because of a lack of physical infrastructure (due in part to lack of access to capital), poor planning, lack of maintenance and poor performance of the power utilities. Based on the continent's huge hydro, solar and geothermal potential, however, he suggests that African countries should establish RE targets, combined with additional incentives for attracting RE investments, in order to overcome these challenges. Lembo and Eleoterio conclude Part II by describing the Latin American experience. They observe that current electricity cooperation and integration projects in the Americas are focused on the development of regional grid interconnections that are supported by bilateral and regional agreements. While acknowledging the efforts of Latin American countries to foster RE electricity trade, they argue that the lack of a common regulatory framework remains a major obstacle to all electricity integration projects in the region. Accordingly, they advocate for new rules to provide coherence to cross-border electricity trade in Latin America.

Part III looks at the main interconnectivity issues raised by the peculiarities of electricity transportation and trade via fixed grids, and their implications for the trade and investment legal framework. Selivanova explores relevant international trade rules disciplining the conditions for access to fixed electricity infrastructure, namely the rules on transit in Article V of the General Agreement on Tariffs and Trade (GATT) and the General Agreement on Trade in Services (GATS) disciplines related to energy services. She finds that Article V does not address, among others, the issue of construction of new transit capacity – for which, she contends, an effective investment framework is needed. She notes that GATS rules do not currently address the anti-competitive practices that are common among incumbents controlling different segments upstream and downstream along the energy value chain. She proposes that additional commitments be negotiated either in an annex to GATS on Energy Services or in a Reference Paper to address competition issues and problems of third-party access to fixed infrastructure. Gudas moves to issues of connection, access and dispatch of electricity to the electricity network, and focuses on the particular question of whether renewables should be given special treatment at the grid operation level. He argues that, instead of priority rules, the regulation of connection to the grid and the operation of the grid should be subject to rules aimed at ensuring neutrality across technologies through grid adaptation and market arrangements. Espa follows with an analysis of the international trade rules on import and export restrictions, which are relevant to addressing the range of practices affecting cross-border electricity (and, in particular, RE electricity) flows. She finds that generic WTO rules are not designed in a way that can accommodate the peculiar needs of (RE) electricity trade, namely cross-border balancing and congestion management. She suggests possible directions to overcome these drawbacks based on a comparative analysis of pertinent electricity-specific rules developed at the regional level incorporating RE promotion goals (i.e. EU secondary law). De Sépibus explores the regulatory challenges posed by the ongoing paradigm shift in electricity regulation, from grid operators to the advent of 'smart grids', focusing on the EU experience. She argues that the current focus on grid tariffs has to give way to a broader view, which encompasses all terms and conditions related to access to the facilities that are essential for smart grid activities. She further contends that the role of the regulator should be revisited so as to reflect the increasing involvement of third parties in the grid business and should be limited to tackling market failures and ensuring the pursuit of public policy goals. Kunze addresses the specific interoperability challenges faced by transmission system operators (TSOs) resulting from the development of decentralised capacity mechanisms and the implementation of 'disruptive' technologies. Building on a case study of Switzerland, he argues that these structural changes call on TSOs to introduce market design structures that allow for the optimisation of flexibility potentials coming from the integration of renewables into the grid. He suggests that such structures have to be defined taking into account political and organisational factors such as the role of the balancing responsible parties and the mechanisms to increase the price signal within a short timeframe at a very early stage of the planning.

Part IV of the book deals with the regulatory challenges arising out of the proliferation of transitional policy instruments aimed at supporting clean energy industries and electricity produced from renewable sources. Rubini addresses the question of the policy space for subsidies in support of RE energy and electricity, in particular under the WTO Agreement on

Subsidies and Countervailing Measures (ASCM). He introduces the concepts of 'point of balance' and 'trade-offs' to challenge ASCM disciplines as they currently stand. Reaching the conclusion that such disciplines, as interpreted by recent case law, are not only unclear and unsound, but also inadequate in the light of the economic, political and legal circumstances of today's world, he stresses the need to create the necessary conditions for starting a process of reform, namely by advancing knowledge and understanding of the current drawbacks, based on consistent and reliable data. Vermulst and Meng examine the most pressing dumping and countervailing duties (CVD) issues that have emerged in the new era of RE-targeted trade disputes, with a focus on the trade defence cases against solar products from China and the issues in the biofuels/biodiesel disputes involving the EU, Argentina and Indonesia at both the EU and the WTO levels. They argue that the proliferation of such disputes has increased the attention paid to certain disputable methodologies applied by some user countries such as the United States and the EU. They posit that such scrutiny will make interested parties better informed as to how subsidies and other support measures are likely to be treated in trade disputes domestically and at the WTO. Holzer, Espa and Payosova explore the legal feasibility of a system of differential taxation on electricity based on its energy sources or its carbon footprint. They argue that imposing higher rates of taxes on electricity produced from fossil fuel and applying lower tax rates or granting exemptions to green electricity can create appropriate incentives for RE production without burdening public resources. They contend that there are options for designing such a scheme in a way that would be compatible with relevant WTO rules under the GATT and the ASCM.

The final three chapters of Part IV explore competition and investment issues related to the promotion of RE electricity, including the role of monopolies, technological transfer and the treatment of regulatory takings. Nartova examines general trends in the effects of competition policy, institutional monopolies and the role of governments in promoting RE electricity. She posits that government incentives and environmental laws are needed to ensure effective development of RE electricity. She goes on to suggest that states should enact appropriate horizontal agreements and practices in their environmental and competition policy and evaluate the environmental impacts of their internal restrictive practices and how they affect competition in the energy sector. Abbott identifies the aspects of technology transfer that may be relevant to long-distance trade in RE electricity and explores whether there are potential obstacles to such trade coming from intellectual property law. He suggests that transfers of technology directed towards establishing a globally interconnected electricity network should begin at a more modest level through bilateral and regional electricity-sharing agreements and programmes. Finally, Matteotti and Payosova examine the tensions emerging in interventional investment law when it comes to finding a reasonable balance between the interests of an investor and a state's legitimate right to regulate the RE sector on its own territory. After considering the most recent RE-related arbitration awards under various investment agreements, they conclude that the current international investment law framework does not seem to offer significant legal security either for investors or for host states. They provide a number of policy recommendations that governments may use when implementing new RE support schemes or amending the existing ones in the RE sector, and which are based on recent interpretations of the fair and equitable treatment standard in investment jurisprudence.

D Conclusions

The contributions to this volume offer interesting and relevant building blocks for use in the quest for law governing international trade and investment in (RE) electricity. They are not exhaustive, and other important areas, such as the implications of the law of the sea for marine electrical cables,[29] are not addressed in this volume, as its scope is limited to trade and investment. Nor have we dealt extensively with the pressing issue of subsidies allocated to fossil fuels and how they inherently influence the regulation of RE sources.[30] A key message of the book is that there is a need for much more constructive dialogue

[29] See D. R. Burnett, R. Beckman and T. M. Davenport (eds.), *Submarine Cables: The Handbook of Law and Policy* (Leiden: Martinus Nijhoff, 2014); T. Davenport, 'The installation of submarine power cables under UNCLOS: legal and policy issues', *German Yearbook of International Law*, 56 (2013), 107; K. Meissner, H. Schabelon, J. Bellebaum and H. Sordyl, *Impacts of Submarine Cables on the Marine Environment* (2007), www.bfn.de/fileadmin/BfN/meeresundkuestenschutz/Dokumente/BfN_Literaturstudie_Effekte_marine_Kabel_2007-02_01.pdf (accessed 15 December 2015).

[30] I. Espa and S. E. Rolland, *Subsidies, Clean Energy and Climate Change*, E15 Initiative (Geneva: International Centre for Trade and Sustainable Development (ICTSD) and World Economic Forum, 2015), http://e15initiative.org/publications/subsidies-clean-energy-and-climate-change/ (accessed 10 March 2016).

among lawyers, economists, political scientists and engineers studying the challenges posed by cross-border electricity trade. Accordingly, this volume aims at stimulating cross-fertilisation across these disciplines by showing how researchers from different fields tackle the same substantive questions and share a common vision for the future of electricity trade.

All contributors agree, in particular, that the unprecedented challenge of global warming requires the decarbonisation of the electricity sector, to be achieved through a rapid and substantial shift to renewables. They also agree that the efficient integration of increasing shares of RE sources and the need to satisfy increasing demand for electricity in many parts of the world will lead to the gradual integration of national, regional and, eventually, inter-regional grids, in parallel with the development of mechanisms designed to cope with decentralised energy production. Finally, they agree that the reinforcement of transmission interconnectors and the extension of grids over larger regions poses a number of regulatory challenges to the trade and investment framework. Market integration prospects depend heavily upon appropriate structures of international governance. Objections raised with respect to the vision of a global grid have thus focused on security and dependency issues, suggesting that such vision will take a long time to evolve in a piecemeal manner. While the proposals formulated and the answers provided differ, all contributors agree that the focus in this phase should be on how to design the process for reform and frame it in the right way. The accent is put on the need to generate knowledge and meaningful discussion among all the different stakeholders on the peculiarities of electricity and, in particular, RE electricity trade and investment as the necessary precondition for possible future negotiations.

In this respect, for all the reasons already mentioned, the aim of this volume is to make a useful contribution to the advancement of knowledge and understanding of the problems raised by the changing landscape of electricity trade, and the challenges this poses to the trade and investment framework. We hope that the volume may help in assessing the way forward within the multilateral trading system and within the emerging regional trade agreements. This will be achieved through identifying relevant building blocks that could inform either the advent of sectorial agreements on trade and investment in electricity or the reform of existing provisions and agreements to render them conducive to achieving the goal of stable and decarbonised supplies of electricity to all humankind.

Bibliography

Barker, C. (2014). Practical management of variable and distributed resources in power grids. In L. E. Jones (ed.), *Renewable Energy Integration: Practical Management of Variability, Uncertainty, and Flexibility in Power Grids*. Amsterdam: Elsevier Academic Publishing, pp. 175–88.

Bielecki, J. and Desta, M. G. (eds.) (2004). *Electricity Trade in Europe: Review of the Economic and Regulatory Challenges*. Alphen aan den Rijn: Kluwer Law International.

Chatzivasileiadis, S., Ernst, D. and Andersson, G. (2013). The Global Grid. *Renewable Energy*, **57**, 372–83.

Cottier, T. (2014). Renewable energy and WTO law: more policy space or enhanced disciplines? *Renewable Energy Law and Policy Review*, **1**, 40–52.

(2015). *Renewable Energy and Process and Production Methods*, E15 Initiative (Geneva: International Centre for Trade and Sustainable Development (ICTSD) and World Economic Forum, 2015), http://e15initiative.org/publications/renewable-energy-and-process-and-production-methods/ (accessed 8 February 2016).

Cottier, T., Malumfashi, G., Matteotti-Berkutova, S., Nartova, O., de Sépibus, J. and Bigdeli, S. Z. (2011). Energy law and policy. In T. Cottier and P. Delimatsis (eds.), *The Prospects of International Trade Regulation: From Fragmentation to Coherence*. Cambridge: Cambridge University Press, pp. 211–44.

Cottier, T., Nartova, O. and Bigdeli, S. Z. (eds.) (2009). *International Trade Regulation and the Mitigation of Climate Change: World Trade Forum*. Cambridge: Cambridge University Press.

Davenport, T. M. (2013). The installation of submarine power cables under UNCLOS: legal and policy issues. *German Yearbook of International Law*, **56**, 107.

Douglas, R., Burnett, D. R., Beckman, R. and Davenport, T. M. (2014). *Submarine Cables: The Handbook of Law and Policy*. Leiden: Martinus Nijhoff.

Economic Consulting Associates (2010). *The Potential of Regional Power Sector Integration: Central American Electric Interconnection System (SIEPAC) Transmission & Trading Case Study*, www.esmap.org/sites/esmap.org/files/BN004-10_REISP-CD_Central%20American%20Electric%20Interconnection%20System-Transmisison%20&%20Trading.pdf (accessed 8 February 2016).

Energy Charter Treaty Secretariat, *Model Intergovernmental and Host Government Agreements for Cross-Border Electricity Projects*, www.energycharter.org/fileadmin/DocumentsMedia/Legal/EMAs_en.pdf (accessed 8 February 2016).

Espa, I. and Rolland, S. E. (2015). *Subsidies, Clean Energy and Climate Change*, E15 Initiative (Geneva: International Centre for Trade and Sustainable Development (ICTSD) and World Economic Forum, 2015), http://e15initiative.org/publications/subsidies-clean-energy-and-climate-change/ (accessed 10 March 2016).

EurElectric. (2015). *Power Statistics and Trends*, www.eurelectric.org/media/249736/power-statistics-and-trends-the-five-dimentions-of-the-energy-union-lr-2015-030-0641-01-e.pdf (accessed 8 February 2016).

(2015). *A Sector in Transformation: Electricity Industry Trends and Figures*, www.eurelectric.org/media/161808/electricityindustrytrendsandfigures2015_lr-2015-030-0064-01-e.pdf (accessed 8 February 2016).

Gönen, T. (2014). *Electrical Power Transmission System Engineering: Analysis and Design*, 3rd edn. Boca Raton, FL: CRC Press.

IEA. (2015). *World Energy Outlook Electricity Database*, www.worldenergyoutlook.org/resources/energydevelopment/energyaccessdatabase/ (accessed 8 February 2016).

(2015). *World Energy Outlook Report*, www.worldenergyoutlook.org/media/weowebsite/2015/WEO2015_Chapter01.pdf (accessed 8 February 2016).

(2015). *World Energy Outlook Special Report: Trade and Climate Change*, www.iea.org/publications/freepublications/publication/WEO2015SpecialReportonEnergyandClimateChange.pdf (accessed 15 December 2015).

Infrastructure Consortium for Africa. (2011). *Regional Power Status in African Power Pools*, African Development Bank, www.icafrica.org/fileadmin/documents/Knowledge/Energy/ICA_RegionalPowerPools_Report.pdf (accessed 8 February 2016).

International Energy Agency. (2013). *Secure and Efficient Electricity Supply during the Transition to Low Carbon Power Systems*, www.iea.org/publications/freepublications/publication/secureandefficientelectricitysupply.pdf (accessed 8 February 2016).

Mäntysaari, P. (2015). *EU Electricity Trade Law: The Legal Tools of Electricity Producers in the Internal Electricity Market*. Heidelberg: Springer International Publishing.

Massachusetts Institute of Technology Energy Initiative. (2011). *The Future of the Electric Grid: An Interdisciplinary MIT Study*. Cambridge, MA: Massachusetts Institute of Technology.

Meissner, K., Schabelon, H., Bellebaum, J. and Sordyl, H. (2007). *Impacts of Submarine Cables on the Marine Environment*, www.bfn.de/fileadmin/BfN/meeresundkuestenschutz/Dokumente/BfN_Literaturstudie_Effekte_marine_Kabel_2007-02_01.pdf (accessed 15 December 2015).

Patterson, W. (1999). *Transforming Electricity*. Earthscan: London.

Pauwelyn, J. (2009). Global challenges at the intersection of trade, energy and the environment: an introduction. In J. Pauwelyn (ed.), *Global Challenges at the Intersection of Trade, Energy and the Environment*. Geneva: Centre for Trade and Economic Integration, pp. 1–8.

Proyeto Mesoamerica, Costa Rica completa línea de transmisión eléctrica SIEPAC, San José, Costa Rica, 16 October 2014, www.proyectomesoamerica.org/

joomla/index.php?option=com_content&view=article&id=731&Itemid=85 (accessed 8 February 2016).

Roggenkamp, M., Redgwell, C., Rønne, A. and del Guayo, I. (eds.). (2007). *Energy Law in Europe: National, EU and International Regulation.* Oxford: Oxford University Press.

Salzman, J. and Wu, M. (2014). The next generation of trade and environment conflicts: the rise of green energy policy. *Northwestern University Law Review*, **108**, 401.

Sauvage, J. and Bahar, H. (2013). 'Cross-border trade in electricity and the development of renewables-based electric power', *OECD Trade and Environment Working Papers* 2013/02, www.oecd-ilibrary.org/trade/cross-border-trade-in-electricity-and-the-development-of-renewables-based-electric-power_5k4869cdwnzr-en?crawler=true (accessed 8 February 2016).

Selivanova, Y. (ed.) (2011). *Regulation of Energy in International Trade Law.* Alphen aan den Rijn: Kluwer Law International.

Slot, P. J. and Skudder, A. (2001). Common features of community law regulation in the network-bound sectors. *Common Market Law Review* **38**, 87.

Talus, K. (ed.) (2015). Special issue on renewable energy disputes. *Oil, Gas & Energy Law*, **3**.

Tchakalov, I., Mitev, T. and Hristov, I. (2013). Bulgarian power relations: the making of a Balkan power hub. In P. Högselius, A. Hommels, A. Kaijser and E. van der Vleuten (eds.), *The Making of Europe's Critical Infrastructure: Common Connections and Shared Vulnerabilities.* Basingstoke: Palgrave Macmillan, pp. 131–56.

PART I

Towards a Global Grid

Foundations and Challenges

2

The State of Play in Cross-Border Electricity Trade and the Challenges towards a Global Electricity Market Environment

SPYRIDON CHATZIVASILEIADIS AND DAMIEN ERNST

A Introduction

Increased environmental awareness has led to concrete actions in the energy sector in recent years. Examples include the European Commission's target of 20 per cent participation of renewable energy sources (RES) in the European Union (EU) energy mix by 2020[1] and California's decision to increase renewable energy in the state's electricity mix to 33 per cent of retail sales, again by 2020.[2] At the same time, several studies have investigated the possibility of having a higher share of renewables in the energy supply system of the future. For instance, the German Energy Agency (dena) assumes 39 per cent RES participation by 2020,[3] while a detailed study from the National Renewable Energy Laboratory (NREL) suggests that meeting US electricity demand in 2050 with 80 per cent RES supply is feasible.[4]

In 2011, Czisch discussed the possibility of having a 100 per cent renewable energy supply system in Europe with interconnections in North Africa and West Asia. Furthermore, Jacobson and Delucchi investigated 'the feasibility of providing worldwide energy for all purposes (electric power, transportation, heating/cooling, etc.) from wind, water,

[1] European Commission, *Impact Assessment on the EU's Objectives on Climate Change and Renewable Energy* (Brussels: European Commission, 2008).

[2] State of California, Senate Bill X1–2 (sbx1 2) (2011).

[3] Deutsche Energie-Agentur (dena), *dena Grid Study II – Integration of Renewable Energy Sources in the German Power Supply System from 2015–2020 with an Outlook to 2025* (Berlin: German Energy Agency, 2010).

[4] M. M. Hand, S. Baldwin, E. DeMeo, J. M. Reilly, T. Mai, D. Arent, G. Porro, M. Meshek and D. Sandor (eds.), *National Renewable Energy Laboratory, Renewable Energy Futures Study*, 4 vols. (NREL/TP-6A20-52409) (Golden, CO: National Renewable Energy Laboratory, 2012), www.nrel.gov/analysis/re_futures/ (accessed 20 June 2012).

and sunlight'. The authors made a detailed analysis and proposed a plan for implementation.[5] They found that the barriers to the realisation of this plan are not technological or economic, but rather social and political.

All these studies suggest that the development of the electricity network will play a crucial role in the efficient integration of increasing shares of renewables, for two main reasons. First, interconnecting RES increases the reliability of supply. Second, long transmission lines can help harvest renewable energy from remote locations with abundant potential and very low production costs. Concrete actions have been taken to exploit the benefits of such interconnections, which will lead to the creation of regional supergrids. EU guidelines already encourage transmission projects such as the Baltic Ring.[6] Projects such as Medgrid[7] and Offshore-Grid[8] have been launched with the aim of interconnecting Mediterranean states with Europe and transferring renewable energy to the major load centres. At the same time, initiatives such as Gobitec[9] in Asia and Atlantic Wind Connection[10] in the United States aim to interconnect the Asian power grids or transmit offshore wind energy to the US east coast.

Expanding the regional transmission grids to supergrids, building new lines and reinforcing existing interconnections are necessary to minimise the need for balancing energy from conventional generators as well as to decrease the need for storage.

In Section B of this chapter, we will discuss the need to extend the power grid to increase the reliability of renewable energy supply and reduce the requirement for balancing power from conventional generators. In Section C, we summarise the analyses in which we have shown that building long-distance and intercontinental interconnections can be economically competitive. We consider the cheap and abundant renewable power in remote locations, as well as the opportunities presented by

[5] M. Z. Jacobson and M. A. Delucchi, 'Providing all global energy with wind, water, and solar power, Part I: technologies, energy resources, quantities and areas of infrastructure, and materials', *Energy Policy*, 39(3) (2011), 1154–69; G. Czisch, *Scenarios for a Future Electricity Supply: Cost-Optimized Variations on Supplying Europe and Its Neighbours with Electricity from Renewable Energies* (Stevenage: Institution of Engineering and Technology (IET), 2011).

[6] A. Boute and P. Willems, 'RUSTEC: greening Europe's energy supply by developing Russia's renewable energy potential', *Energy Policy*, 51 (2012), 618–29.

[7] www.medgrid-psm.com (accessed 20 December 2016).

[8] www.offshoregrid.eu (accessed 20 December 2016).

[9] www.gobitec.org (accessed 20 December 2016).

[10] www.atlanticwindconnection.com (accessed 20 December 2016).

the time-zone difference for energy arbitrage between the continents. In that respect, we also introduce the vision of the Global Grid and a global electricity market environment. To facilitate our discussion on cross-border electricity trade, in Section D, we describe merchant transmission investment, an investment scheme for financing the building of new lines, and how this depends on the market structure of the regions connected by such lines. In Section E, we outline the market operating schemes in different regions and then summarise the mechanisms for cross-border electricity trade. Finally, in Section F, we focus on the challenges that arise in an effort to couple different regional markets for cross-border electricity trade. Section G concludes.

B The Need to Extend Transmission Infrastructure to Balance Demand and Supply from Renewables

Power systems started to develop at a regional level in order to supply larger areas' load demand from single generators. Interconnections among several countries emerged in the second half of the twentieth century.[11] The main reason for interconnecting different power systems was to increase reliability and security of supply. If one area experienced a disturbance, neighbouring areas could provide the missing power for a given amount of time, until the fault was cleared. In the past two decades, however, interconnections have become the platform for ever-increasing transfer of electricity and for electricity trade. With the increased integration of fluctuating energy sources, the role of interconnections is becoming more important.

Several studies investigating very high RES penetration in electric power systems have been carried out in Europe and the United States, and also in Asia. All these studies conclude that storage and increased interconnection capacity are major factors contributing to the balancing of RES variability and load.

Becker et al. investigated the transmission needs in Europe for a fully renewable pan-European electricity supply.[12] They considered high-resolution wind and solar data collected in Europe over eight consecutive

[11] http://theinstitute.ieee.org/technology-focus/technology-history/the-star-that-sparked-europes-power-grid513 (accessed 20 December 2016).

[12] S. Becker, R. A. Rodriguez, G. B. Andresen, S. Schramm and M. Greiner, 'Transmission grid extensions during the build-up of a fully renewable pan-European electricity supply', *Energy*, 64 (2014a), 404–18.

years. They found that quadrupling today's net transfer capacity (NTC) of the European interconnections leads to a 33 per cent reduction in the need for balancing energy. Wind power benefits more from such interconnections than solar, as 'larger interconnected areas eliminate to a significant degree weather correlations that lead to generation patterns'.[13]

Similar studies have been carried out for the United States. In a second study, Becker et al. investigated the optimal mix of wind and solar for a fully renewable US electricity system based on thirty-two years of data for the whole contiguous United States.[14] They determined the optimal mix for each of the ten Federal Energy Regulatory Commission (FERC) regions independently, as well as for the whole of the United States. Among other results, they showed that interconnecting the ten FERC regions reduces the need for balancing power by up to 30 per cent compared with each region having to balance power supply and demand independently.

Mills and Wiser focused on the economic value of increased PV penetration in the United States.[15] They showed that absent any mitigation measure, more PV cannot help with the evening load (after the sun is down), which leads to a decrease in the marginal benefits of PV. One suggested remedy is to overdimension PV installations, namely by installing PV capacity that exceeds the electricity demand during peak times. Despite this leading to a curtailment of PV production during peak hours, it will also lead to a flatter power production curve during the day.

A pan-Asian power grid has been suggested by Taggart et al., based on electricity market simulations.[16] They found that the total cost of electricity could be reduced by up to USD 6.55 billion *per year* by interconnecting wind farms in China and solar PV installations in Australia.

Aboumahboub et al. investigated the need for building additional 'peaking' gas power plants for balancing renewable energy.[17] In their

[13] Ibid.

[14] S. Becker, B. A. Frew, G. B. Andresen, T. Zeyer, S. Schramm, M. Greiner and M. Z. Jacobson, 'Features of a fully renewable US electricity system: optimized mixes of wind and solar PV and transmission grid extensions', *Energy*, 72 (2014b), 443–58.

[15] A. D. Mills and R. H. Wiser, 'Changes in the economic value of photovoltaic generation at high penetration levels: a pilot case study of California', *IEEE Journal of Photovoltaics*, 3(4) (2013), 1394–402.

[16] S. Taggart, G. James, D. Zhaoyan and C. Russell, 'The future of renewables linked by a transnational Asian Grid', *Proceedings of the IEEE*, 100(2) (2012), 348–59.

[17] T. Aboumahboub, K. Schaber, P. Tzscheutschler and T. Hamacher, 'Optimization of the utilization of renewable energy sources in the electricity sector', *Proceedings of the 5th IASME/WSEAS International Conference on Energy & Environment* (EE '10) (2010), pp. 196–204.

study they compared the number of necessary conventional power plants for balancing power in the case that there are interconnecting lines between regions and in the case that there are not. Their results for both the European and a potential Global Grid showed that by building a sufficient number of interconnecting lines, the need for dispatchable conventional power plants could be reduced by between half and one eighth of the present level.

All these studies conclude that the benefit from interconnections is much more significant in the presence of a high proportion of wind in the renewable mix. On the other hand, if the renewable mix is mostly based on PV solar power, storage plays a significant role, as this can provide the missing energy during the night.

In an electricity system supplied by renewable sources, storage is equally as important as a strong transmission grid and should be considered either as an alternative – when the majority of the renewable sources are small-scale plants such as rooftop solar PVs – or as a complement to long interconnections, especially when most of the renewable supply comes from remote wind farms and solar parks. Transmission-grid studies already incorporate storage options in their projections for the future.[18] Bulk storage will be necessary for absorbing non-transmissible power and relieving congestion. Besides pump-hydro power plants, which are currently the most widespread type of bulk energy storage, new technologies, such as compressed-air, hydrogen, flywheel farms or batteries,[19] can also be used in the future as bulk storage systems.

An additional factor, however, that needs to be taken into account when considering bulk storage is the necessity for grid reinforcements. An investigation of the integration of additional RES[20] showed that even if there were storage able to absorb 100 per cent of the excess power, 65 per cent of the proposed grid reinforcements would still be necessary. In this respect, it should also be noted that, comparing bulk storage efficiency with transmission losses, a 6,000 km long high-voltage, direct current (HVDC) line has a greater efficiency presently than hydrogen or battery storage (ultra HVDC at 800 kV exhibits 3 per cent losses per 1,000 km, in comparison with the 50 per cent efficiency of hydrogen storage).

[18] See, e.g., Deutsche Energie-Agentur (dena), *dena Grid Study II.*

[19] For example, Tesla Motors is building a battery 'gigafactory': http://fortune.com/2014/10/28/tesla-closes-on-free-nevada-land-for-gigafactory/ (accessed 20 December 2016).

[20] Deutsche Energie-Agentur (dena), *dena Grid Study II.*

C Long-Distance Interconnections: Towards a Global Electricity Environment

As we have seen in the previous sections, if we are envisioning an electricity system supplied 100 per cent from renewables, it is necessary to build strong transmission grids that extend over large regions.

In earlier work we have proposed the next natural step in introducing the vision of a globally interconnected power network: the Global Grid.[21] The Global Grid aims at interconnecting the regional supergrids into one global electricity network. High-capacity long transmission lines will interconnect wind farms and solar power plants, supplying load centres with green power over long distances. Besides introducing the concept, our work further highlighted the multiple opportunities that the Global Grid offers, and discussed possible investment mechanisms and operating schemes. Through cost–benefit analyses we showed that long-distance interconnections forming a Global Grid can be technologically feasible and economically competitive.

Recently, support for the concept of a global transmission grid powered by renewables has come from Zhenya Liu, the chief executive officer of the China State Grid Corporation. The China State Grid Corporation is one of the largest transmission system operators in the world, serving 128 million customers.[22,]

Our analyses were based on real market price data, results from other studies and realistic assumptions. In this section we provide a summary of these results.

I Stages in Creating a Global Power Grid

We identify three main drivers for the creation of a globally interconnected network. First, the need to harvest remote renewable energy resources, either further offshore or in deserts, will lead to continuously expanding regional supergrids. Second, taking advantage of the shift in

[21] See S. Chatzivasileiadis, D. Ernst and G. Andersson, 'The Global Grid', *Renewable Energy*, 57 (2013), 372–83; S. Chatzivasileiadis, D. Ernst and G. Andersson, 'Global power grids for harnessing world renewable energy' in L. E. Jones (ed.), *Renewable Energy Integration: Practical Management of Variability, Uncertainty and Flexibility in Power Grids* (London: Elsevier Academic Press, 2014), pp. 175–88.

[22] J. Riesz, 'China leading the way to a global renewable grid', http://reneweconomy.com.au/2014/china-leading-way-global-renewable-grid-2-84674 (accessed 19 September 2014); Z. Liu, *Global Energy Interconnection* (London: Elsevier, 2015).

peak demand periods between continents owing to time-zone differences, remote RES plants located at similar distances from two regions can connect and sell their power always at peak price. Third, the time-zone difference between the continents creates opportunities for electricity arbitrage, which can lead to profitable electricity trade. Based on these drivers, the next paragraphs present the three main stages we envision as leading to the development of a Global Grid environment.

II Harvesting RES from Remote Locations

In this section we present two examples of harvesting renewable energy potential from remote locations.

In an earlier publication we described how tapping the renewable potential in Greenland can be a realistic example of progress towards global interconnections.[23] Greenland was selected as a representative example for three reasons. First, it has a significant wind and hydro potential.[24] Second, it is close to Iceland, and Hammons et al. have already shown that the Iceland–UK interconnection is a viable option.[25] The two governments have already been discussing its possible realisation.[26] Third, all interconnecting sections along this route have lengths or sea depth comparable to currently existing projects.[27] Therefore, we expect that Europe can benefit from importing wind and hydro power harvested in Greenland. Assuming that a wind farm in Greenland would be feasible, in the next sections we will present a more detailed analysis of how intercontinental trade could be a viable option in a transmission route over Greenland.

Similar to the case of Greenland is that of a wind farm on the Kerguelen islands.[28] The Kerguelen islands are located in the Indian

[23] Chatzivasileiadis, Ernst and Andersson, 'The Global Grid'.

[24] See Jacobson and Delucchi, 'Providing all global energy with wind, water, and solar power, Part I'; R. Partl, *Power from Glaciers: The Hydropower Potential of Greenland's Glacial Waters*, IIASA Research Report RR-77-020 (1977).

[25] T. Hammons, A. Olsen, P. Kacejko and C. Leung, 'Proposed Iceland/United Kingdom power link – an in-depth analysis of issues and returns', *IEEE Transactions on Energy Conversion*, 8(3) (1993), 566–75.

[26] D. Carrington, 'Iceland's volcanoes may power UK', *The Guardian* (11 April 2012), www.guardian.co.uk/environment/2012/apr/11/iceland-volcano-green-power/ (accessed 16 September 2014).

[27] For more details, see Chatzivasileiadis, Ernst and Andersson, 'The Global Grid'.

[28] Chatzivasileiadis, Ernst and Andersson, 'Global power grids for harnessing world renewable energy'.

Ocean, approximately equidistant from South Africa and Australia (~4,100 km), and they experience high and continuous winds. Based on the published wind data, onshore wind farms on Kerguelen Islands are expected to have a capacity factor of 60–70 per cent. In our analysis we have accounted for three different wind capacity factors – 40, 50 and 60 per cent – and two different submarine-cable cost projections. Assuming that the wind farm revenues are generated by selling its power to South Africa and receiving the feed-in tariffs determined by the South African government,[29] in our study we determine the maximum wind farm production costs so that the wind farm, including the submarine-cable investment, remains profitable. We find that for a capacity factor of 60 per cent, if its production cost is below 0.085 USD per kWh, the wind farm on Kerguelen Islands can compete with wind farms located in South Africa. Currently, such production costs can be achieved by large wind farms. Assuming a lower capacity factor of 40 per cent, the wind production costs must not exceed 0.033–0.055 USD/kWh. Taking into account that Delucchi and Jacobson project the production costs for onshore wind farms to start below 0.040 USD/kWh,[30] values such as 0.033/kWh could be highly probable by 2020 and beyond.

III Interconnecting Two Continents over Remote RES Locations

In our earlier analysis, we assumed that a 3 GW wind farm off the eastern shores of Greenland was feasible and noted that some investors have decided to connect this wind farm with a 3 GW line to Europe through Iceland and the Faroe Islands.[31] We investigated whether a connection to North America would be profitable, taking into account two effects. First, owing to the time-zone difference, the wind farm will always be able to sell the power it produces at peak prices, e.g., 50 per cent of the time to Europe and 50 per cent of the time to the United States. Second, by creating a link between Europe and the United States over Greenland, opportunities for electricity trade between the continents emerge. As the

29 Wikipedia, 'Energy in South Africa' (2013), http://en.wikipedia.org/wiki/Energy_in_South_Africa (accessed 17 December 2013).

30 M. A. Delucchi and M. Z. Jacobson, 'Providing all global energy with wind, water, and solar power, Part II: reliability, system and transmission costs, and policies', *Energy Policy*, 39(3) (2011), 1170–90.

31 See Chatzivasileiadis, Ernst and Andersson, 'The Global Grid'.

wind farm can produce power only for a limited time (we assume a capacity factor of 40 per cent), the cable capacity can be used for electricity trade for the remaining hours.

1 Offering RES at Peak Prices

In our earlier analysis, we estimated that a 3 GW cable can deliver about 20 TWh per year, taking into account the transmission losses.[32] Of this, about 10 TWh is allocated to wind farm production. This means that for about 50 per cent of the time the cable capacity is available for electricity trade.

We found that the costs per delivered kWh for the wind farm by building the Greenland–USA transmission route would increase by 21–25 per cent due to additional investment costs. On the other hand, if off-peak prices are half of peak prices, revenues will increase by 31–33 per cent if the wind farm always sells the power it produces at peak price – 50 per cent of the time to Europe and 50 per cent of the time to the United States. This results in additional profits of 7–12 per cent for the wind farm, as shown in Table 2.1.

2 Intercontinental Electricity Trade

Besides selling the wind power produced, 10 TWh of the cable capacity is available for electricity trade between the continents. Our analysis of the electricity trade opportunities is based on real price data for 2012. We obtained the hourly spot prices from the European Power Exchange (EPEX)[33] in Germany and the PJM Interconnection[34] in the United States. Owing to the time-zone difference between the two continents, the two electricity markets experience their peak and lowest prices at different times.[35] We mainly examine two levels of utilisation: 30 per cent and 50 per cent. Our investigations show that, through the revenues generated from the electricity trade, the route between Greenland and North America can be amortised within 10–12 years if the line is used 50 per cent of the time, and within 14–17 years if the cable is used for electricity trade only 30 per cent of the time. Translating these results into profits, we find that by conducting trade in electricity for 30 per cent

[32] Ibid.

[33] European Power Exchange, www.epexspot.com (accessed 20 December 2016).

[34] PJM Interconnection LLC, www.pjm.com (accessed 20 December 2016).

[35] For more details, see Chatzivasileiadis, Ernst and Andersson, 'Global power grids for harnessing world renewable energy'.

Table 2.1 *Wind farm in Greenland: Summary of the cost–benefit analysis of connecting the wind farm to the United States*

Transmission route: Europe–USA over Greenland; total cable energy capacity: 20 TWh per year			
	Wind farm production	Electricity trade	
Utilisation (% of total time)	~10 TWh (40%)[a]	~6 TWh (30%)	~10 TWh (50%)
Profit increase	7–12%	24–27%	39–42%

[a] The wind farm is located in the middle of the Europe–United States path. As a result, it incurs only half of the transmission losses. That means that the same amount of power, e.g., 10 TWh, can be delivered over a shorter distance, resulting in a lower utilisation factor of the transmission path.

of the time, the net profits[36] will increase by 24–27 per cent compared with what the wind farm can achieve from selling its wind power only to the United Kingdom. If electricity trade is carried out for 50 per cent of the time, the profits may increase by up to 42 per cent. Table 2.1 summarises our results.

In conclusion, it seems that a connection to both continents would be a profitable strategy for the wind farm in Greenland. As detailed in this section, we investigated two possible sources of income created by building a transmission route from Europe to the United States over a wind farm in Greenland. The first was always selling the wind power produced at peak prices, either in Europe or in the United States, and the second was trading electricity between the continents. Although we expect that the transmission route will be used for both options at the same time, we have shown that both options result *independently* in the profitable operation of the wind farm.

IV Intercontinental Interconnections by Direct Lines

In the section, we examine the direct connection of two continents by transmission lines. In an earlier analysis we considered the expected

[36] The investment costs of the additional cable have been deducted from the electricity trade revenues.

transmission costs per delivered kWh for a direct submarine cable between Europe and the United States.[37] We estimated the cost of a 5,500 km, 3 GW submarine cable to be in the range of 0.0166 euros to 0.0251 euros per kWh delivered. We ascertained that, except for the most expensive RES generators, it would apparently be more economical for the United States to import RES power from Europe than to operate its own fossil-fuel power plants.[38]

We then extended our analysis to find out whether the cost for a long submarine cable could be amortised through the revenues arising from the electricity trade between the two continents.[39] Again, we used the hourly spot prices for 2012 provided by PJM in the United States and EPEX in Germany. In our calculations we accounted for the transmission losses incurred by an 8,000 km-long corridor connecting the United States with Germany. We estimate that for an 80 per cent utilisation of the cable (i.e. the cable is used only 80 per cent of the time), the amortisation period ranges from 18 to 28 years. In the less favourable case where cable utilisation is 50 per cent, the amortisation period increases to about 23–35 years. Although such amortisation periods might not be very appealing for private investors, these results highlight that from a social welfare point of view, such a cable is beneficial to society.

V *The Global Grid: An Illustration*

Intercontinental interconnections, such as those discussed above, can lead to a globally interconnected network. We envision that the power supply of the Global Grid will depend on RES. Large renewable potential exists in remote locations, such as in deserts and offshore. Long HVDC lines will constitute the main corridors in a Global Grid environment, transmitting bulk quantities of power over long distances. HVDC lines are superior to AC technologies for long-distance transmission, as they

37 Chatzivasileiadis, Ernst and Andersson, 'The Global Grid'.

38 The result is based on electricity generation from conventional sources estimated by Delucchi and Jacobson, 'Providing all global energy with wind, water, and solar power, Part II'. Fuel cost projections are based on 2009 US Energy Information Administration data. Unconventional sources of oil and gas such as oil sands and natural gas shales are also considered in the US Energy Information Administration's projections of the same year.

39 See Chatzivasileiadis, Ernst and Andersson, 'Global power grids for harnessing world renewable energy'.

exhibit lower losses, provide active and reactive power support and can connect non-synchronous grids.[40]

D Merchant Transmission Investment

In the previous sections we discussed the need for substantial reinforcements of transmission lines to decrease the need for balancing power, while also presenting results for the economic feasibility of new lines connecting different markets over long distances, e.g., Europe with the United States. Before moving on to the different market operation structures for electricity trade, we discuss the investment schemes for transmission infrastructure, as their profits can depend on these market structures. Besides regulated investment, which is the most common type of transmission investment, here we focus on merchant transmission investment (MTI). MTI profits derive mainly from electricity trading, and therefore the line's profitability is dependent on the market-operation structure of the points connected by the line.

MTI, unlike regulated investments, does not need to be coordinated by a central authority and is not subject to regulation, and its profits derive mainly from trading electricity between the two areas that the transmission line connects. As a result, any interested party, including the transmission system operator (TSO) or transmission owners, can undertake a merchant investment as long as the difference in the electricity prices between the two areas allows for a profit margin.[41]

A comparison of transmission investment schemes concluded that different transmission investment models successfully tackle different inefficiencies but also have their own shortcomings.[42] The literature concludes that developing good regulatory mechanisms that will also provide opportunities for merchant investors to develop projects seems to be a good approach, but at the same time poses a significant research challenge. As Brunekreeft suggests,[43] controllable flows, namely HVDC lines and flexible AC transmission systems (FACTS), can operate

40 More details on power generation and transmission through the Global Grid are published in Chatzivasileiadis, Ernst and Andersson, 'The Global Grid'.

41 For more details, see S. Chatzivasileiadis, *Transmission Investments in Deregulated Electricity Markets*, Technical Report, ETH Zurich (2012), and the references therein.

42 Ibid.

43 G. Brunekreeft, 'Market-based investment in electricity transmission networks: controllable flow', *Utilities Policy*, 12(4) (2004), 269–81; P. Joskow and J. Tirole, 'Merchant transmission investment', *The Journal of Industrial Economics*, 53(2) (2005), 233–64.

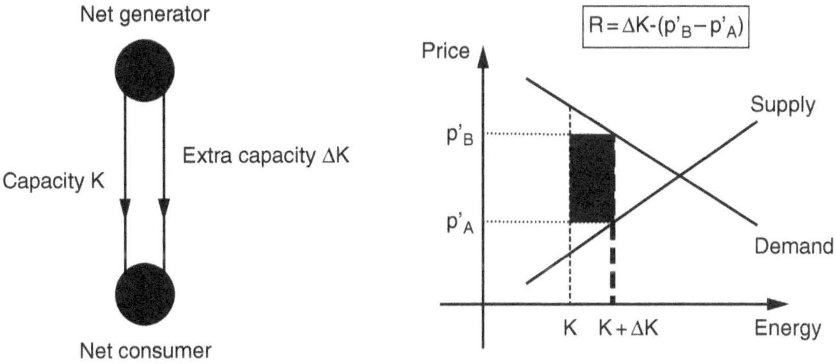

Figure 2.1 Illustration of the revenue generation for the merchant transmission investment scheme

sufficiently well under an MTI regime, overcoming several of its inefficiencies. For example, BritNed and NorNed[44] are DC interconnections that successfully follow the MTI model.

Concerning merchant transmission investments, owing to the benefits that interconnections offer, more such projects are expected to follow in the coming years, connecting different regions and electricity markets. Owing to the size of the regional electricity markets, it could be envisioned that different owners will construct parallel lines, thus facilitating competition not only between the areas, but also between the lines.[45] For example, assuming controllable flows, if a contracted energy transfer between points A and B can flow along two different paths, the corresponding line owners could compete over which path the energy follows.

E Cross-Border Electricity Trade: State of Play

Cross-border electricity trading depends to a certain extent on the market structure of the individual regions that are connected with inter-tie lines. In this section we provide a summary of the most common electricity-market operating schemes, which can be implemented on a

44 www.tennet.eu/nl/grid-projects/international-projects/norned.html (accessed 20 December 2016); www.britned.com/ (accessed 20 December 2016).

45 Peak electricity demand is in the range of several hundred GW for each region (e.g., Europe, North America or China), while the capacity of an interconnector is expected to be around a couple of GWs.

regional basis. We then describe the cross-border capacity allocation mechanisms that lead to the coupling of the regional markets.

I Bilateral Trading

Bilateral trading takes place in most electricity markets, e.g., in the United States and Europe. Under this scheme a seller and a buyer agree on the exchange of a certain amount of energy for a certain amount of time at a certain price. Bilateral contracts can be hourly, weekly, monthly or longer-term, or they can be for peak or off-peak periods during a day. The advantage of bilateral trading is that it does not need a central market authority. Currently, in California, there are also proposals to abolish the day-ahead energy market, keeping only the fifteen-minute real-time unit commitment and the five-minute real-time balancing market (economic dispatch), and allowing day-ahead transactions to be carried out through bilateral trading.

II Power Exchange

In a power-exchange market structure, sellers and buyers participate in bidding over price and the quantity they can offer (or that they need, in the case of buyers). The market operator clears the market, ensuring that the demand can be covered by the supply, dispatching the generators with the lowest bids. Although a power-pool market structure provides for electricity consumers' participation, market regulations usually allow only large consumers to bid in the market. Small consumers usually pay a pre-agreed price and might be represented by load-serving entities. Network constraints are not considered in a power exchange, but should be taken into account by the system operator. Where the market outcome violates network constraints, re-dispatching is necessary. Power pools are currently operating, for example, in Australia.[46] Market clearing in Australia takes place every five minutes, and six dispatch prices are averaged every half-hour to determine the spot price for each trading interval for each of the regions of the Australian National Electricity Market: Queensland, New South Wales, Tasmania, Victoria and South Australia.

[46] http://en.wikipedia.org/wiki/National_Electricity_Market (accessed 20 December 2016).

III Locational Marginal Pricing

In locational marginal pricing (or nodal pricing), the market operator receives bids (pairs of quantity and price) from power sellers and buyers and clears the market, taking into account network constraints. The price that a generator receives varies according to its location and is also dependent on transmission losses and transmission congestion. The advantage of locational marginal pricing over a power pool is that it takes into account network constraints. As a result, the market outcome satisfies the need for reliable operation of the grid, while proper price signals are generated, taking into account congested transmission lines. On the other hand, the algorithm to solve such an optimisation problem is more complex than the economic dispatch taking place in the power exchange, as it needs a full representation of the network. Furthermore, locational marginal pricing allows generators that are located downstream of congested lines to attain market power. Nodal pricing is currently implemented in all six organised markets in the United States,[47] as well as in Singapore, among others. Poland is also in the process of implementing a nodal pricing market.

The market clearing varies in each market. In Singapore it is every thirty minutes. In California (the California Independent System Operator, or CAISO) there is a day-ahead market as well as a real-time market which clears every fifteen minutes for the unit commitment and every five minutes for real-time balancing (economic dispatch). In most US regions where an organised market exists, the market clearing mechanism co-optimises the dispatch of energy and reserves.

IV Zonal Pricing

Zonal pricing – in accordance with nodal pricing – establishes different electricity prices for different locations in the network.[48] In contrast to nodal pricing, where, in the event of congestion, prices might differ for every node, for zonal pricing a group of nodes is aggregated to one zone. These zones are mostly defined a priori as the concept focuses on certain flowgates, which may be subject to congestion. An example is the

[47] ISONE, NYISO, PJM, MISO, SPP, CAISO.

[48] T. Krause, 'Evaluating congestion management schemes in liberalized electricity markets applying agent-based computational economics', PhD thesis, ETH Zurich, Switzerland (2007).

Norwegian system, where the system operator splits the national transmission system into two zones (north and south) in the event of congestion. If the demand for transmission services does not exceed system capabilities, different network zones are not established, and thus there is only one clearing price for the whole network. The European electricity grid also has a zonal-pricing market structure.

V *Ancillary Services/Reactive Power Markets*

Besides energy markets, there also exist ancillary services markets. Regulating energy, spinning and non-spinning reserves are traded in day-ahead and real-time markets in both the European and US electricity markets. Switzerland and Austria serve as an example of a coupled ancillary services market, as each region is eligible to bid for ancillary services from the neighbouring region.

Switzerland also operates a reactive power market, allowing hydro generators in the distribution grid to offer reactive power and receive compensation for these services in order to keep the voltage level at certain buses in the transmission grid within secure limits.[49]

VI *Cross-Border Transmission Capacity-Allocation Mechanisms*

Special transmission capacity-allocation mechanisms have been developed to couple neighbouring markets and reserve transmission capacity in the inter-ties, also known as cross-border lines. In Sections VII and IX we distinguish between two main capacity-allocation mechanisms.

VII *Explicit Auctions*

In explicit transmission-capacity auctions, bids are accepted to reserve transmission capacity on the inter-ties. To make use of this capacity, however, the bidder also needs to bid in an energy market, e.g. a power exchange. Thus, the electricity trade following this scheme requires two steps: first, reserving the necessary transmission capacity in the interface to transmit power, and second, participating in an energy market to buy or sell the necessary amount of energy.

[49] M. Zerva and M. Geidl, 'Contribution of active distribution grids to the coordinated voltage control of the Swiss transmission system', 18th Power Systems Computation Conference (PSCC), Wroclaw, Poland, 18–22 August (2014).

VIII Implicit Auctions: Flow-Based Allocation

The flow-based allocation mechanism is currently in use in central Europe (e.g. the Central West European Region, or CWE). This method couples the electricity trade with the physical flows resulting from it. A centralised auction office for the whole flow-based allocation (FBA) region is established, which maintains a simplified model of the power system under its responsibility. All interested parties submit bids to this office. FBA has several advantages, as it takes into account the loopflows induced from the trade (in the uncontrollable alternating current, or AC, lines), allows for inter-regional trade, allocates energy contracts and transmission capacity at the same time and supports the integration of regional power markets towards a single market.[50]

In the European context, each zone is usually a single country or a TSO control area. For the cross-border flow-based allocation method, no detailed modelling of the network inside each zone is necessary. However, TSOs derive power-transfer distribution factors in order to represent the network on the interfaces, that is, the inter-zonal interconnections. This information is then maintained by the central auction office.

IX Regions with No Electricity Markets, Or with Electricity Markets under Development

Not all world regions have developed electricity markets. Arab states belonging to the Gulf Cooperation Council[51] (GCC) plan to form a common wholesale market. The GCC Interconnection Authority will act as a gateway towards achieving a regional and pan-Arab power pool, becoming a broker between regional and international utilities and regulators.[52] Similarly, regions in the United States, such as the US Pacific Northwest region – which is under the responsibility of Bonneville Power Authority – have no organised market.

[50] More information on how flow-based allocation avoids loopflows is provided in Chatzivasileiadis, *Transmission Investments in Deregulated Electricity Markets*, and the references therein.

[51] GCC includes Saudi Arabia, Kuwait, Bahrain, Qatar, the United Arab Emirates, and the Sultanate of Oman.

[52] H. K. Al-Asaad, 'Electricity power sector reform in the GCC region', *The Electricity Journal*, 22(9) (2009), 58–64.

F Challenges in Cross-Border Electricity Trade

As we have seen in previous sections, transmission reinforcements will be necessary to smooth out the variability in the supply of renewables and in electricity demand. In our analysis, we also showed that interconnecting regions over longer distances can be technically feasible and an economically competitive option. At the same time, interconnecting different regions requires the coupling of the respective electricity markets to facilitate electricity trade and power exchange.

Several challenges arise from coupling regional electricity markets. We distinguish below between the political and regulatory challenges, and the challenges pertaining to the different market structures and investment schemes.[53]

I Political and Regulatory Challenges

By extending the markets over wider regions, load centres are expected to become increasingly dependent on imported energy. This raises at least two questions. The first concerns the security of supply. What is the impact of importing energy, for instance, from Russia or the African deserts? Desertec, which proposed building concentrated solar-power plants in the Sahara region and transporting cheap renewable energy to Europe, showed that Europe could gain significant economic benefit from such an investment. However, following the unrest in the Arab countries, the concept has lost support. This problem is similar to those facing other energy carriers, such as oil or gas. The difference with electricity is, however, that it cannot be stored.[54] This has both benefits and drawbacks. On the one hand, the importing country cannot be energy-independent unless it invests in its own generation capacity or in different interconnection paths in order to diversify the pool of its energy suppliers. On the other, from the viewpoint of the exporting country, renewable energy that cannot be stored will be lost, resulting in substantial economic loss. A regulating mechanism needs to be found that will share the risks of an energy 'embargo' between an exporting and an importing country.

[53] The technical challenges are discussed in Chatzivasileiadis, Ernst and Andersson, 'The Global Grid'.

[54] At least for the moment, there is no technology that can store large amounts of electricity in an efficient way.

A further challenge relates to the cost of imported electricity. Assuming that different countries follow different policies with respect to environmental taxes, e.g., the carbon tax, the cost of energy will differ. To illustrate this challenge let us assume that cross-border electricity trade takes place between Russia and the EU countries. We further assume that the EU has imposed taxes on all conventional generation, so that production from renewable technologies in Europe is cheaper than gas, nuclear and coal generation. On the other hand, we assume that Russia has low-cost gas-fired power plants, and does not impose any environmental taxes on CO_2 or other emissions. In such a case, importing electricity from Russia may be cheaper than producing electricity from renewables in power plants in Europe. Several questions arise. Will this imported energy count towards meeting the environmental targets set by the EU? Should there be an import tax, similar to the tolls that are imposed for the import of goods? Lower-cost energy would be beneficial for the European economy, but what is the economic impact of the environmental damage in Russia or globally?

An additional challenge relates to the development of microgrids. Drivers such as a more resilient grid and decreased distribution losses lead to the installation of more microgrids. Local generation can act both as an alternative and as a complement to large centralised generation and bulk power transmission. Regulation plays a significant role in this case too. How will the costs of rooftop PV systems and batteries and the pricing scheme for distribution tariffs evolve? If local generation becomes abundant, the energy flowing on the distribution network will significantly decrease, resulting in a significant drop in revenue for the distribution companies. Depending on how the transmission and distribution tariffs are regulated, microgrids might opt to disconnect and operate off-grid in an effort to avoid the grid tariffs.[55] How will this affect the transmission network and cross-border electricity trade? And what will be the impact on the available budget – and need – for transmission investments? On the other hand, as long as the microgrids remain

[55] At the moment the cost of battery storage is significant for large energy capacities. Given that solar irradiation in central Europe is about three times higher during the six sunniest months (April–September) than in the six least sunny ones (October–March), the necessary battery capacity for seasonal storage and the associated cost is very high. This will, therefore, be the main reason at present for deciding against disconnecting the microgrids; however, technological (or regulatory) developments might change this in the future.

connected, proper regulating mechanisms will be necessary for electricity trade where microgrids can also participate as entities.

The widespread development of microgrids will also depend on the costs of installing new distributed-generation power plants. Although the cost of distributed generation has significantly fallen in recent years, Dinica[56] suggests that in the long run a flattening-out or a U-shaped curve for production costs would be more likely. For wind generation, she notes that '[in Spain] the increases in technological performances and installed capacity per turbine did not compensate for the decrease in resource quality and availability', and this resulted in an increase of the wind power production costs from 5.1–6.5 euro cents per kWh in 2005 to an estimated 5.6–7.0 euro cents per kWh in 2010. On the other hand, this trend is not currently observed for solar PV panels, owing to economies of scale and significant technological advances. As long as the trend that Dinica suggests holds for both wind and solar PV in the future, we expect that local generation will act as a complement rather than as a complete alternative to bulk renewable generation. Both will be necessary so that small residential and large industrial and commercial loads can be equally supplied with 'green' energy.

Last but not least, coordination of activities between the regional regulators and a common coordinating body will be necessary. For example, in North America, the North American Electric Reliability Council (NERC) is a regulatory authority whose mission is to ensure the reliability of the bulk power system in Canada, the United States and part of Baja California in Mexico.[57] Similarly, efforts towards an internal European market have led to the Agency for the Cooperation of Energy Regulators (ACER) recently being formed in Europe.[58] As long as regions extend beyond the central European or the North American boundaries, e.g., with interconnections to Asia or Latin America respectively, these bodies should also be extended.

II Challenges in Market Operation

Considering cross-border flows and market structures, the first question is whether we need a more centralised or decentralised market operation scheme. Europe seems to be moving to a more centralised structure with

[56] V. Dinica, 'Renewable electricity production costs – a framework to assist policy-makers' decisions on price support', *Energy Policy*, 39 (2011), 4153–67.

[57] www.nerc.com (accessed 20 December 2016).

[58] www.acer.europa.eu (accessed 20 December 2016).

flow-based allocation mechanisms. California, by contrast, has recently discussed maintaining only bilateral trading for day-ahead trading in the future.

Taking into account that an HVDC overlay grid connecting different market regions may exist in the future, if we opt for more centralised coordination of the electricity trade, various options are available. One possible option is a hierarchical market model, equivalent to the schemes currently proposed for a flow-based allocation mechanism. In this case we distinguish the DC grid from the underlying regional AC networks. An interregional market can be formed in which each regional market participates as an individual player.[59] Here, considerations should be made in a fair and transparent way so that the HVDC flows can be calculated. The advantage of such a scheme is that the structure of each regional market probably does not need to be significantly modified.

Another alternative is a more 'horizontal' operation of the interregional market. For this option, two possibilities should be distinguished. First, an operation similar to the current power exchanges could be envisioned. The second possibility is more complex, in which the interregional market operator, taking into account the whole grid and local grid constraints, clears the market and determines the power flows. Here, as in the case of the hierarchical model, a proper mechanism for calculating the flows must be designed. Such a scheme, however, introduces significant complexity and raises questions, such as whether to opt for a nodal pricing scheme or a zonal scheme in which several nodes are aggregated to a node.

A solution determined through an optimal power flow maximises social welfare and determines the dispatch of the generators and the flow on the HVDC lines.[60] The pricing of the multi-terminal HVDC nodes, however, remains an open question. The locational marginal pricing concept does not apply as well to HVDC grids, given that the HVDC flows are controllable.

Challenges arise also from the introduction of new investment schemes. As already mentioned, the MTI model applies well to HVDC lines. Given

[59] Individual players need not be the size of regional markets. They can also comprise a certain group of entities (above a certain MW level of production or consumption) that belong to the same energy market.

[60] For information on technical implementation of an optimal power flow in multi-area AC and HVDC grids under different operation schemes, the interested reader can refer to E. Iggland, R. Wiget, S. Chatzivasileiadis and G. Andersson, 'Multi-area DC-OPF for HVAC and HVDC grids', *IEEE Transactions on Power Systems*, 30(5) (2015), 2450–9.

the numerous advantages of such lines for long-distance transmission, we expect that in the future several HVDC lines will form an HVDC grid, and some of these HVDC lines could be under the MTI model. In such an environment, several operational and regulatory questions arise. One could envision competition between parallel HVDC lines. Do merchant transmission investments need to be regulated, and if so, how? Additionally, assuming that an HVDC line connects a nodal pricing market with a zonal market, how can the revenues of the line be calculated? And what happens if an HVDC line extends over a whole country? For example, if country A is adjacent to country B, and country B is adjacent to country C, what happens if we build a line that directly connects country A with country C passing through country B? With the HVDC technology, in order for country B to have access to the transmitted power, a multi-terminal converter is necessary. If no such converter is available and no trade with country B takes place, is country B somehow eligible for part of the profits generated by the HVDC line just because it offers rights of way?

G Discussion and Conclusions

Aiming at a fully renewable system, as we shift from fully dispatchable conventional generators to fluctuating RES, it is necessary to rethink how to design, plan and operate the grid. Efficient storage and transmission reinforcements are two of the most important factors required to balance renewable supply and demand and maintain a reliable electricity supply while avoiding carbon emissions. Reinforcing transmission interconnections and extending the grid over larger regions reduces the necessary balancing power, especially for wind generators, that should otherwise be provided from conventional generators. At the same time, grid expansion allows the harvesting of renewable potential abundant in remote locations, such as offshore or in deserts. As larger regions become interconnected, mechanisms for transparent and efficient cross-border energy trading should be put in place to facilitate the power exchange between regions.

In this chapter, we first summarised our previous work on the feasibility of long-distance interconnections from an economic perspective. We envisioned that such interconnections could form the basis of a Global Grid.[61] We showed that connecting Europe with the United States over a wind farm in Greenland can be profitable in two ways: first, by always selling the wind power at peak prices, half of the time to Europe and half of

[61] See Chatzivasileiadis, Ernst and Andersson, 'The Global Grid'; Chatzivasileiadis, Ernst and Andersson, 'Global power grids for harnessing world renewable energy'.

the time to the United States, and second, through electricity trade between the regions. We found that the wind farm's profits could increase by 7–12 per cent in the first case and by up to 42 per cent in the second case. We further investigated direct interconnections between Europe and the United States through submarine cables. Our analysis was based on real price data from Germany and PJM in the United States. We found that the amortisation period for such a cable would lie between eighteen and thirty-five years, assuming 50–80 per cent utilisation of the cable.

Next, we focused on the challenges emerging from cross-border electricity trade when interconnecting larger regions. To facilitate this, first we provided a short discussion on investment schemes for new transmission lines, focusing on the merchant transmission investment model. This is a scheme that can apply well to the installation of new HVDC lines, and which is dependent on the market structure of the regions connected by the line. Subsequently, we focused on the main market-operating schemes for electricity trade in different regions, as well as the market-coupling schemes and the transmission capacity-allocation mechanisms for cross-border lines. We summarised the explicit capacity-allocation and flow-based allocation schemes.

We then elaborated the numerous challenges that emerge from long-distance interconnections and the coupling of several regional markets. We distinguished between political and market operation challenges. We discussed regulatory matters, such as the cost of imported electricity if the exporting and importing regions follow different policies with respect to environmental taxes. We further mentioned challenges pertaining to the market operation in the presence of an HVDC overlay grid and the coupling of nodal and zonal markets, as well as the regulatory and operational framework of an MTI coupling different markets.

In conclusion, we note that coupling several regional markets with the long-term vision of creating a global electricity market environment presents several challenges. Identifying these challenges and working on solutions is the first step towards the creation of larger electricity markets, leading to increased competition, lower electricity prices and potentially greater security of supply.

Bibliography

Aboumahboub, T., Schaber, K., Tzscheutschler, P. and Hamacher, T. (2010). 'Optimization of the Utilization of Renewable Energy Sources in the Electricity Sector', in *Proceedings of the 5th IASME/WSEAS International Conference on Energy & Environment* (EE '10).

Al-Asaad, H. K. (2009). Electricity power sector reform in the GCC region. *The Electricity Journal*, **22**(9), 58–64.

Becker, S., Rodriguez, R. A., Andresen, G. B., Schramm, S. and Greiner, M. (2014a). Transmission grid extensions during the build-up of a fully renewable pan-European electricity supply. *Energy*, **64**, 404–18.

Becker, S., Frew, B. A., Andresen, G. B., Zeyer, T., Schramm, S., Greiner, M. and Jacobson, M. Z. (2014b). Features of a fully renewable US electricity system: optimized mixes of wind and solar PV and transmission grid extensions. *Energy* **72**, 443–58.

Boute, A. and Willems, P. (2012). RUSTEC: greening Europe's energy supply by developing Russia's renewable energy potential. *Energy Policy*, **51**, 618–29.

Brunekreeft, G. (2004). Market-based investment in electricity transmission networks: controllable flow. *Utilities Policy*, **12**(4), 269–81.

Carrington, D. (2012). 'Iceland's volcanoes may power UK', *The Guardian*, 11 April, www.guardian.co.uk/environment/2012/apr/11/iceland-volcano-green-power (accessed 16 September 2014).

Chatzivasileiadis, S. 2012. *Transmission Investments in Deregulated Electricity Markets*, Technical Report, ETH Zurich.

Chatzivasileiadis, S., Ernst, D. and Andersson, G. (2013). The Global Grid. *Renewable Energy*, **57**, 372–83.

(2014). Global power grids for harnessing world renewable energy. In L. E. Jones, ed., *Renewable Energy Integration: Practical Management of Variability, Uncertainty and Flexibility in Power Grids*. London: Elsevier Academic Press, pp. 175–88.

Czisch, G. (2011). *Scenarios for a Future Electricity Supply: Cost-Optimized Variations on Supplying Europe and Its Neighbours with Electricity from Renewable Energies*. Stevenage: Institution of Engineering and Technology (IET).

Delucchi, M. A. and Jacobson, M. Z. (2011). Providing all global energy with wind, water, and solar power, Part II: reliability, system and transmission costs, and policies. *Energy Policy*, **39**(3), 1170–90.

Deutsche Energie-Agentur (dena). (2010). *dena Grid Study II – Integration of Renewable Energy Sources in the German Power Supply System from 2015–2020 with an Outlook to 2025*. Berlin: German Energy Agency.

Dinica, V. (2011). Renewable electricity production costs – a framework to assist policy-makers' decisions on price support. *Energy Policy*, **39**, 4153–67.

European Commission. (2008). *Impact Assessment on the EU's Objectives on Climate Change and Renewable Energy*. Brussels: European Commission.

Hammons, T., Olsen, A., Kacejko, P. and Leung, C. (1993). Proposed Iceland/United Kingdom power link – an in-depth analysis of issues and returns. *IEEE Transactions on Energy Conversion*, **8**(3), 566–75.

Hand, M. M., Baldwin, S., DeMeo, E., Reilly, J. M., Mai, T., Arent, D., Porro, G., Meshek, M. and Sandor, D. (eds.) (2012). National Renewable Energy

Laboratory, Renewable Energy Futures Study, *4 vols.* (NREL/TP-6A20-52409). Golden, CO: National Renewable Energy Laboratory, www.nrel.gov/analysis/re_futures/ (accessed 20 June 2012).

Iggland, E., Wiget, R., Chatzivasileiadis, S. and Andersson G. (2015). Multi-area DC-OPF for HVAC and HVDC grids. *IEEE Transactions on Power Systems*, **30**(5), 2450–9.

Jacobson, M. Z. and Delucchi, M. A. (2011). 'Providing all global energy with wind, water, and solar power, Part I: technologies, energy resources, quantities and areas of infrastructure, and materials. *Energy Policy*, **39**(3), 1154–69.

Joskow, P. and Tirole, J. 2005. Merchant transmission investment. *The Journal of Industrial Economics*, **53**(2), 233–64.

Krause, T. 2007. *Evaluating congestion management schemes in liberalized electricity markets applying agent-based computational economics.* Ph.D. thesis, ETH Zurich, Switzerland.

Liu, Z. (2015). *Global Energy Interconnection.* London: Elsevier Academic Press.

Mills, A. D. and Wiser, R. H. (2013). Changes in the economic value of photovoltaic generation at high penetration levels: a pilot case study of California. *IEEE Journal of Photovoltaics*, **3**(4), 1394–1402.

Partl, R. (1977). Power from Glaciers: The Hydropower Potential of Greenland's Glacial Waters, IIASA Research Report RR-77-020.

State of California. (2011). Senate Bill X1-2 (sbx1 2).

Taggart, S., James, G., Zhaoyan, D. and Russell, C. (2012). The future of renewables linked by a transnational Asian Grid. *Proceedings of the IEEE*, **100**(2), 348–59.

Wikipedia. (2013). 'Energy in South Africa', http://en.wikipedia.org/wiki/Energy_in_South_Africa (accessed 17 December 2013).

Zerva, M. and Geidl, M. (2014). 'Contribution of active distribution grids to the coordinated voltage control of the Swiss transmission system', 18th Power Systems Computation Conference (PSCC), Wroclaw, Poland, 18–22 August.

3

Technological Challenges and New Frontiers in International Electricity Trade

ANDERS PLEJDRUP HOUMØLLER

A Introduction: Why a Greening Economy Requires Improved International Trade in Electricity

In 2013, electricity produced by wind turbines constituted 34 per cent of Denmark's total electricity production. During the first six months of 2014, wind energy's share in total Danish electricity production was 43 per cent. These percentages are the highest in the world.

Denmark is divided into two asynchronous electricity areas: western Denmark and eastern Denmark. A 600 MW direct current (DC) link connects the two areas. Table 3.1 gives data for western Denmark, where most of the Danish wind turbines are situated.

As can be inferred from Table 1, during periods of strong wind, Denmark – and especially western Denmark – is forced to export electricity. Normally, the cross-border energy flow will go towards Denmark during periods in which there is little wind.

The Danish experience is reflected in much of Europe: a high percentage of renewable energy requires a well-functioning, multinational electricity market. This gives us a flexible electricity system able to handle intermittent energy production.

It may be possible to create a more integrated electricity market by building more grid. However, building more grid is of little value unless the organisation of the cross-border energy trade is efficient. This can be

Table 3.1 *Approximate loads and wind turbine capacity for western Denmark (2014)*

Approximate maximum load (MW)	Approximate minimum load (MW)	Installed wind turbine capacity as of 1 January 2014 (MW)
3,600	1,200	3,788

illustrated by the following example: on Christmas Day 2012, wind turbines created a big production surplus in western Denmark. This production surplus was foreseen in the day-ahead plans. Nevertheless, the day-ahead plans created a huge flow of electricity from Germany towards western Denmark. This was caused by an error in the system used to make the day-ahead plans.

The remainder of this chapter will focus on day-ahead congestion management. This is because the Danish experience has amply illustrated that the organisation of cross-border electricity trading is a crucial component of a 'green' electricity supply system.

The chapter goes on to describe the European day-ahead congestion management method known as 'market coupling'. With market coupling, the electricity exchanges' day-ahead prices are used to set the day-ahead plans for cross-border energy flows. Market coupling is an important step towards an integrated electricity market. However, as we shall see, the current European market coupling is neither cost-efficient nor customer-oriented. This is a problem for the planned greening of the European electricity supply system.

B The European Concept of a Spot Market

In the European electricity supply business, a *spot market* is a market operated by an exchange, in which electrical energy is traded on a day-ahead basis. A *spot exchange* is an exchange operating a spot market. Via the Internet, players who want to trade on the spot market send their bids to the spot exchange at the latest by noon on the day before the day of operation. The day of operation is the day on which the electrical energy is produced and consumed.

When the spot exchange's prices are calculated, the purchase bids are aggregated to a demand curve. The sale offers are aggregated to a supply curve.

At noon, calculation of the day-ahead prices starts. When the calculation is finished, the exchange prices are published. These exchange prices are called *spot prices*. Figure 3.1 shows an example of spot prices.

C Price Zones

In Europe, a price zone is a geographical area within which the players can trade electrical energy on a day-ahead basis without considering grid bottlenecks (i.e. no need to consider bottlenecks). A spot price is

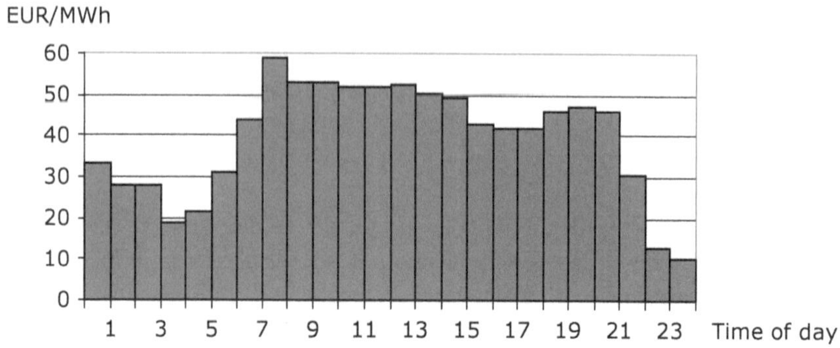

Figure 3.1 Spot prices in eastern Denmark, Monday 6 August 2012
Source: Author

calculated for each price zone for each hour of the following day. For example, Figure 3.1 shows the twenty-four spot prices for eastern Denmark for 6 August 2012.

Owing to grid bottlenecks, the Baltic–Nordic area is divided into a number of price zones. Figure 3.2 shows the Baltic–Nordic price zones as of October 2013.

D Day-Ahead Grid Congestion Management: Market Coupling

Apart from calculating day-ahead prices, the spot market is also used to carry out day-ahead congestion management in much of Europe. This day-ahead congestion management is called market coupling.

To explain market coupling, let us consider a grid bottleneck with a capacity of 600 MW. We will consider one given hour of the following day. Assume that the price-calculating computer during the calculation of the spot prices finds there will be different prices on the two sides of the bottleneck during this hour. In this case the computer will automatically insert an extra purchase of 600 MWh in the low-price zone and an extra sale of 600 MWh in the high-price zone, and then carry on with the price calculation.

The extra sale and the extra purchase are made the day before the day of operation. The next day, when the given hour is reached, the extra purchase in the low-price zone will cause a production surplus of 600 MWh: in the low-price zone, there are producers who will produce the extra 600 MWh, as they have sold this to the exchange. However, in the low-price zone, there is no corresponding local consumption. Owing to the production surplus, electricity must flow out of the low-price zone.

Figure 3.2 The Baltic–Nordic price zones as of October 2013
Source: Author

Likewise, the extra sale in the high-price zone will lead to a production deficit of 600 MWh: in the high-price zone, there are consumers who will consume the 600 MWh. However, in the high-price zone there is no corresponding local production. Owing to the production deficit, electricity must flow towards the high-price zone (Figure 3.3).

E Calculation of Spot Prices

Figure 3.4 illustrates the spot price calculation for a simple example with only two price zones.

The spot exchange reduces the price differences between the zones. If the grid capacity between the two zones is high, the exchange can level out the prices and create a common spot price for the two zones.

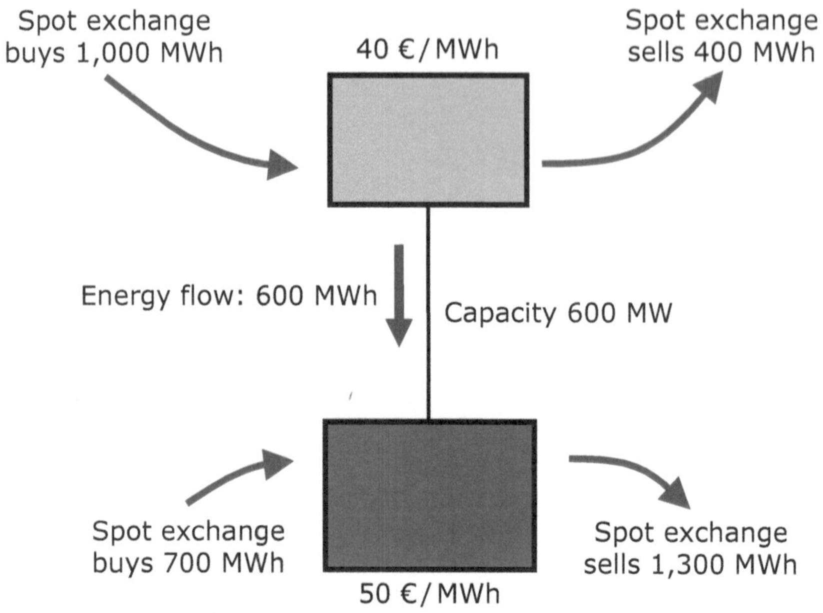

Figure 3.3 Market coupling: an example for one hour of operation
Source: Author

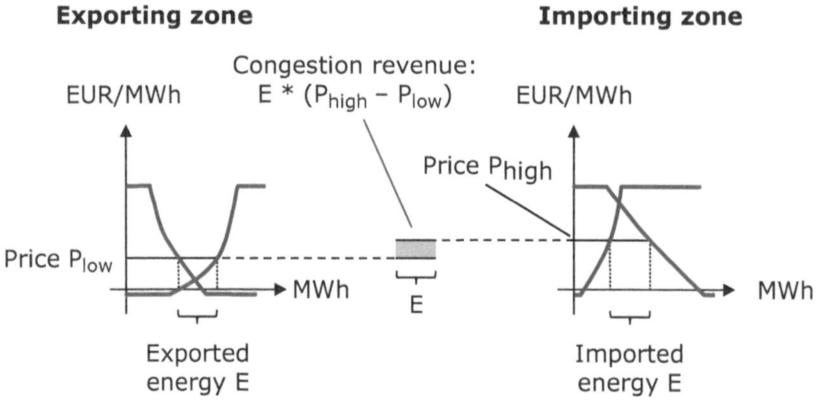

Figure 3.4 Spot price calculation when there is market coupling
Source: Author

However, Figure 3.4 illustrates a case where the grid capacity is too low for this: there is a spot price difference ($P_{high} - P_{low}$) between the two zones. This creates an arbitrage revenue, as the exchange is buying the transferred energy E at price P_{low} and selling E at price P_{high}.

This arbitrage revenue is called the congestion revenue (or congestion rent). The congestion revenue is given to the owners of the cross-border capacity (normally transmission system operators).

F European Day-Ahead Congestion Management

Unfortunately, the spot exchanges become monopolies when we introduce market coupling.[1] The consequences of this have not yet been taken into account; there is a severe lack of regulation of the European spot exchanges, which has led to heavy losses being inflicted on market players and societies.

As a case in point, consider the market coupling in Northern Europe from November 2010 to February 2014, which was operated by the European Market Coupling Company (EMCC).

The operation of the market coupling ran as follows: every day, perfectly reliable spot prices for the whole of northern Europe were calculated by EMCC. Unfortunately, for purely political reasons, the spot exchanges insisted on carrying out local re-calculations of the local spot prices. The re-calculated spot prices were used in the settlement of the market players' spot trading.

The local re-calculations crashed repeatedly. As an example of this, we may consider the spot prices for the Baltic–Nordic area for 13 August 2012. As usual, EMCC calculated reliable spot prices, including for the Baltic–Nordic area. However, the re-calculation of the Baltic–Nordic spot exchange, Nord Pool Spot, crashed. The loss for sellers of electricity in the Baltic–Nordic area was EUR 895,000, as the unreliable spot prices were used in the spot trading settlement. Polish buyers of electricity suffered a similar loss.

Figure 3.5 illustrates the spot prices for Southern Sweden on that day.

G PCR Market Coupling in Europe

In February 2014, EMCC market coupling was replaced with so-called PCR ('price coupling regions') market coupling. Following the experiences with EMCC market coupling, PCR market coupling does not include local re-calculations of spot prices. However, neither have the participating spot exchanges been able to agree on the establishment of

[1] See 'Single spot exchange for the single electricity market', www.houmollerconsulting.dk/facts-findings/ (accessed 1 November 2015).

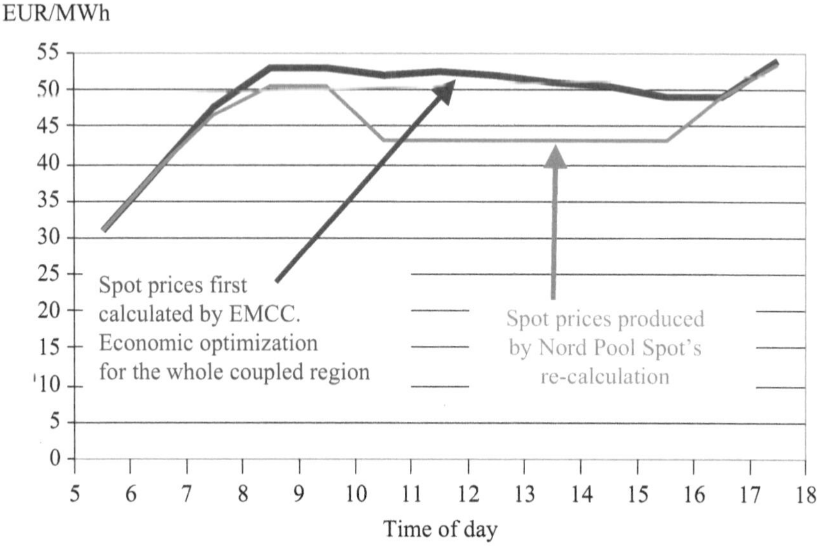

Figure 3.5 Spot prices for southern Sweden, 13 August 2012
Source: Author

a single entity that will calculate the spot prices. Hence, all the participating spot exchanges calculate all the spot prices, using the same software and the same information on all spot bids for the whole coupled region. Consequently, the number of administrations involved in the calculations equals the number of participating spot exchanges. At the launch of PCR in February 2014, this amounted to four spot exchanges. As market coupling expands to cover more of Europe, the number of administrations involved in the calculations will expand accordingly, and the number of computers, software installations and spot trading floors involved in the calculations should equal the number of participating spot exchanges.

Not surprisingly, even the establishment of PCR market coupling has proven to be very cumbersome and extremely expensive. Moreover, the PCR development process was complicated by bickering between the spot exchanges. For obvious reasons, the operation of PCR is also unnecessarily expensive, as there are many redundant staff, computers and software installations involved in the calculations. Hence, we conclude it would be strongly preferable to have one regulated spot exchange for the entire coupled area.

H Conclusion from the Experiences of Market Coupling in Europe

The lessons from Europe are very clear: when you establish market coupling, it is much simple to have one spot exchange for the whole coupled area rather than several spot exchanges. Politically, with one spot exchange, the countries in the coupled area can each have a share in the ownership of the spot exchange. The political issues can be sorted out in meetings of the spot exchange's board. Having one administration operating the market coupling avoids the risky and costly redundancy associated with PCR. It is also advantageous to separate the discussion of political issues from that of technical issues: the political issues can be solved in the single spot exchange's board meetings, whereas the technical issues can be solved by the single administration operating the market coupling.

The regulators in the coupled region must form a forum, which is tasked with establishing common rules for the regulation of the single spot exchange.

With a single spot exchange, Europe can have a reliable, cost-efficient day-ahead congestion management system. In turn, this greatly supports the introduction of more renewable energy in the electricity supply system. For example, at times when some parts of Europe have strong winds, the congestion management system will ensure the energy surplus is exported to the areas that have the highest demand for electricity, thereby optimising the socio-economic value of the renewable energy.

4

Economic and Geopolitical Determinants of Trade in Electricity

DAVID ROBINSON*

A Introduction

In this chapter, I consider some of the economic and geopolitical determinants of trade in conventional and renewable electricity, with an emphasis on the latter. The central question is whether electricity interconnections and trade promote economic welfare and geopolitical stability, as argued by proponents of a Global Grid, or if the causality is principally in the other direction.

Some clarification is required about what I mean by 'economic and geopolitical determinants of electricity trade'. I use 'economic determinants' to refer both to the capital costs associated with building interconnectors and to the short-term operating costs and losses related to transporting electricity over that infrastructure. I also refer to the distributional consequences – the sharing of the costs and the rents – of integrating electricity systems.

With regard to geopolitical determinants, I discuss two kinds of relationship. The first is the impact of geography and resource endowments on the politics and economics of electricity trade, as in the availability of hydroelectricity potential for exports from Quebec to the United States. The second is the impact of domestic and international politics on the electricity sector, for instance the consequences of an EU decision to promote renewable power for the need for interconnections, or of political instability in North Africa for the prospects for exports of renewable power to Europe.

The argument is as follows. First, in some cases there are sound economic reasons to build interconnectors and promote electricity trade

* The contents of this paper are the author's sole responsibility. They do not necessarily represent the views of any of the organizations with which the author is affiliated, or any of the members of those organizations.

involving renewable energy. For instance, interconnection and trade may be necessary to achieve scale economies on projects that would only make economic sense at a large scale, such as a major hydro project in a small country like Laos or a major solar project in a North African country. Economies of scale may also lower the cost of meeting climate policy objectives, where some 'non-market' revenues are required to justify the investment. For example, new interconnectors and trade could lower the costs of developing offshore wind resources in the North Sea and mitigate the problems of intermittency associated with wind and solar power.

Often, however, interconnection and trade does not make economic sense, in particular owing to the high fixed costs and losses associated with transporting electricity over long distances, especially where the risk and related cost of capital are substantial. Moreover, trends towards distributed generation (e.g. solar photovoltaic (PV) systems), energy efficiency, green credit trading and the relocation of data centres all *reduce* the need for international electricity trade. In other words, although new interconnections and trade will sometimes favour the efficient development of renewable energy, this is not always true.

Third, geopolitics sometimes favours electricity trade, but often not. On the one hand, electricity interconnection may assist individual countries or political groupings, like the EU, to achieve geopolitical aims at the least cost. On the other, geopolitical considerations – in particular the risk of depending on politically unstable countries for electricity supply, and the influence of domestic politics on the energy mix – discourage new electricity interconnections and trade. There is also growing public and political opposition to building high-voltage networks. In view of these limits, it is not surprising that projects like Desertec – intended to bring renewable power from North Africa to central Europe – have run into problems.

Finally, in response to the central question of this chapter, I argue that favourable geopolitical and economic conditions drive electricity interconnection and trade. Although there may be examples where the causality goes in the direction suggested by proponents of the global grid – that is, that electricity interconnections promote geopolitical stability and economic welfare – this is not generally true. I think policy should focus on creating the geopolitical and economic conditions that favour trade.

The remainder of this chapter is divided into five parts. In Section B, I introduce the concept of the global grid. In Sections C, D and E, I address the following three questions, before drawing conclusions in Section F.

- What are the economic and geopolitical determinants of trade in electricity?
- How does renewable energy change these determinants, if at all?
- What reforms are required to increase efficient international trade in electricity?

B The Global Grid

The original idea of the 'Global Grid', which was proposed by Buckminster Fuller in 1969, includes an economic argument as well as a geopolitical and social one: interconnection of electricity systems around the world would lower costs for people in all the connected systems, would make war more difficult to wage because of physical interdependence related to energy and would generally improve social welfare. The initial formulation did not emphasise renewable energy sources, but the more recent ones do.

The website of GENI (the Global Energy Network Institute) describes its mission as follows, emphasising the role of integrating renewable energy through the physical interconnection of electric power networks:

> GENI's mission is to conduct research and to educate world leaders and the public about the critical viability of the interconnection of electric power networks between nations and continents, with an emphasis on tapping abundant renewable energy resources, what we call 'the GENI Initiative.' Our research shows that linking renewables between all nations will mollify conflicts, grow economies and increase the quality of life and health for all. This is a strategy rooted in the highest priority of the World Game simulation developed by Dr. Buckminster Fuller three decades ago.[1]

In short, the idea is that the Global Grid will lower costs and improve political stability and social welfare. If we look around the world, it is hard to prove this thesis. Some countries, for instance Thailand and Cambodia, have continued to trade electricity in spite of political differences or even war. On the other hand, there are instances where energy grids that have been built are used as political weapons, as is the case with Russia's gas exports to neighbouring countries, or where contracts were rescinded overnight, such as the gas contract between Argentina and Chile. Even where political relations are amicable and the institutional

[1] www.geni.org/globalenergy/about_us/mission/index.shtml (accessed 20 October 2015).

relations for trade are well established, there are often domestic political reasons why electricity interconnections are not built (France–Spain), not used (Central America) or used in an opportunistic way (Spain–Portugal). I explore some of these examples below. Suffice it to say here that political conditions seem to drive trade and interconnection more often than the other way around, and that economic viability is probably a necessary, but not sufficient, condition.

C What Are the Economic and Geopolitical Determinants of Electricity Trade?

I Economic Determinants

1 Economic Benefits

Lower costs and improved social welfare. Most economists would probably agree that interconnection and trade in electricity is desirable when it lowers the costs of providing an electricity service. Since electricity is generally not storable, system stability requires immediate access to reserve capacity, which may be most efficiently provided from a distant resource connected to the same network. A related economic benefit of interconnection is the maintenance of a constant voltage. Early forms of interconnection were primarily aimed at providing additional security, but each system was largely self-supplied and had different electricity prices reflecting the costs of the system.

However, this was a timid form of integration. Economic efficiency and social welfare improves for the combination of the two systems if they are treated as a single system. Investment in generation will occur where the costs are lower and operations will use the plants with the lowest operating costs, regardless of their location. As a result of this more advanced form of integration, electricity prices on either side of the interconnector converge, with overall costs falling and with economic benefits to share. This has happened, for instance, between Spain and Portugal following the integration of the two markets in the Iberian Electricity Market (MIBEL); prices are the same on both sides of the border more than 90 per cent of the time, with the percentage up significantly on the period prior to the creation of MIBEL.

Interconnection and trade in electricity between systems has grown mainly because there are economic gains to be made from trade, as described above. The logic applies equally within countries and between them. This explains the creation of power pools, the expansion of networks within most advanced countries and interconnectors between regional or national systems.

Scale economies. Interconnection sometimes supports investment in larger and more efficient power stations than would otherwise be economically justified by the domestic market. An example is in southeast Asia in the Grand Mekong system, which includes Thailand, Cambodia, Laos, Vietnam, Myanmar and the south of China. Laos is the country with the smallest population and electricity demand, but it has significant hydro potential. Laos benefits from exporting electricity through the taxes it levies, as well as by providing labour opportunities and electricity to its remote regions. The other countries benefit from cheaper electricity. In this case, geography encourages trade because the countries share long borders, while investment by the (foreign) electricity buyers provides strong incentives for successful projects.

2 Economic Limits

Capital costs and losses. There are economic limits to integration and trade. These include the capital cost of interconnectors and network reinforcement, and the losses incurred in transport. The longer the distances covered, the more inhospitable the terrain and the greater the environmental sensitivity to overland lines, the higher these costs are. For instance, the new interconnector currently being built between Spain and France in the Eastern Pyrenees will add 1.4 GW of transmission capacity (3 per cent of Spain's peak demand).[2] It is 60 km long, passes underground through difficult mountain terrain and will cost about €700 million. If the interconnector were much shorter, the terrain were easier to cross and there were no environmental objections, the cost would be significantly reduced, to well below €100 million, according to credible sources. These sources also claim that without financial assistance of approximately €200 million from the EU, the interconnector would probably not have been built.

Supply security. Because electricity cannot be stored, system operators must always have flexible reserves available to cope with fluctuations in demand and supply. While it may be economically optimal to rely on reserves from another system and to treat the systems as if they were one, system operators are often reluctant to rely heavily on the adjacent system for their security of supply. This may be due to geopolitical

[2] For a general introduction to Spain's electricity interconnections, see www.ree.es/sites/default/files/downloadable/electricity_interconnections.pdf (accessed 20 October 2015).

concerns, but could also reflect the transaction costs of trade and the economic costs of potential blackouts or brownouts that could result from an inability to control reserves.

Distributional issues. Integration of systems implies convergence of prices. This affects consumers and producers on both sides of the interconnector, and has distributional consequences. For example, let us assume that wholesale prices were initially lower in Spain than in Portugal. Integration of the two markets would initially benefit Portuguese consumers through lower prices, but would negatively affect generators in Portugal by lowering their prices and market share. On the other side of the border, the increased demand from Portugal would initially raise Spanish prices, benefiting generators in Spain but not final customers. While integration should benefit both systems in the longer term, there are distributional effects that need to be taken into account.

Competition issues. Interconnectors bring new competitors into the connected markets. Inevitably, incumbents will resist new entry, even more so when the market is already facing conditions of excess generation capacity.

These economic limits to trade exist before geopolitical considerations are taken on board. Let me now turn to some of the geopolitical determinants of electricity trade.

II Geopolitical Determinants

1 Geopolitical Benefits

Necessary but not sufficient conditions. Favourable economic conditions are usually a necessary but not sufficient condition for investment in interconnectors. Another necessary but not sufficient condition is political commitment and confidence among trading partners; this is important because electricity interconnectors and trade will always involve at least two governments making decisions related to investment, regulating the trading parties, ensuring security of supply and sharing the costs and the rents.

Furthering common policy objectives. Political and/or economic union may create a set of shared policy objectives that could be furthered by integrating electricity grids. The EU single market is a good example of a broader political framework that should favour electricity market integration in the interest of lowering costs and prices, thereby improving the competitiveness of certain of its industries in international markets. The North American Free Trade Agreement (NAFTA) is another example, where a free trade area should facilitate integration of electricity systems. Even Central America has established an institutional framework to encourage economic integration and trade, including the building of interconnections and a regime to support electricity trade, the Central American Electrical Interconnection System (*Sistema de Interconexión Eléctrica de los Países de América Central*; SIEPAC).

Exploiting locational advantages of individual countries. Some individual countries or regions see political and economic advantages to encouraging trade. There are at least three types of trading opportunity: those related to the potential to export or import base-load electricity from systems with lower electricity costs and/or excess capacity to systems where prices are higher (for instance from France to Italy, or Quebec to New York); those where there is potential for trade of electricity at peak periods (for instance from Spain and Switzerland to France); and those for transit countries (for instance Switzerland).

2 Geopolitical Limits

Domestic political objectives may limit trade. Even where the economics are favourable and there is a political framework that might appear to favour trade, domestic politics often get in the way. These include national control over energy security, as well as support for national fuel suppliers and other industries. EU legislation specifically allows each Member State to determine its own fuel mix and to favour specific domestic fuels (e.g. domestic coal in Spain), ostensibly to achieve security of supply.

Sometimes, domestic political limits can be overcome by pressure from a supranational governing entity such as the EU.[3] A proven case is Sweden. Northern Sweden has a surplus of electricity but cannot, because of transmission congestion, always get it to southern Sweden. At such times of congestion, the Swedish transmission system operator (TSO) used to curtail southern Sweden's power exports to Denmark, in order to prevent the price in southern Sweden rising above that in the north. Following concerns expressed by Brussels, the Swedish TSO decided to split the Swedish power market into several bidding zones, producing a lower price in the north than in the south. This has reduced congestion inside Sweden. It has also produced a different geographical pattern to price convergence, particularly between southern Sweden and Denmark.

[3] Thanks to David Buchan of the Oxford Institute for Energy Studies (OIES) for this example.

Self-reliance. One of the main geopolitical reasons for not building an interconnector is the risk of relying on imports. This risk is minimal among advanced countries where energy is largely depoliticised and where there are shared cultural, political and economic traditions, for instance among the Nordic countries. On the other hand, in SIEPAC, trade is very limited in spite of the existence of physical interconnection and an institutional agreement among the member countries. In that case, the reluctance to trade is because governments are unwilling to agree long-term firm contracts: sellers want to maintain the right to restrict exports when supplies are tight, and buyers are reluctant to sign contracts that can be interrupted. Even within European markets, some countries close their markets to imports, alleging technical reasons that are questionable.

Environmental objections. It is increasingly difficult to obtain authorisation to build high-voltage interconnectors or indeed networks within a country. This contributes not only to the higher cost of building them – for instance, of laying cable underground rather than installing overhead cables – but also sometimes to the impossibility of obtaining authorisation to build. This certainly has contributed to France's resistance against building further interconnector capacity over the Pyrenees.

High risk associated with interconnectors. Another geopolitical barrier to trade concerns the risk of opportunistic behaviour related to interconnectors after they have been built. This is an extreme case of sunk costs and the related risk of post-contractual opportunistic behaviour. The problem becomes more acute when there are multiple parties involved in the relationship, especially when one or more of the interconnected countries are politically unstable. The concern may be about physical interruptions in supplier or transit countries. But more generally the risk is that the buyer will not pay or that the seller will not deliver. Private investors are used to taking risks related to oil and gas development in overseas countries because the returns often outweigh the risks. But they are less likely to take risks related to building networks and interconnectors in foreign countries, because returns will probably not compensate the related risks.

Implications for the Global Grid. The building of a Global Grid runs into problems because of the economic and geopolitical limits to integration. Very long distance (e.g. transatlantic) interconnectors imply significant capital costs and technical losses. Even for projects where the economics appear to make sense, distributional issues arise over sharing the costs and benefits over a long period. Furthermore, investment in interconnectors entails high risk associated with opportunistic behaviour and changes in market conditions, making it prohibitively expensive. If we add a lack of political stability or of mutual confidence, the likelihood of investment usually vanishes.

D How Does Renewable Energy Change These Determinants?

I Some Ways in which Renewable Energy Is Different

Many of the same economic and geopolitical drivers of conventional electricity trade apply also to renewable electricity. However, renewables are different in some important ways that change the nature of electricity trade, the need for it and the relative importance of its determinants. Let me summarise the main differences before examining how they affect trade in electricity.

- *Subsidies.* To date, most new renewable electricity resources (especially wind and solar energy) have been subsidised, either by customers or by taxpayers, largely because of a political decision to promote de-carbonisation through electricity. This reinforces the idea that politics – not markets – are driving investment and operating decisions in the sector, including decisions to build and use interconnectors.
- *The siting issue.* A wind or solar farm should be situated where the optimal wind or solar resource is found.[4] On the one hand, this could lead to international trade, as for instance with large hydro projects, such as the Mekong hydro system. However, sometimes this siting issue undermines the case for trade; indeed, most countries favour renewables because they are indigenous resources.
- *The relationship of fixed to variable costs.* Renewable electricity typically has high fixed costs and low marginal costs. Penetration of renewable power changes the electricity merit order because conventional fossil fuel plants have relatively high marginal costs and are displaced by renewables.[5] This tends to lower market-determined electricity prices. Conventional plants consequently find themselves selling less output at lower

[4] That there are so many rooftop solar PV systems in Germany and so few in Spain is a reflection of regulation and subsidies, not economic logic.

[5] 'Merit order' refers to the order in which electricity-generation plants operate, giving preference to the plants with lower marginal costs.

market prices. Meanwhile, power generated from renewables in most countries often receives additional revenues outside the electricity market.

- Wind and solar energy are intermittent sources of power, which means that production is determined by the weather. Intermittency leads to the need for flexible energy resources to respond to increases and decreases in renewable output, since large-scale storage technologies – apart from hydro – are still not commercially available. The flexible resources for the most part come from the same conventional power stations (e.g. gas-fired plants) that are earning less as a result of the penetration of renewables.

Some of these characteristics will encourage greater trade in renewable electricity and others will have the opposite effect. In other words, the relationship between growing penetration of renewables and international trade is case-specific.

II Economic Determinants of Trade in Renewable Electricity

1 Economic Benefits

Some of the benefits of trading renewable electricity fall into the same categories as for conventional electricity; this is true in particular for cost reduction through realising scale economies. However, other benefits from trade are related to the intermittent nature of some renewables.

Lower costs of investment. Interconnection and trade may reduce the unit cost (per kWh) of building large-scale renewable projects. For instance, the North Seas Countries Offshore Grid Initiative (NSCOGI) involves collaboration between EU Member States and Norway to create an integrated offshore energy grid linking wind farms and other renewable energy sources. According to its website, 'It seeks to maximize the efficient and economic use of the renewable energy resources as well as infrastructure investments'.[6] On the other hand, these North Sea projects receive subsidies as part of an EU-wide decarbonisation strategy and they are more expensive than alternative ways of lowering emissions, so one should not exaggerate the cost savings. It is better to think of them as a means of reducing the cost of achieving a wider range of policy objectives, including security of supply and industrial innovation as well as carbon dioxide (CO_2) emission reduction.

[6] See the NSCOGI website: www.benelux.int/nl/kernthemas/energie/nscogi-2012-report (accessed 20 October 2015).

Reducing average cost by raising capacity factors of intermittent renewable projects and thereby lowering CO_2 emissions. Other benefits of expanded interconnection involve improving capacity factors for intermittent renewable power stations. For instance, Spain's wind power capacity is 22 GW, with a maximum generation of about 17 GW, compared to an off-peak demand of about 22 GW. Once nuclear, cogeneration and other must-run generation is taken into account, Spain at times finds itself with an excess of wind-generated electricity. In these circumstances, owing to the shortage of interconnection capacity with France, Spain must temporarily shut down some of its wind stations. If wind power is displaced by carbon-emitting generation, this raises CO_2 emissions. It also raises the price of electricity in Spain and France. Increased interconnector capacity would avoid these additional short-run costs and emissions. On the other hand, increasing interconnector capacity is expensive, so a cost–benefit analysis is needed to determine whether the investment is economically justified.

Lower balancing costs. In some circumstances, trade can help to lower balancing costs associated with renewables. For instance, intermittent renewables over a wider geographical area may reduce the combined variability if climate patterns differ or if there are complementary renewables connected to the network. Supporters of the NOCSGI initiative in the North Sea argue that it could make the system less susceptible to changing weather; Norway's hydroelectric power plants could act as a 'giant battery', storing the power produced and releasing it at peak times, or when wind strength is low. Cross-border trade in electricity may also assist in balancing the system if the renewable energy is dispatchable, as it is for stored hydro, geothermal and biomass.[7]

Lowering the cost of green certificates. Green certificates (or renewable energy certificates, RECs, in the United States) are tradable commodities whose underlying energy source is renewable. They are typically used to meet regulatory obligations on the buyer to contract for a certain amount of renewable energy. Trade in green certificates does not necessarily require the building of networks to transport renewable energy to the buyer. The buyer of the green certificate (e.g. in Germany) may pay less to meet its

[7] See H. Bahar and J. Sauvage, 'Cross-border trade in electricity and the development of renewables-based electric power: lessons from Europe', *OECD Trade and Environment Paper* No. 2013/02 (2013), p. 9, www.oecd.org/officialdocuments/publicdisplaydocumentpdf/?cote=COM/TAD/ENV/JWPTE(2012)20/FINAL&docLanguage=En (accessed 21 September 2014).

renewable energy obligations if the energy were produced in a country where the costs of renewable energy were lower (e.g. in Morocco). This makes economic sense from the final customers' perspective of cost minimisation. However, this is not really trade in electricity; rather, it is a way of reducing the costs of meeting renewable energy obligations.

2 Economic Limits

Some of the limits associated with interconnectors and trade for renewable electricity are similar to those for conventional electricity. For instance, this is the case for the impact of distance on costs. Other limits are more specific to renewables, in particular the opposition to trade as a result of the influence of domestic subsidies.

High costs. As with conventional electricity trade, the longer the distance and the more investment required for interconnectors, the greater the fixed costs and risks (which in turn raise the cost of capital); line losses also rise with increasing distance. However, investment in high-voltage direct current (DC) lines can reduce costs. Furthermore, sometimes the economics favour long-distance projects because the cost of distant renewables plus the network costs is lower than the cost of renewable power in the buyer's own region. As explained later in the chapter, it is often other geopolitical factors that stop long-distance renewable energy projects from proceeding.

Complexity, intermittency and unwanted flows. The complexity of system operations increases along with the share of intermittent renewables. Since wind and solar power are intermittent sources, they require the system operator to respond quickly to changes in output from these plants. In countries that have not invested enough in flexibility to absorb these changes, the result can be unwanted flows of energy into surrounding countries. An example is the difficulty faced as the result of German wind power in the north of the country and related grid congestion. Germany has inadequate internal transmission to fully transport wind-power electricity from the north to the centres of demand in the south. The result is unwanted 'loop flows' of north German wind-based electricity into the neighbouring Polish, Czech and Dutch markets. This sometimes leads the bordering countries to restrict imports from Germany.

Supply security. The growth of renewable power in any country brings with it the need to provide backup. This usually means that conventional generation is required to provide flexibility (ramping up and down). However, conventional plants may close because prices have fallen as a result of the penetration of renewables. Trading partners may wish to take advantage of low-price electricity across the border, but this can create a problem of supply security in the importing country. For instance, the Netherlands imports low-cost wind power from northern Germany, but these imports reduce average wholesale power prices in the Netherlands and could undermine supply security if backup plant were not available owing to lower prices. One should not exaggerate this problem, however, since the solution is to provide adequate returns for flexible plants.

Limits to trade due to distributed energy resources (DER). We are witnessing a trend towards DER: distributed generation (DG, such as rooftop solar), energy efficiency, conservation, the use of batteries and demand response. All of these energy resources originate inside consumer premises and to a large extent they reduce the need to purchase energy from the grid. For instance, millions of premises now generate their own electricity in Germany, the United States and Australia. DER reduce total demand for electricity generated from conventional plants, and also the need for international trade of conventional electricity.

III Geopolitical Determinants of Trade in Renewable Electricity

Some of the geopolitical benefits and costs of trading renewable electricity are similar to those that affect trade in conventional electricity, for instance those related to a country's physical resources and location. However, the geopolitical determinants are somewhat different for trade in renewables, mainly because renewable power is so heavily influenced by government policy decisions on subsidisation.

1 Geopolitical Benefits

Reduced cost of achieving regional energy policy. The EU sees the development of renewable power as part of its geopolitical response to the challenges of global climate change, energy security (reliance on imported oil and gas) and economic competitiveness. While the United States develops shale resources as part of its overall geopolitical strategy, the EU has made a commitment to developing renewable power. In some cases, additional electricity interconnection and trade could lower the costs of achieving the EU objective.

The promotion of renewables in the EU is seen as a means of reducing reliance on imported fossil fuels, notably natural gas from Russia. This may make sense for geopolitical and environmental reasons, but it could also constitute a barrier to international trade. I am not sure how the

WTO would interpret decisions to favour domestic sources of renewable energy at the expense of competing imports of fossil fuels.

Geopolitical benefits for individual countries or groupings. Countries with significant solar energy potential (e.g. in the Mediterranean area) could potentially provide renewable electricity to distant countries of northern and central Europe. Countries with geothermal and hydro resources (e.g. Iceland or Norway) have limited domestic markets and could take advantage of scale economies by exporting their electricity. And countries with significant offshore wind resources (e.g. in the North Sea) see an opportunity to share infrastructure costs to bring their resources to market. In some cases, the economics may support this strategy; in others, they may not.

2 Geopolitical Limits

Domestic political objectives. Most governments that have supported the development of renewable power to date have done so as part of an industrial strategy aimed at promoting new technologies, local employment and the development of businesses with global markets.[8] This typically involves 'out-of-market' payments, or subsidies, usually borne by residential customers. This tends to weaken domestic political support for trade in green electricity for two reasons. First, imports benefit industrial development and employment in other countries, not in the country where subsidies are paid. Second, as long as a country's consumers are paying the subsidies, they are often reluctant to finance exports, especially when the regulatory regime is based on feed-in tariffs related to output.

This domestic policy focus is actually supported by EU legislation, which places responsibility for the design of renewable energy regulations with Member States. The European Court of Justice's judgment in the case of Ålands Vindkraft confirms this in supporting Sweden's right to deny support for renewable electricity to Finnish producers.[9]

[8] One could argue that these industrial policies have failed, at least in the case of solar power systems, since most European solar PV manufacturers have gone out of business as a result of Chinese imports. Nevertheless, governments continue to pursue and protect local economic advantages, even at the expense of raising the cost to final consumers, for instance through the imposition of trade barriers on Chinese solar panels. Furthermore, even if some industrial processes are carried out in other countries, there are many 'soft' services (e.g. installation, software, financing) that are carried out locally.

[9] For an analysis, see E. Durand and M. Keay, 'National support for renewable electricity and the single market in Europe: the Ålands Vindkraft case' (2014), www.oxfordenergy.org/2014/08/national-support-for-renewable-electricity-and-the-single-market-in-europe-the-alands-vindkraft-case/ (accessed 5 January 2016).

High risk associated with interconnectors. Interconnection to support trade of renewable power over long distances involves significant risk, in much the same way as for conventional electricity sources. Where there is a high degree of confidence, political commitment and political stability, for instance in Norway and the United Kingdom, the risk associated with a dedicated interconnector between them may be acceptable. Otherwise, interconnection projects are very unlikely to proceed, especially when one or more of the countries are politically unstable.

Desertec. This project reflects the difficulties of building major interconnectors to trade renewable energy. The combination of high capital costs, environmental opposition to new grids, excess Spanish generation capacity and political instability in North Africa has contributed to the failure (so far, at least) of the project. The initial aim of the project was to transport renewable energy from the Maghreb to Europe and to meet up to 20 per cent of Europe's electricity demand in 2050.[10] One version of this project involved building a Super Grid across Spain, through France and into Germany. According to reliable sources, this project makes economic sense on paper. However, it was politically unrealistic given the environmental opposition to new Super Grids in Europe. A second version involved expanding interconnector capacity between Morocco and Spain, and between Spain and France, while using existing national networks. Although less expensive and less likely to cause environmental opposition, the timing for this more modest plan was unfortunate. Spain had excess generation capacity and was not interested in adding to it, and France did not appear to want to expand the interconnector capacity with Spain. But it was political instability in North Africa that was the nail in the coffin for this project. Until that instability passes, it is hard to imagine Desertec proceeding.

E What Can and Should Be Done to Support Renewables Trade?

The rest of this chapter offers some thoughts on how to overcome some of the restrictions on integration and trade in electricity when the economic conditions are favourable.

I Socialising the Costs of Interconnectors

The main reason to build interconnectors and to facilitate trade in electricity is to improve social welfare through lower costs of electricity.

[10] www.euractiv.com/energy/desertec-abandons-sahara-solar-p-news-528151 (accessed 20 October 2015).

However, the distributional effects of converging prices can discourage economically rational interconnection and trade. Furthermore, the high risks associated with these investments make them unlikely to proceed without strong political support and commitment from the governments involved. This suggests that interconnectors should be built as part of a multi-country planning process and that the costs and benefits of the interconnectors should be socialised – in other words, shared – according to a set of principles agreed in advance.

II Political and Economic Integration

Even when the interconnector has been built, trade may be restricted or distorted by political decisions, as we see in Central America and in Europe. This suggests the need for closer economic and political integration, or other confidence-building measures.

III Redesigning and Regionalising Wholesale Markets

Wholesale electricity markets were designed in the 1990s in the USA and most of the EU with conventional (fossil fuel) plant in mind. In the EU, most countries have 'energy only' markets, where conventional power stations are supposed to recover their fixed costs during periods of high prices and without extra capacity payments. With the increasing penetration of low marginal cost renewable power, these traditional wholesale markets are unsustainable. They do not permit the recovery of conventional power stations' fixed costs. Nor do they provide the basis for recovering costs of renewable power in the absence of subsidies. These markets are also not designed to integrate distributed generation or to incorporate demand response that is growing as a result of smart grids and smart meters. There are many different models being considered for the sector, including the introduction of new capacity and flexibility markets. The point is that a redesign is urgently needed in order to make liberalisation and decarbonisation compatible.[11]

In order to cope with the increasing importance of renewables and their impact on trade flows (e.g. German loop flows affecting Poland and the Czech Republic), the new wholesale markets should at least be

[11] M. Keay, J. Rhys and D. Robinson, 'Decarbonization of the electricity sector: is there still a place for markets?', *Oxford Institute for Energy Studies* (2012), www.oxfordenergy.org/2012/11/decarbonisation-of-the-electricity-sector-is-there-still-a-place-for-markets/ (accessed 20 October 2015).

regional. Otherwise, there will be a growing reluctance to open borders because of the additional complexity, security and other implications outlined in this chapter.

IV *Regional or Global Green Certificate Markets*

One of the goals of EU policy has been to encourage the development of renewable energy, initially set at 20 per cent for 2020 (implying close to 40 per cent for renewable electricity). However, this has been implemented at a national level, with very different targets and regulatory systems and at a very high cost, resulting in significant increases in final electricity prices. In order to lower the cost of achieving any given renewables target, it makes sense to establish an EU-wide regime that encourages the development of renewables where the costs are lowest.[12]

For instance, an EU-wide regime would be likely to encourage solar generation where solar conditions are optimal, in the south of Europe rather than in the north. It would support the development of renewables generation in Iceland owing to the availability of hydro and geothermal energy, even though markets would be elsewhere. In some cases, this would involve additional physical trade in electricity and would require new interconnectors. In other cases, it would increase trade in financial instruments (green certificates), with the renewable power being consumed where it is generated.

At least in principle, the recent decision to adopt an EU-wide renewable target of 27 per cent for 2030 is consistent with a more economically rational development of renewable power. Furthermore, the EU's decision to expand interconnector capacity between Spain and France is consistent with the expansion of trade in renewables (from Spain to France). However, the decision is very short on specifics with respect to how this 'binding' EU-wide target will be met.[13]

There are different ways in which this EU-wide regime might be organised, including the adoption of a green certificates scheme, along the lines of the UK's Renewable Obligation Certificates (ROCs). Under

[12] This recommendation assumes that the EU will have a renewables target. I would actually prefer an effective carbon tax regime or cap-and-trade regime that provides long-term signals for investors, with no renewables target.

[13] D. Buchan, M. Keay and D. Robinson, 'Energy and climate targets for 2030: Europe takes its foot off the pedal', *Energy Comment*, Oxford Institute for Energy Studies (2014), www.oxfordenergy.org/wpcms/wp-content/uploads/2014/10/Energy-and-climate-targets-for-2013-Europe-takes-its-foot-off-the-pedal1.pdf (accessed 20 October 2015).

this scheme, all electricity retail suppliers are obliged to purchase a certain percentage of their power from renewable sources, or to buy ROCs. The Commission has proposed a green certificate scheme, but it was rejected by the Member States. With the costs of meeting renewables targets having reached unsustainable levels, this idea should be revisited.

V Trade in Information by Relocating Data Centres

The Internet has created a new and growing source of demand for electricity, namely, to power data centres, which now account for about 3 per cent of electricity demand in developed countries. Provided that data can be moved internationally (digitally) at very low cost, it is often sensible to locate data centres near to sources of renewable energy, even if the source is not connected to a high-voltage grid. This avoids the need for more expensive (or CO_2-emitting) electricity generation, as well as networks, in countries where the data is consumed. And it avoids the need to build expensive transmission networks to bring renewables to traditional markets. Indeed, it means that previously isolated sources of renewable energy can be exploited. Iceland and other Nordic countries are an example, especially owing to the availability of cooling water. But the latest data centres apparently need less cooling, so it is now feasible to have major data centres supplied by solar energy. This would be a particularly interesting opportunity for solar-rich countries and regions to consider.

VI Free Trade in Renewable Energy Equipment

Probably the most important contribution of EU (especially German) renewable energy policy has been to create a market for new renewable technologies. This in turn has driven down prices for equipment such as solar panels and wind turbines, often produced in China and other emerging countries. The International Renewable Energy Agency (IRENA) recently reported that the cost of solar panels had fallen by 80 per cent in the preceding five years, and the cost of wind turbines by 30 per cent. This lowers the cost of achieving a given penetration of renewables and increases the potential penetration for a given budget. It is therefore a serious concern when governments introduce quantitative restrictions or tariff barriers on imports of this sort of equipment. Both the United States and the EU have introduced such barriers. This is not the place to debate the details of these cases. Suffice it to say that import barriers are usually erected to protect domestic producers (e.g. panel manufacturers), at the expense of consumers and the commercial interests that serve them (e.g. the importers and installers of those panels, finance and software companies and other service providers who integrate those panels into consumer premises). The burden of proof should be high before imposing these barriers.

F Conclusion

The case for the Global Grid rests on a fundamental geopolitical principle: that physical integration of world electricity grids will lower costs and make the world a safer place. However, it is far from clear that physical integration of networks achieves this goal. Indeed, integration of energy grids has sometimes been a source of geopolitical tension and economic risk.

I argue instead that sound economics, political stability and shared values are the drivers for electricity integration. They are the necessary, if not sufficient, conditions for building interconnectors and for overcoming the distributional and other limits to trade. Any reform process aiming to favour welfare-improving trade of electricity should therefore start by establishing the basis for political stability and a shared long-term vision for the peoples of the different countries involved.

When these conditions exist, there will be opportunities for mutual benefit from trading electricity, both renewable and conventional. However, realising those benefits requires well-designed markets and regulations that provide efficient incentives to invest and to trade. The challenge is to obtain political support for electricity interconnection and trading, while allowing markets to allocate resources efficiently.

Bibliography

Bahar, H. and Sauvage, J. (2013). 'Cross-border trade in electricity and the development of renewables-based electric power: lessons from Europe', *OECD Trade and Environment Paper* No. 2013/02, www.oecd.org/officialdocuments/publicdisplaydocumentpdf/?cote=COM/TAD/ENV/JWPTE(2012)20/FINAL&docLanguage=En (accessed 21 September 2014).

Buchan, D., Keay, M. and Robinson, D. (2014). 'Energy and climate targets for 2030: Europe takes its foot off the pedal', *Energy Comment, Oxford Institute for Energy Studies*, www.oxfordenergy.org/wpcms/wp-content/uploads/

2014/10/Energy-and-climate-targets-for-2013-Europe-takes-its-foot-off-the-pedal1.pdf (accessed 20 October 2015).

Durand, E. and Keay, M. (2014). *National Support for Renewable Electricity and the Single Market in Europe: The Ålands Vindkraft Case*, www.oxfordenergy.org/2014/08/national-support-for-renewable-electricity-and-the-single-market-in-europe-the-alands-vindkraft-case/ (accessed 5 January 2016).

Keay, M., Rhys, J. and Robinson, D. (2012). 'Decarbonization of the electricity sector: is there still a place for markets?' *Oxford Institute for Energy Studies*, www.oxfordenergy.org/2012/11/decarbonisation-of-the-electricity-sector-is-there-still-a-place-for-markets (accessed 20 October 2015).

5

Good Governance for a Global Electricity Grid: What Are the Ingredients?

STEIVAN DEFILLA*

A Introduction

Although electricity is a widely used energy form, especially in developed countries, its specificities are rarely understood by non-experts. This misunderstanding often reflects public opinion, in particular when the discussion relates to the extension and development of new electricity infrastructures. Electricity is a high-technology product with both important similarities and differences to products developed by the information technology (IT) industry. This chapter will specifically show that technology is sufficiently important in the electricity sector to make it the major driver of future evolution. The chapter first describes the most important technological drivers in the electricity sector (Section B). With a basis in those technological drivers, Section C describes the various regulatory challenges that exist and responses to these challenges. Section D provides an overview of the existing institutional setting which governs the electricity grids around the world, placing a special focus on the Energy Charter Treaty (ECT). Section E concludes. The research concludes that the ECT, as a unique multilateral and legally binding framework dealing exclusively with energy issues, could play a key role in the world's energy governance, including governance for a global electricity grid, provided that renewables receive sufficient priority and that a specific instrument for electricity can be developed.

* The author is grateful to Iryna De Meyer for her support.

B Technological Drivers

I Demand Trends

According to the report on Energy Technology Perspectives prepared by the International Energy Agency (IEA) in 2014, electricity will play a key role in the developed economies' energy sectors in the twenty-first century.[1] This observation is based upon long-term trends that show the double role played by electricity, driving not only the world's economic growth but also the transition towards low-carbon energy systems. According to the 2015 Key World Energy Statistics, electricity consumption increased from 9.4 per cent of total final consumption in 1973 to 18.0 per cent in 2013, while the consumption share of other fuels, such as oil and coal, decreased.[2]

A vision of the electricity system of the twenty-first century can be built upon the following technological elements.

Electricity is the most polyvalent and the most user-friendly energy product, and its global share in the energy consumption basket is therefore likely to grow further. The polyvalence of electricity can be illustrated through its multiple uses: everyday experience demonstrates that electricity can be used to satisfy practically all of society's various energy needs. Moreover, electricity is the only energy form currently able to drive our information technology (IT) systems.[3] In modern society, electricity is also by far the most widely used form of energy for lighting. Furthermore, electricity is increasingly used in land transportation systems, such as electric trains and, more recently, electric vehicles. Two pioneering efforts – both Swiss – show that photovoltaic electricity can be used as an exclusive power source for ships and planes.[4] Since the launch of the *USS Nautilus*, the first nuclear submarine, in 1954, electricity has proven to be the best energy form used for submarines.[5] In the twentieth century, electricity

[1] IEA, press release on the report on Energy Technology Perspectives 2014, 'Taking on the challenges of an increasingly electrified world', www.iea.org/newsroom/news/2014/may/taking-on-the-challenges-of-an-increasingly-electrified-world-.html (accessed 9 December 2016).

[2] IEA, *2015 Key World Energy Statistics* (2015), p. 28, www.iea.org/newsroom/news/2015/november/key-world-energy-statistics-2015-available-for-download-in-time-for-climate-talk.html (accessed 9 December 2016).

[3] Wherever other energy forms (e.g. oil, gas, coal and uranium) play a role, this role is in the generation of electricity.

[4] PlanetSolar ship project, www.planetsolar.org (accessed 9 December 2016); Solar Impulse, solar-powered aircraft project, www.solarimpulse.com (accessed 9 December 2016).

[5] Submarine Forces Museum, www.ussnautilus.org (accessed 9 December 2016).

gradually entered most industrial processes and took a greater role in households. Electricity has been seen as a factor of economic growth.[6] Among the areas in which electricity has not yet achieved full penetration are transport and heating/cooling. The present section of this chapter postulates that in the twenty-first century the role of electricity will increase in these areas as new technologies become available, and that the technologies which will eventually be most widely used are those that enhance energy efficiency while also empowering consumers.

Future electricity growth will be strong, especially in developing countries and emerging economies. Owing to higher rates of growth in electricity consumption than is the case for other energy forms, investment in electricity infrastructure needs to be greater than investment in other energy systems. In its 2014 *World Energy Investment Outlook*, the IEA states that investments in the power sector in non-Organisation for Economic Co-operation and Development (OECD) countries were made mainly to satisfy a rapidly growing demand for electricity which, in the preceding ten years, had seen an average increase of 6.5 per cent.[7] In monetary terms, power sector investments in non-OECD countries grew from US$ 135 billion in 2000 to US$ 360 billion in 2011.[8] The majority of the incremental investment was in construction of new power plants, whereas 35 per cent of investment was in transport and distribution grids. Around 60 per cent of investment in non-OECD countries went to China.[9]

II E-mobility

Conventional internal combustion engines have an energy efficiency level of approximately one fifth. In other words, four fifths of the input energy is lost in the form of unused heat. Obviously, new technologies such as electric vehicles will improve this situation. One of the demand drivers of de-carbonisation is the electrification of individual mobility. Bloomberg New Energy Finance forecast that the cost of electric vehicles would be

[6] National Research Council; Division on Engineering and Physical Sciences; Energy Engineering Board; Commission on Engineering and Technical Systems; Committee on Electricity in Economic Growth, *Electricity in Economic Growth* (Washington, DC: National Academy Press, 1986), pp. 16, 40, www.nap.edu/catalog/900/electricity-in-economic-growth (accessed 9 December 2016).

[7] IEA, *World Energy Investment Outlook* (2014), p. 92, www.iea.org/publications/freepublications/publication/WEIO2014.pdf (accessed 9 December 2016).

[8] Ibid. [9] Ibid.

equal to the cost of internal combustion vehicles as of the middle of the current decade and thereafter will further decrease owing to learning curves that are still far from reaching a plateau.[10] Electric mobility refers to 'vehicles which rely on plug-in electricity for their primary energy, whether or not they have an auxiliary internal combustion engine for range extension or for keeping the battery charged up', and is described as the most important technological development in the automotive industry in recent years.[11] E-mobility includes not only cars but also other vehicles, such as bicycles, collective passenger transport, delivery vans, scooters and so on.[12] Electric vehicles do not work on the basis of low-efficiency internal combustion and are, furthermore, able to generate, store and, hence, recycle energy each time they brake. This will substantially increase their energy efficiency, making the progress of electric vehicle development likely to become unstoppable worldwide.

Currently, the assumption is that electric cars require charging by cable. This could change in the future as the options for wireless charging might rapidly increase. In a first step, stationary wireless charging will be made available for parked cars in the form of a plate fixed to the ground in garages and parking places.[13] In a second phase, dynamic wireless charging might allow for batteries to be charged while driving. This will require chargers to be implanted in the roadbed, a process that could start in large cities after maturation of the technology, perhaps in a decade's time. A further development would be for roads themselves to be used to produce photovoltaic electricity. Such solar roads, built by replacing conventional road surfaces with solar tiles, could not only generate electricity but could also incorporate dynamic wireless charging, warming strips to melt ice on the road and other features such as light-emitting diodes (LEDs) to replace conventional road signs and lane

[10] Bloomberg New Energy Finance, press release, 'Electric vehicle battery prices down 14% year on year' (16 April 2014), http://about.newenergyfinance.com/about/press-releases/electric-vehicle-battery-prices-down-14-year-on-year/ (accessed 9 December 2016).

[11] Fédération Internationale de l'Automobile, 'Towards e-mobility: the challenges ahead' (2011), p. 4, www.fiaregion1.com/download/emobility/emobility_full_text_fia2011.pdf (accessed 9 December 2016).

[12] D. Sabadka, 'Expectations and trends in individual mobility', *Transfer inovácií*, 24 (2012), http://blog.narotama.ac.id/wp-content/uploads/2014/11/EXPECTATIONS-AND-TRENDS-IN-INDIVIDUAL-MOBILITY.pdf (accessed 9 December 2016).

[13] R. Kaufman, 'Recharge and roll: electric carmakers plan to cut the cord', *Scientific American* (28 October 2011), www.scientificamerican.com/article/wireless-electric-car-charging/ (accessed 9 December 2016).

markers.[14] Multifunctional smart roads would allow, *inter alia*, for smart traffic management. A revolution will also be triggered if Google manages to make self-driving cars available to the public by 2020.[15] This development would presumably require new legislation worldwide.

III District Heating and Cooling by Means of Heat Pumps

Another technological driver marking changes in the electricity sector is the substitution of fossil-fuelled heating with electricity-driven heat pumps. This could trigger a technological change at the level of district heating in countries where the latter form of heating currently prevails. Efficiency gains can be expected because a heat pump uses only about one-third of the energy used in a traditional heating system.[16] In hot countries where inefficient decentralised cooling prevails, district cooling networks would bring efficiency gains. Most OECD countries have temperate climates, but district cooling is likely to penetrate into hotter countries outside the OECD, provided that these countries put in place the requisite regulatory framework. District cooling is likely to be more energy-efficient than individual room-by-room cooling owing to economies of scale and better management.

IV Greater Synergy with the IT Industry

Electricity needs instantaneous adaptation of supply to demand. In alternating current (AC) grids (which are the norm worldwide), second-by-second primary regulation (i.e. short-term fine-tuning of supply to demand) is done automatically by spinning reserves used for frequency regulation. Secondary regulation (i.e. coarse adaptation of production to daily or longer-term demand fluctuations) requires manual intervention and programming, making use of information related to producers and consumers. Smart technologies to manage production and consumption will be more widely used for this purpose in the future. New electricity switches will include a large amount of

[14] *The Economist*, 'On the not so sunny side of the street' (5 June 2014), www.economist.com/blogs/schumpeter/2014/06/crowdfunding-solar-roadways (accessed 9 December 2016).

[15] T. Halleck, 'Google Inc. Says Self-Driving Car Will Be Ready By 2020', *International Business Times* (14 January 2015), www.ibtimes.com/google-inc-says-self-driving-car-will-be-ready-2020-1784150 (accessed 9 December 2016).

[16] Depending on the coefficient of performance (COP) – e.g. if a heat pump delivers three units of heat for one unit of electricity consumed, its COP equals three.

information on consumer behaviour and might suggest immediate action to be taken by the consumer. If the consumer so desires, he or she will be able to programme the switches to take action automatically. Synergies between electricity and IT companies are therefore likely to emerge. IT companies will have an increasing role to play in operating smart grids, allowing improvements in efficiency, reliability, economics and sustainability of the production and distribution of electricity. It remains to be seen where new common infrastructures for IT and electricity might also be envisaged.

V Improvements in Electricity Storage

At the producer end, carbon-free production is more often intermittent than carbon-based production. This increases the requirement for backup capacities (battery storage, pump storage or gas turbine backup). Major technological progress is expected in the area of accumulator storage. As Miller noted in 2013, 'although hydropower pumped storage is the only industrial available power storage option available today, future developments are possible and are definitely needed for significantly expanding variable wind and solar power penetrations into existing power grids. Possible power storage options such as various thermal or chemical energy conversion, capacitor/battery, static potential energy, compressed air, dynamic mechanical, etc. must be developed.'[17]

Battery storage in homes received a major breakthrough with Tesla's launch of a US$ 3,000 home battery in Los Angeles on 30 April 2015.[18] This has the potential to become the silver bullet for integrating intermittent renewable energy. The batteries in electric vehicles can also play an important role in storage. Two models are in use. The first is the battery swap model, where the battery to be changed is replaced by a charged one. This model is used, for instance, for electric buses in the Chinese city of Qingdao, where the entirely automated swap process takes barely a few minutes.[19] The other model is the plug-in system, which is used elsewhere, and which has the disadvantage that charging takes several hours.

[17] J. Miller, 'Why expanded alternative energy increases the need for natural gas', *The Energy Collective* (29 January 2013), http://theenergycollective.com/jemillerep/178096/expanded-wind-and-solar-power-increase-need-natural-gas (accessed 9 December 2016).

[18] www.teslamotors.com/powerwall (accessed 9 December 2016).

[19] Phoenix Contact, 'Battery swapping for electric buses in Qingdao' (2013), www.phoenixcontact.com/assets/downloads_ed/local_gb/web_dwl_promotion/5733e.pdf (accessed 9 December 2016).

In the interest of achieving a good balance, both systems should be present. Electricity system operators prefer the battery swap system as it gives them better access to storage capacity, that is, to the charged batteries before they are inserted into vehicles. However, vehicle manufacturers might prefer the plug-in system as this gives them more choice in the type of battery used, and thereby influences the added value of the battery in the car.

VI Decoupling Economic Growth from Emissions

Economic growth provides populations with jobs and other goods. A simple model developed by the author shows that 2 per cent of economic growth worldwide until the end of the twenty-first century would increase global GDP by a factor of five compared to 2007 figures. It should be noted, however, that the fifth assessment report of the Intergovernmental Panel on Climate Change (IPCC), released in 2013, points out the importance of limiting the global temperature increase to a maximum of 2° C. The decisive factor is cumulative greenhouse gas (GHG) emissions since the beginning of the Industrial Revolution. The only scenario that is by and large in line with a temperature increase of 2° C is the one limiting CO_2 emissions to 421 ppm and emissions of other GHG gases to 475 ppm equivalent CO_2 by 2100.[20] The author's model translates this into 1 ton of CO_2 per person per year, that is, four times less than today's average global per capita emission per year. As GDP grows by a factor of five and CO_2 emissions should decrease by a factor of four, GDP carbon intensity therefore must decrease by a factor of twenty by 2100. Hence a decisive decoupling of economic growth from emissions is necessary. According to the IEA estimates, despite certain achievements in the field of renewable energy, the carbon intensity of the global energy supply has hardly improved in the past few decades.[21]

These calculations demonstrate the important role played by renewable energies, as they offer the best option for decoupling emissions from GDP. However, the role of energy efficiency also remains critical, because

[20] Representative Concentration Pathway RCP2.6; cf., e.g., IPCC, Working Group I, 'Summary for policymakers' in IPCC Fifth Assessment Report, *Climate Change 2013: The Physical Science Basis* (Cambridge: Cambridge University Press, 2013), p. 29, www.ipcc.ch/report/ar5/wg1/ (accessed 9 December 2016).

[21] IEA, 'Taking on the challenges of an increasingly electrified world'.

energy efficiency is currently the cheapest way to mitigate CO_2 emissions. Under the hypothesis that energy efficiency policies would act on global energy demand and stabilise it at the 2 kW per person level, as postulated by Novatlantis of the Polytechnic School of Zürich,[22] the OECD countries, as well as some oil-producing countries, would have to improve their energy efficiency by cutting their per capita consumption by a factor of 4–20, depending on the case.[23] This would allow developing countries and emerging economies such as India to catch up in terms of development as well as per capita energy consumption. At the global level, practically all future increases in energy demand would be compensated by a corresponding increase in energy efficiency, resulting in a stabilisation of global energy demand at a level by and large near the present consumption level. For that to happen, conceptual and regulatory frameworks would have to be set up which would allow energy efficiency to be considered as a genuine investment with corresponding returns and bankability. This may mean setting up a legislative framework favouring the development of energy service companies (ESCOs) with a business model based on the objective of making a profit on energy savings.

Two more recent developments need to be mentioned in this context. In September 2015, the UN General Assembly adopted a number of Sustainable Development Goals and targets.[24] For the first time, an energy-related goal has been included: Goal 7, which aims at ensuring access to affordable, reliable, sustainable and modern energy for all. It contains three targets: (7.1) to ensure universal access to modern energy services; (7.2) to increase substantially the role of renewable energy; and (7.3) to double the global rate of improvement in energy efficiency. These (not legally binding) goals point in the same direction as other developments described in this section, namely towards a stronger role for electricity, more renewables and more energy efficiency.

The second recent development to be mentioned in this context is Chinese President Xi Jinping's announcement, at the adoption of the UN Sustainable Development Goals, of China's proposal for discussion on the establishment of a global energy network to facilitate efforts to meet global

[22] www.2000watt.ch/ (accessed 9 December 2016).

[23] The only table listing per capita country consumption in watts, referring to 2003, can be found at http://en.wikipedia.org/wiki/List_of_countries_by_energy_consumption_per_capita (accessed 9 December 2016).

[24] United Nations Sustainable Development Goals, https://sustainabledevelopment.un.org/sdgs (accessed 9 November 2016).

power demand through clean and green alternatives.[25] This echoed a July 2014 proposal by Zhenya Liu, Chairman of the State Grid Corporation of China, to establish a Global Energy Internet (GEI).[26] The GEI strategy is to advance clean energy by replacing fossil-fuel energy with renewables and to replace coal and oil used for electricity production by promoting the use of electric vehicles and electric heating and cooling.[27]

C Regulatory Challenges and Responses

I Permit Granting Procedures for Electricity Infrastructures

In the developed world, procedures for building electricity lines are most often perceived as taking too long. In most developed countries, even after the use of all possible means to speed up the process, lines require near to ten years to build, if they can be built at all. It has been proposed that best practice would be to create one-stop shops for granting electricity permits. One-stop-shop measures are often supported by the population and the local authorities, and one of the main benefits of this approach would be a significant decrease in the number of permissions required and other administrative burdens. It would provide an opportunity to coordinate and publish environmental and other data.[28] The main challenge is that this would require different public administration departments to work together.

Still, this measure alone is insufficient to accelerate the relevant procedures. The time-consuming element is not so much the environmental impact assessment, which can be carried out according to publicly available schemes before the procedure is started and submitted with the project at the moment the procedure starts, but local opposition – which is also an uncontrollable element. Projects can be challenged at all

[25] X. Jinping, address at the adoption of the UN Sustainable Development Goals, September 2015. For the English translation see www.fmprc.gov.cn/mfa_eng/zxxx_662805/t1302359.shtml (accessed 9 December 2016).

[26] Z. Liu, *Global Energy Interconnection* (London: Elsevier, 2015).

[27] L. L. Lai, 'Global energy internet and interconnection', *IEEE Smart Grid Newsletter* (October 2015), http://smartgrid.ieee.org/newsletters/october-2015/global-energy-internet-and-interconnection (accessed 9 December 2016).

[28] European Commission, '"Impact assessment" – Accompanying the document "Proposal for a Regulation of the European Parliament and of the Council on guidelines for trans-European energy infrastructure and repealing Decision No 1364/2006/EC"' (2011), pp. 59–60, http://ec.europa.eu/smart-regulation/impact/ia_carried_out/docs/ia_2011/sec_2011_1233_en.pdf (accessed 9 December 2016).

court levels. These procedures cannot be shortened unless general civil rights are diminished.

As best practice, project proponents are advised to enter into dialogue with local communities and evaluate, as objectively as possible, which concrete aspects of the project impinge on local communities, and in what way. If disutility which affects local communities can be objectively identified, project proponents are advised to propose economic compensation. The recipients of this compensation should be carefully identified and should sign a compensation agreement specifying what type of disutility is being compensated for.[29]

In developing countries and emerging economies, the problem of permit granting is very different. In these countries, one of the major obstacles to investment in electricity is that electricity prices are often heavily subsidised, and hence too low to offer sufficient incentives to investors. Furthermore, many countries have quasi-institutionalised corrupt practices. Where these obstacles can be overcome, construction can be very rapid. China's investments in developing regions such as Africa are examples of rapidly constructed infrastructures.[30]

II Long-Term Stability

Another area in which regulatory measures are needed is long-term stability. In some developed countries, especially in situations where the grid is not sufficiently meshed and the market is not sufficiently liquid, regulatory frameworks for building electricity infrastructure should allow for longer-term contracts as a means to compensate for lack of market stability. Where grids are highly meshed and markets are liquid, long-term financial contracts, that is, long-term options and futures, can suffice. In this situation, the contracts might be able to ensure that suppliers can sell electricity and receive a return on their investments as well as insure buyers against regulatory reforms and, consequently,

[29] See for example Swiss Federal Office for Energy, *Sectoral Plan for Deep Geological Repositories [of nuclear waste], Conceptual Part* (2008), pp. 22, 35, www.bfe.admin.ch/radioaktiveabfaelle/01375/04389/index.html?lang=en (accessed 9 December 2016).

[30] M. Davies, *How China Is Influencing Africa's Development*, Background paper (Paris: OECD Development Centre, 2010), p. 21, www.oecd.org/development/pgd/45068325.pdf (accessed 9 November 2016).

against price changes.[31] The duration of the contract between electricity consumers and generators can be set as equal to the projected lifetime of the generation or transmission equipment.[32]

Political risks – such as currency inconvertibility, transfer restrictions, expropriation, terrorism, civil disturbance and failure to honour financial obligations, among others – threaten the design, building, execution and operation of long-term contracts related to new electricity infrastructure. Therefore these types of contractual schemes may need regulations providing for guarantee mechanisms, which would work as mitigation tools with respect to such events. The schemes could make use of instruments such as investment protection, investment guarantees, traditional private insurance policies, partial risk guarantees or guarantees conceded by export credit agencies, and therefore influence the price of the debt and other financial structures required in this type of project.[33]

Moreover, there can be risk associated with the operation of the electricity infrastructure. To avoid this risk, it is necessary to ensure third-party access to the electricity grid for existing lines as well as transparent tariff fixation for transmission activities.[34] Access to the grid is vital for private generation developers, and transparent conditions for access will attract investments in generation.[35] Therefore, guaranteed access to the transmission grid should be provided for independent power producers in a clear and non-discriminatory manner. To provide legal certainty to investors and private developers, guaranteed access to the grid and the conditions for receiving it should be covered by national regulations.[36]

III Transit Tariffs

Electricity market regulation is a permanent challenge worldwide. This is because the marginal electricity loss from grids depends on the square of the energy transmitted in the same direction, and it is diminished by transmitting electricity in the opposite direction. It is impossible to reproduce these physical characteristics in a grid-pricing model without complete information about all producers and consumers. Resolving this issue is the key to regional electricity cooperation. No perfect model has

[31] Dii, *Desert Power: Getting Started – The Manual for Renewable Electricity in MENA* (2012), p. 209, www.db.com/cr/de/docs/Desert-Power-Getting-Started-Full-Report-English-Screen(1).pdf (accessed 9 December 2016).
[32] Ibid. [33] Ibid., p. 150. [34] Ibid., p. 127. [35] Ibid., p. 134. [36] Ibid., p. 134.

as yet been implemented. To date, regulators and transmission system operators have used second-best solutions – for example, market coupling and market splitting – as best practice. Smart grids may improve this situation, although they might require smart regulations in order to work properly.

IV Compatibility with WTO

The EU's transition to decarbonised energy forms was supposed to be incentivised primarily by emissions trading. In reality, nationally fixed feed-in tariffs have taken the lead, and have been very successful in bringing about a strong increase in decarbonised production.

Feed-in tariffs could, however, be challenged by future WTO panels, either because of issues related to the 'made in Ontario' policy[37] or based on governmental policies regarding electricity prices. In the latter case, the government should bear in mind the national treatment obligation, according to which imported energy products should not be treated less favourably than local ones. Given their design, it is nearly impossible for imported renewable energy products to benefit from feed-in tariffs. Therefore, if brought before the WTO, feed-in tariffs might prove to be incompatible with WTO principles.

Neither feed-in tariffs nor other measures in support of low-carbon electricity (or that penalise carbon-intensive electricity) can readily be applied to imports at the border. This is one of the main drivers of carbon leakage, which occurs when the price of production of a good in a country that internalises its GHG emissions is higher than that in a country that does not internalise them.[38] This creates disadvantages for countries that do internalise the costs of GHG emissions. This is particularly problematic for electricity production. If, at the border, electricity is imported as a mix of high-carbon and carbon-free electricity, no specific measure can be taken against imports of high-carbon electricity. A best practice recommended to address this issue is electricity labelling,

[37] M. Wilke, 'Feed-in tariffs for renewable energy and WTO subsidy rules: an initial legal review', *ICTSD Programme on Trade and Environment: Trade and Sustainable Energy Series 4* (2011), www.ictsd.org/downloads/2011/11/feed-in-tariffs-for-renewable-energy-and-wto-subsidy-rules.pdf (accessed 9 December 2016).

[38] See for example, Swiss Federal Office of Energy, *Electricity Labelling and Certificate of Origin for Electricity* (2013), www.bfe.admin.ch/themen/00612/00614/?lang=en (accessed 9 December 2016).

which would specify the production method and origin of electricity. Labelling would also address this issue for all other goods.

V Synchronous Operation

Another regulatory challenge in the electricity sector concerns economies of scale and synchronous operation. For instance, the EU and Russia have made attempts at negotiating on synchronous interconnection – the first in the late 1980s and the second in the early 2000s – but they were not able to agree on modalities and conditions before 2009. Russia's entry into the WTO has changed the framework conditions for this negotiation. The General Agreement on Tariffs and Trade (GATT) and the Agreement on Technical Barriers to Trade (TBT) reverse the burden of proof. By virtue of the most-favoured-nation principle (MFN), the EU could in principle be obliged to treat Russia in the same way as third parties, examples of which include Switzerland and Turkey. The EU concluded a free trade agreement (FTA) with Switzerland in 1972, which applies also to electricity (Harmonized Commodity Description and Coding Systems (HS) code 27.16).[39] Switzerland is therefore not a benchmark for EU–Russia relations in the electricity field. The EU also has an FTA and a customs union with Turkey, but this agreement seems not to apply to electricity.[40] Yet the EU and Turkey are in synchronous operation. Russia could therefore ask for synchronous operation with the EU, using Turkey as the MFN benchmark. The EU would have to supply proof that AC–DC–AC converter stations are the least costly option to address the legitimate concern of keeping the frequency stable.

The concrete threat of Russia asking the EU to grant synchronous operation at the WTO has been reduced by the introduction of capacity payments into the Russian electricity market, making Russian electricity less competitive in Western Europe.[41] Capacity mechanisms, understood

[39] Article 2, Agreement between the European Economic Community and the Swiss Confederation, Brussels, 22 July 1972, in force 1 January 1973, *Official Journal* L 300, P. 0189–0280, http://eur-lex.europa.eu/LexUriServ/LexUriServ.do?uri=CELEX:21972A0722(03):EN:HTML (accessed 9 December 2016).

[40] Decision No 1/95 of the EC-Turkey Association Council of 22 December 1995 on implementing the final phase of the Customs Union (96/142/EC), *Official Journal of the European Communities* L35/14, http://eur-lex.europa.eu/legal-content/EN/TXT/PDF/?uri=OJ:L:1996:035:FULL&from=EN (accessed 9 December 2016).

[41] O. Barkin, *The Russian Capacity Market: Main Principles and Pricing*, Council of European Energy Regulators (CEER) Conference (February 2011), www.ceer.eu/portal/

as 'additional mechanisms influencing the volume of installed generation capacity to ensure short-term reliability and long-term generation adequacy', are being discussed and re-introduced in several countries.[42] As mentioned above, Russia has already made electricity markedly more expensive; in particular, 'the Russian capacity mechanism has affected the use of interconnections between the Nordic and Russian markets, leading to inefficient use of the transmission link between Finland and Russia'.[43] Russian electricity's comparative disadvantage would however be diminished once more if Western European states also begin to introduce capacity mechanisms that make their electricity more expensive.

VI Standards for Batteries and Plugs

As mentioned previously, electric vehicles raise the problem of standards for batteries and plugs. For instance, in electric passenger vehicles, the battery is charged inside the vehicle as needed, using DC or AC. In commercial vehicles, battery-swapping may have advantages – as shown by the City of Qingdao, which has adopted this system.[44] As mentioned above, regulation should enable both systems to coexist.

Of lesser importance, but still creating an obstacle to trade, is the multiplicity of electricity plugs, which in a certain sense shows the art of differentiation. Four different types of electricity are used in mains grids in final consumption worldwide:

1. 220–240 V at 50 Hz: used in Eurasia, Africa, Australia, Argentina and Chile (except Saudi Arabia, Japan, Korea, Philippines and Madagascar);
2. 110–127 V at 60 Hz: used in the Americas (except Peru, Argentina, Chile and parts of Brazil), Liberia, Saudi Arabia and southern Japan;
3. 220–240 V at 60 Hz: used in Peru, Korea, the Philippines and parts of Brazil;
4. 110–127 V at 50 Hz: used in Madagascar and northern Japan.

page/portal/EER_HOME/EER_INTERNATIONAL/CEER-FTS/Workshop_on_Capacity_Markets/Mr%20Barkin%20NP%20MC.pdf (accessed 9 December 2016).

[42] KU Leuven Energy Institute, *EI Fact Sheet: Capacity Mechanisms* (2013), www.mech.kuleuven.be/en/tme/research/energy_environment/Pdf/ei-factsheet3-eng-def.pdf (accessed 9 December 2016).

[43] S. Viljainen, O. Gore and M. Makkonen, *Linking together Electricity Markets with Different Market Designs – Case Study of the Future Use of the Finnish–Russian Interconnection* (Stockholm: Elforsk rapport, 2014), p. 1, www.elforsk.se/Documents/Market%20Design/projects/ER_14_05.pdf (accessed 9 December 2016).

[44] See note 20.

In the sense of technical necessity, the number of different plugs used could be diminished to four. The situation would also justify four different plug types being available to consumers. However, there are in fact fifteen different types of electricity plug, ranging from Type A to Type O.[45] Some types of electricity plug see widespread use, whereas others are used exclusively by one country. International regulation might work to simplify the situation, for example, by introducing multi-plug wall sockets.

VII Cybercriminality

In periods of heightened international tension, electricity grids, together with financial enterprises, are prime targets for cybercriminality. Even small grids are subject to millions of attacks per week. Part of a comprehensive solution may lie in increased international cybersecurity standards. Governments take very diverse approaches in this respect and there is little coordination between them. Not all EU countries allow additional cybersecurity costs to be financed by increasing the regulated costs of network use. International regulation could be necessary to make progress and avoid cyber-attacks which could create longer blackouts.

VIII Climate Adaptation

Electricity networks are heavily exposed to the effects of climate change, and in particular extreme climate events such as storms. This warrants the prompt construction of supplementary lines and the realisation of other measures, such as heating specific parts of the grid to protect it against ice formation. Also, decentralised electricity storage can help improve the local security of supply. Electricity should be the prime target sector for climate adaptation measures in all countries.

D Institutional Setting – the Place of the Energy Charter Treaty

The institutional approach towards worldwide electricity regulation can be understood by looking at the different global non-governmental

[45] World Standards, *Plug & Socket Types* (2014), www.worldstandards.eu/electricity/plugs-and-sockets/ (accessed 9 December 2016).

organisations (NGOs) and intergovernmental organisations that play a role in the electricity industry.

I European Network of Transmission System Operators for Electricity (ENTSO-E)

Within the EU framework, transmission operators are represented in ENTSO-E, an organisation with non-governmental status.[46] Although ENTSO-E is open to transmission system operators from non-EU countries, the focus is very much on EU law. EU law, in spite of being regional in its geographical scope, plays a pioneering role in experimenting with closer cooperation, such as integration between countries. ENTSO-E is therefore a very important institution, but it cannot be expected to take on a global role, as this is not its mission.

II Friends of the Supergrid

One of the newer organisations addressing electricity issues is Friends of the Supergrid, an industry association promoting and influencing the policy agenda and the regulatory framework for a European Supergrid. It includes businesses representing the entire supply chain: companies that will bring in the high-voltage direct current infrastructure and corresponding technology, and others involved in developing, installing, owning and operating the infrastructure.[47] Friends of the Supergrid could play a role beyond European borders, depending on how well it manages to integrate members from outside the EU.

III GO15

GO15 is a voluntary initiative comprising seventeen of the largest power grid operators.[48] The operators of the initiative supply electricity to nearly 3.5 billion customers, represent about two-thirds of the world's

[46] For more information see the ENTSO-E website, www.entsoe.eu (accessed 9 December 2016).

[47] For more information sec the Friends of the Supergrid website, www.friendsofthesupergrid.eu (accessed 9 December 2016).

[48] For more information see the GO15 website, www.go15.org (accessed 9 December 2016).

electricity demand and are 'responsible for integration of 2 518 GW of generation capacity into the grid, of which 21% is from renewable energy sources'.[49] Its vision is to act as a flagship and to promote 'the transition of the electric power industry to the power grid of the 21st century'. GO15 sees its mission as reaching a high level of consensus or unanimity with respect to the power grid operators' strategic questions, elaborating a mutual perception of the related technologies and best examples and promoting the realisation of this common perception.[50]

IV *International Electrotechnical Commission (IEC)*

The IEC is the most important global standard-setting organisation in the electricity sector.[51] It represents the national electricity standard-setting bodies of most of the countries of the world. These standards are voluntary, but pursuant to the TBT Agreement of the WTO, governments must refer to them in relevant national regulations in order to comply with Articles 2.2 and 2.8 of the TBT agreement.[52] More than 100 standards have already been defined for smart grids.[53]

V *World Trade Organization (WTO)*

On the global level, the WTO is the main regulatory organisation. It is now geographically a quasi-universal organisation dealing with practically every aspect of trade. The general rules are complemented by a set of more specific (and also more restrictive) rules on agriculture. Energy is totally integrated into the general provisions. Calls for an energy agreement in the WTO therefore have little chance of being realised. In spite of considerable development of the WTO rulebook in the early 1990s, the WTO still deals mainly with general aspects of non-discrimination, as anchored in its main principles of MFN and national treatment (not yet generalised in services). An important substantive provision, however, is the general elimination of quantitative restrictions (GATT Article XI),

[49] Ibid. [50] Ibid.

[51] International Electrotechnical Commission, www.iec.ch/ (accessed 9 December 2016).

[52] WTO, *Technical Information on TBTs*, www.wto.org/english/tratop_e/tbt_e/tbt_info_e.htm#agree1 (accessed 9 December 2016).

[53] International Electrotechnical Commission, *Core IEC Standards*, www.iec.ch/smartgrid/standards/ (accessed 9 December 2016).

which in the General Agreement on Trade in Services (GATS) is called market access (Article XVI). The Doha Round, which began in 2001, was supposed to address the question of access of investments,[54] as well as competition issues, which would have completed the trade-relevant rulebook of the WTO. So far, however, no agreement has been reached in any of the areas of the Doha Round. Given the slow progress of the Doha negotiations as far as deepening of substance is concerned, the WTO has in the past fifteen years progressed in widening the scope of its membership (e.g. Russia became a WTO Member in 2012), and FTAs among its Members have mushroomed. Such FTAs, however, are negotiated not in the WTO Secretariat, but directly among the WTO Members concerned.

VI United Nations Conference on Trade and Development (UNCTAD)

UNCTAD is another institution with global outreach. This is the main UN body dealing with international trade and development.[55] UNCTAD is most of all an economic forum for the leaders of its more than 180 member countries. Its general mandate is to improve the well-being of the population by helping developing countries and countries in transition in their development. UNCTAD organizes the annual high-level World Investment Forum, which analyses and discusses trends in international investment policies.

VII Energy Charter

Among the organisations with global potential is the Energy Charter, which was created in 1994 as a baby of the post-Cold War period under the assumption that energy would be a cornerstone of cooperation between the Commonwealth of Independent States and the rest of the world. For this cooperation to be effective, a legal framework was created. The Energy Charter Treaty (ECT) covers investment protection, transit, trade and energy efficiency. It is an important part of an international effort to build a legal foundation for energy security and to strengthen the rule of law on energy issues. It is aimed at creating a level playing field of rules to be observed by all participating governments, mitigating risks

[54] Also known as the making of investments or pre-investment stage.

[55] For more information see the UNCTAD website, www.unctad.org (accessed 9 December 2016).

associated with energy-related investment and trade.[56] The ECT is based on the principles of open, competitive markets and sustainable development, and the provision of legal rules for investments, trade and transit.[57]

Strong protection for investments is a cornerstone of the Energy Charter; this translates energy security into investment security. Investment provisions are fully operational, as evidenced by the recent Yukos investor–state arbitration case, which applied the Energy Charter provisions to Russia. Investors have called for further development of regulation in the energy sector. Effective protection mechanisms may be a positive factor allowing a country to attract foreign investments. The problem with the Energy Charter is that it does not give preferential treatment to clean energies, meaning that, at a time of increased public pressure to fight climate change, the Treaty might become obsolete. The Treaty's perceived protection of investments in fracking in European countries was criticised in a public letter addressed to the former President of the EU Commission in January 2014.[58] Furthermore, the UNCITRAL convention on transparency in investor–state dispute settlement will only apply to treaties concluded after this UNCITRAL convention has entered into force, and hence not to the ECT.

The Energy Charter also applies to trade in energy. Trade provisions are based on the WTO; however, the Charter has a special mandate to negotiate a plurilateral agreement within the Energy Charter constituency with the aim of introducing a binding customs standstill regime on energy materials and products, as well as on energy-related equipment used in the energy sector. In this area, the Energy Charter cooperates with WTO members to provide technical support for the Environmental Goods Agreement negotiations, which started in July 2014.[59] In the area of transit, the ECT goes slightly beyond the corresponding GATT

[56] See the list of investor–state dispute settlement cases initiated under the ECT, including the Yukos arbitral award, at www.energycharter.org/what-we-do/dispute-settlement/all-investment-dispute-settlement-cases/ (accessed 9 December 2016).

[57] For more information see the Energy Charter website, www.energycharter.org/ (accessed 9 December 2016).

[58] Joint open letter to EU institutions, 16 January 2014 (updated 26 January 2014), http://frackingfreeireland.org/2014/01/16/joint-open-letter-to-eu-institutions/ (accessed 9 December 2016).

[59] *WTO: 2014 News Items*, 'Azevêdo welcomes launch of plurilateral environmental goods negotiations' (8 July 2014), www.wto.org/english/news_e/news14_e/envir_08jul14_e.htm (accessed 9 December 2016).

provisions, in the sense that indirect national treatment for transit has been agreed upon.

The third element of the Energy Charter is energy efficiency. A separate Protocol on Energy Efficiency and Related Environmental Aspects has been drawn up. The role of energy efficiency should not be underestimated. As mentioned previously, energy efficiency is currently the cheapest form of energy use. Unfortunately, energy efficiency is still not sufficiently often considered as an investment. Instead, many economic actors and writers consider energy efficiency projects as merely imposing costs, without yielding returns. By promoting ways to support the consideration of energy efficiency projects as investments, they can become bankable. There is a need to introduce the necessary legislative framework to help energy efficiency play its role.

The Energy Charter has the potential to be a global player. On 20 May 2015, more than seventy states and two regional economic integration organisations adopted and signed the International Energy Charter in The Hague. This Charter updates the European Energy Charter of 1991, originally designed as a regional East–West forum for cooperation with transition countries, to a global North–South forum for cooperation with developing countries. The Energy Charter is expected to widen its geographical scope and to strengthen its authority within the Energy Charter constituency and in the international arena.[60] However, if the ECT is to be useful to future developments of electricity as outlined in this chapter, it should be complemented by an electricity instrument (a Treaty or a Protocol).

The impossibility of starting negotiations on an electricity protocol triggered regional cooperation in the Energy Charter. In 2007, the Task Force for Regional Electricity Cooperation was created. Since then it has held annual meetings in Central Asian countries, at which the topics addressed include the transit tariff methodology to be applied among the countries of the region.

Regional cooperation has also begun in north-east Asia around the Gobitec and Asia Supergrid project, where no regional forum exists. This cooperation aims at clean electricity production from renewable energy (wind and solar) in the Gobi desert (Gobitec) and its transport

[60] For the updated list of members and observers see www.energycharter.org/who we-are/members-observers/ (accessed 9 December 2016).

to China, South Korea and Japan (Asia Supergrid).[61] Another of its objectives is to strengthen relations between the participants in the initiative, and thereby to contribute to political stabilisation in the region and cooperation between the countries involved.

In 2015, the Energy Charter also began to play a role in other areas where no regional cooperation exists (e.g. in the Caucasus).

E Conclusions

In its examination of the options for good governance for the future electricity grid, this chapter has set out the main technological drivers, the most important regulatory challenges that result from them and an overview of the existing international electricity governance. Governance is understood in this chapter as a set of legally binding and enforceable rules. The chapter has shown that the role of electricity is bound to grow as a result of technological progress. The two areas in which such growth is envisaged are mobility and heating and cooling. The regulatory challenges are enormous, but the integration of intermittent renewable energy, the (indirect) storage of electricity and the challenge of electric mobility are certainly among the most important. Electricity will in future be more strongly linked with IT and will become smarter, to the extent that the cooperation between electricity and IT approaches the level of a merger.

Among the (potentially) global organisations existing at present, the WTO, the Energy Charter and UNCTAD would have a role to play here. However, none of these organisations is specifically designed to do so. The WTO lacks the depth needed for addressing electricity issues, the Energy Charter lacks a specific instrument for electricity and UNCTAD concentrates on investments rather than on electricity. None of the regional organisations can aspire to becoming global. Friends of the Supergrid, GO15 and the IEC are NGOs that do not have a mandate to negotiate or enforce binding rules, albeit that they could have a very important role to play once binding rules have been addressed at an intergovernmental level.

[61] For more information see the Gobitec Initiative website, www.gobitec.org (accessed 9 December 2016). See S. Mano, B. Ovgor, Z. Samadov, M. Pudlik, V. Jülch, G. Sokolov and J. Y. Yoon, *Gobitec and the Asian Super Grid for Renewable Energies in Northeast Asia* (Brussels: Energy Charter, 2014), p. 10, www.energycharter.org/what-we-do/investment/investment-thematic-reports/gobitec-and-the-asian-supergrid-for-renewable-energy-sources-in-northeast-asia-2014/ (accessed 9 December 2016).

Bibliography

Agreement between the European Economic Community and the Swiss Confederation, Brussels, 22 July 1972, in force 1 January 1973, *Official Journal* L 300, 31/12/1972 P. 0189–0280, http://eur-lex.europa.eu/LexUriServ/LexUriServ.do?uri=CELEX:21972A0722(03):EN:HTML (accessed 9 December 2016).

Barkin, O. (2011). *The Russian Capacity Market: Main Principles and Pricing*, Council of European Energy Regulators (CEER) Conference, February, www.ceer.eu/portal/page/portal/EER_HOME/EER_INTERNATIONAL/CEER-FTS/Workshop_on_Capacity_Markets/Mr%20Barkin%20NP%20MC.pdf (accessed 9 December 2016).

Bloomberg New Energy Finance, press release, *Electric Vehicle Battery Prices Down 14% Year on Year*, 16 April 2012, http://about.newenergyfinance.com/about/press-releases/electric-vehicle-battery-prices-down-14-year-on-year/ (accessed 9 December 2016).

Davies, M. (2010). *How China Is Influencing Africa's Development*. Background paper. Paris: OECD Development Centre, April, www.oecd.org/development/pgd/45068325.pdf (accessed 9 December 2016).

Decision No 1/95 of the EC-Turkey Association Council of 22 December 1995 on implementing the final phase of the Customs Union (96/142/EC), *Official Journal of the European Communities* L35/14, http://eur-lex.europa.eu/legal-content/EN/TXT/PDF/?uri=OJ:L:1996:035:FULL&from=EN (accessed 9 December 2016).

Dii. (2012). *Desert Power: Getting Started – The Manual for Renewable Electricity in MENA*, www.db.com/cr/de/docs/Desert-Power-Getting-Started-Full-Report-English-Screen(1).pdf (accessed 9 December 2016).

The Economist. (2014). On the not so sunny side of the street, 5 June, www.economist.com/blogs/schumpeter/2014/06/crowdfunding-solar-roadways (accessed 9 December 2016).

European Commission. (2011). 'Impact assessment' – Accompanying the document 'Proposal for a Regulation of the European Parliament and of the Council on guidelines for trans-European energy infrastructure and repealing Decision No 1364/2006/EC', http://ec.europa.eu/smart-regulation/impact/ia_carried_out/docs/ia_2011/sec_2011_1233_en.pdf (accessed 9 December 2016).

Fédération Internationale de L'Automobile. (2011). *Towards E-mobility: The Challenges Ahead*, www.lowcvp.org.uk/assets/reports/emobility_full_text_fia.pdf (accessed 9 December 2016).

IEA. (2014a). Taking on the challenges of an increasingly electrified world, press release on the report on *Energy Technology Perspectives 2014*, 12 May, www.iea.org/newsroomandevents/pressreleases/2014/may/taking-on-the-challenges-of-an-increasingly-electrified-world-.html (accessed 9 December 2016).

IEA. (2014b). *World Energy Investment Outlook*, www.iea.org/publications/freepublications/publication/WEIO2014.pdf (accessed 9 December 2016).

IEA. (2015). *Key World Energy Statistics*, www.iea.org/publications/freepublications/publication/key-world-energy-statistics-2015.html (accessed 9 December 2016).

IPCC. (2013). 'Working Group I, Summary for policymakers', in *IPCC Fifth Assessment Report, Climate Change 2013: The Physical Science Basis*. Cambridge: Cambridge University Press, www.ipcc.ch/report/ar5/wg1/ (accessed 9 December 2016).

Fracking Free Ireland. (2014). Joint open letter to EU institutions, 16 January 2014 (updated 26 January 2014), http://frackingfreeireland.org/2014/01/16/joint-open-letter-to-eu-institutions/ (accessed 9 December 2016).

Kaufman, R. (2011). Recharge and roll: electric carmakers plan to cut the cord, *Scientific American*, 28 October, www.scientificamerican.com/article/wireless-electric-car-charging/ (accessed 9 December 2016).

KU Leuven Energy Institute. (2013). *EI Fact Sheet: Capacity Mechanisms*, 2013–03, www.mech.kuleuven.be/en/tme/research/energy_environment/Pdf/ei-factsheet3-eng-def.pdf (accessed 9 December 2016).

Liu, Z. (2013). *Global Energy Interconnection*. London: Elsevier.

Lai, L. L. (2015). Global energy internet and interconnection, *IEEE Smart Grid Newsletter*, October, http://smartgrid.ieee.org/newsletters/october-2015/global-energy-internet-and-interconnection (accessed 9 December 2016).

Mano, S., Ovgor, B., Samadov Z., Pudlik M., Jülch V., Sokolov G. and Yoon, J. Y. (2014). *Gobitec and the Asian Super Grid for Renewable Energies in Northeast Asia* (Brussels: Energy Charter), www.energycharter.org/what-we-do/investment/investment-thematic-reports/gobitec-and-the-asian-supergrid-for-renewable-energy-sources-in-northeast-asia-2014/ (accessed 9 December 2016).

Miller, J. (2013). Why expanded alternative energy increases the need for natural gas, *The Energy Collective*, 29 January, http://theenergycollective.com/jemillerep/178096/expanded-wind-and-solar-power-increase-need-natural-gas (accessed 9 December 2016).

National Research Council; Division on Engineering and Physical Sciences; Energy Engineering Board; Commission on Engineering and Technical Systems; Committee on Electricity in Economic Growth. (1986). *Electricity in Economic Growth*. Washington, DC: National Academy Press, www.nap.edu/catalog/900/electricity-in-economic-growth (accessed 9 December 2016).

Phoenix Contact. (2013). *Battery Swapping for Electric Buses in Qingdao*, www.phoenixcontact.com/assets/downloads_ed/local_gb/web_dwl_promotion/5733e.pdf (accessed 9 December 2016).

Sabadka, D. (2012). Expectations and trends in individual mobility, *Transfer inovácií*, 24/2012, pp. 223–5, www.sjf.tuke.sk/transferinovacii/pages/archiv/transfer/24-2012/pdf/223-225.pdf (accessed 9 December 2016).

United Nations Sustainable Development Goals. (n.d.). https://sustainabledevelopment.un.org/sdgs (accessed 9 December 2016).

Swiss Federal Office of Energy. (2013). *Electricity Labelling and Certificate of Origin for Electricity*, www.bfe.admin.ch/themen/00612/00614/?lang=en (accessed 9 December 2016).

Swiss Federal Office for Energy. (2008). *Sectoral Plan for Deep Geological Repositories [of nuclear waste], Conceptual Part*, www.bfe.admin.ch/radioaktiveabfaelle/01375/04389/index.html?lang=en (accessed 9 December 2016).

Tam, D. (2012). Google's Sergey Brin: you'll ride in robot cars within 5 years. *C|NET*, 25 September, www.cnet.com/news/googles-sergey-brin-youll-ride-in-robot-cars-within-5-years/ (accessed 9 December 2016).

Tesla Motors. (n.d.) www.teslamotors.com/powerwall (accessed 9 December 2016).

Thomas, H. (2015). 'Google Inc. says self-driving car will be ready by 2020', *International Business Times*, 14 January, www.ibtimes.com/google-inc-says-self-driving-car-will-be-ready-2020-1784150 (accessed 9 December 2016).

Viljainen, S., Gore, O. and Makkonen, M. (2014). *Linking Together Electricity Markets with Different Market Designs – Case Study of the Future Use of the Finnish–Russian Interconnection.* Stockholm: Elforsk rapport, www.elforsk.se/Documents/Market%20Design/projects/ER_14_05.pdf (accessed 9 December 2016).

Wikipedia. (n.d.) List of countries by energy consumption per capita, http://en.wikipedia.org/wiki/List_of_countries_by_energy_consumption_per_capita (accessed 9 December 2016).

Wilke, M. (2011). Feed-in tariffs for renewable energy and WTO subsidy rules: an initial legal review. *ICTSD Programme on Trade and Environment; Trade and Sustainable Energy Series*, 4, www10.iadb.org/intal/intalcdi/PE/2011/08721.pdf (accessed 9 December 2016).

World Standards. (2014). Plug & socket types, www.worldstandards.eu/electricity/plugs-and-sockets/ (accessed 9 December 2016).

WTO. (n.d.) Technical information on TBTs, www.wto.org/english/tratop_e/tbt_e/tbt_info_e.htm#agree1 (accessed 9 December 2016).

Xi Jinping (2015). Address at the adoption of the UN Sustainable Development Goals, September. For the English translation see www.fmprc.gov.cn/mfa_eng/zxxx_662805/t1302359.shtml (accessed 9 December 2016).

PART II

Regional Experiences in Promoting (Green) Electricity Trade

6

Towards a New Deal for the Integration of Renewable Power Generation in the Internal Energy Market: A Regulatory Perspective on European Climate and Energy Policy

JÉRÔME LE PAGE

A Introduction

In November 2014, Johannes Teyssen, CEO of the European energy giant E.ON, announced the separation of his company's declining power generation business from more lucrative and regulated activities such as renewable power generation and network management. Following this, in September 2015, David Crane, CEO of the US utility NRG Energy Inc., announced plans to distance the company from its historic power generation business to focus on renewable power generation. The shift, in both cases, is intended to dissociate the companies from what is perceived as a doomed, unprofitable business and concentrate instead on activities identified as being the future of the electricity sector. Though not often referred to in official press statements, the generous subsidies and grid privileges awarded to renewable energy sources are likely to have played a determining role in the decisions of these two CEOs. While these choices may make sense economically in these two cases – largely because of political inflexions in Germany in the case of E.ON, and for probably more strategic reasons in the case of NRG – one cannot help wondering about the future of the electricity sector were many more traditional utilities to move in the same direction.

The intermittency of renewable power generation means that as long as demand remains largely inelastic, and in the absence of scalable electricity storage solutions, the need for traditional power generation will continue. And in the short term, traditional power generation is still necessary to cover more than just the periods of low production of electricity from renewable sources. Yet it seems increasingly difficult for these two segments of the electricity sector, the old and the new, to coexist. While the economic downturn in Europe may partly explain the situation, regulatory interventions in the market bear a great share of responsibility for letting traditional utilities fall into unprofitability – and for preventing the emergence of sizeable demand-response and storage capabilities. Realising the incoherence of a number of policies decided at the level of the European Union (EU) and the lack of coordination in their implementation at the national level is key to addressing the problem. Because efforts to liberalise the electricity market and decarbonise the electricity sector have borne fruit and should not be abandoned, fundamental reform is needed to ensure the coherence of these policies and the emergence of a truly European, undistorted market for electricity in the region. Elements of this reform are underway; some of these consist only of minor regulatory changes and others require strong political commitment and will, but only in this way will the energy sector as a community be able to deliver sustainable, competitive and secure electricity to European consumers in the decades to come.

B Renewable Energy Promotion and Energy Market Liberalisation: Lessons Learnt from the 1990–2015 Period

From the early 1990s to the adoption of the Third Energy Package in 2009, the advent of energy market liberalisation profoundly transformed the electricity sector in Europe, allowing electricity to flow, according to market prices and across borders, to where it is most needed. In a concomitant effort, the EU and its Member States have set out on the path towards decarbonisation, with a particular focus on the power-generation industry. As these two policy streams continue to be pursued in parallel, growing concerns arise as to the compatibility of the policy instruments used to reach their objectives.

I The Genesis of Energy Market Liberalisation

The process of liberalisation of the electricity markets started in the early 1990s, in Great Britain and Scandinavia. Well before the first energy package of the EU, these Member States saw the need to ensure security of supply at the lowest cost by dismantling incumbent monopolies. The unbundling of power generation and supply businesses from transmission system operation allowed markets to develop and competition to emerge in a traditionally state-run environment.

Electricity markets in Great Britain and Scandinavia started developing as organised pools. By the mid-1990s, wholesale markets had started developing in continental Europe as well, in the form of spontaneous day-ahead markets formed by the growing mass of bilateral day-ahead spot transactions, often facilitated by traders.

By the end of the 1990s, the EU had decided to set the first basic rules translating the fundamental elements of the EU Treaty for the electricity sector, and most specifically the principles of competition and non-discrimination. The first energy package, adopted in 1996, ensured that wholesale electricity markets throughout Europe followed these basic principles, and paved the way for a series of legislative and regulatory measures in the years to follow.[1]

Following the second and third energy packages,[2] the Security of Supply Directive,[3] and the Transparency Regulation,[4] the framework guaranteeing free competition and non-discrimination in electricity markets became extensive. Detailed rules are still being developed and implemented in the form of network codes and binding guidelines developed by the European Network of Transmission System Operators for Electricity (ENTSO-E) under the supervision of the Agency for the Cooperation of Energy Regulators (ACER) and the European Commission to ensure convergence and harmonisation in operational, connection and market rules in all EU Member States.

[1] Directive 96/92/EC of 19 December 1996 concerning common rules for the internal market in electricity, http://eur-lex.europa.eu/LexUriServ/LexUriServ.do?uri=CELEX:31996L0092:EN:HTML. All weblinks in footnotes were accessed on 1 December 2015, unless indicated otherwise.

[2] Directive 2003/54/EC of 26 June 2003 concerning common rules for the internal market in electricity and repealing Directive 96/92/EC, http://eur-lex.europa.eu/legal-content/EN/TXT/?uri=celex:32003L0054, and Regulation (EC) 1228/2003 of 26 June 2003 on conditions for access to the network for cross-border exchanges in electricity, http://eur-lex.europa.eu/legal-content/en/ALL/?uri=CELEX:32003R1228; Directive 2009/72/EC of 13 July 2009 concerning common rules for the internal market in electricity and repealing Directive 2003/54/EC, http://eur-lex.europa.eu/LexUriServ/LexUriServ.do?uri=OJ:L:2009:211:0055:0093:EN:PDF, and Regulation (EC) 714/2009 of 13 July 2009 on conditions for access to the network for cross-border exchanges in electricity and repealing Regulation (EC) No 1228/2003, http://eur-lex.europa.eu/legal-content/EN/ALL/?uri=CELEX:32009R0714.

[3] Directive 2005/89/EC of 18 January 2006 concerning measures to safeguard security of electricity supply and infrastructure investment, http://eur-lex.europa.eu/legal-content/EN/TXT/?uri=CELEX:32005L0089.

[4] Regulation (EU) 543/2013 of 14 June 2013 on submission and publication of data in electricity markets and amending Annex I to Regulation (EC) 714/2009, http://eur-lex.europa.eu/legal-content/EN/TXT/?uri=CELEX:32013R0543.

Market arrangements still differ in Europe, from fundamental aspects such as the centralisation or decentralisation of dispatch decisions to detailed features such as imbalance settlement periods – all elements that affect market functioning, hedging strategies and investment decisions. Nonetheless, all markets in Europe share a common feature, that is, the emergence of an energy price when demand meets supply, based on a merit order of power-generation units dispatched according to their marginal cost of generation.

This fundamental feature of the European energy markets was soon to become the heart of the most fundamental challenge since the start of the liberalisation process: the significant penetration of electricity from renewable energy sources (RES-E), which, by the end of the 2000s, had turned the traditional merit order of power-generation units on its head.

II The Development of Climate Action Policies

At the end of the 1990s, while energy market liberalisation was gaining speed, growing concern in the population and among the world's political leaders – particularly in Europe – about climate change led to the development of a number of instruments to curb carbon emissions. With the Kyoto Protocol, adopted in 1997, signatory countries were incentivised to put in place instruments to lower the emissions of their most polluting industries, chief among which was power generation. This led to the decision to set up the EU Emissions Trading System (EU ETS) under the Clean Development Mechanism/Joint Initiative (CDM/JI).[5] This market-based cap-and-trade system was designed based on the exchange across borders in Europe of EU allowances (EUAs) to emit certain volumes of carbon dioxide (CO_2) and other greenhouse gases based on National Action Plans (NAPs), whereby the amount of EUAs freely allocated to polluting industries would gradually be replaced by auctions.

Meanwhile, certain EU Member States, initially Germany and Denmark, started promoting the development of carbon-free power generation in the form of RES-E. Promotion of RES-E took the form of both financial support and grid access/dispatch privileges. This move was

[5] More information on the CDM/JI is provided in Article 12 of the Kyoto Protocol, http://unfccc.int/resource/docs/convkp/kpeng.pdf.

sanctioned by the EU in the first Renewable Energy Directive of 2001,[6] enabling a fast development of RES-E power generation throughout the region. This first Directive provided a framework for non-mandatory RES-E consumption targets for each Member State and opened the door for the establishment of national support schemes for RES-E.

The second Renewable Energy Directive, of 2009,[7] confirmed this policy orientation on the part of the Union with the establishment of the 20-20-20 targets: 20 per cent absolute CO_2 equivalent (CO2e) emission reduction from a 1990 baseline, 20 per cent of RES in the Union's final energy consumption and a 20 per cent increase in the Union's energy efficiency, to be reached by 2020. This translated into mandatory RES-E deployment targets for each Member State as a percentage of final energy consumption. By the end of the 2000s, all Member States had adopted national standards for the promotion of RES-E, both for technical aspects (grid access and dispatch) and for financial support.

As of 2015, EU Member States seemed generally on track to reach their mandatory RES deployment targets. In October 2014, the EU Council approved a new set of targets for the 2020–30 period, pledging to reduce CO2e emissions by 40 per cent, setting a Europe-wide mandatory target of 27 per cent RES in the Union's final energy consumption and establishing an indicative target of increasing energy efficiency in the region by 27 per cent.[8]

III The Unveiling of the EU's Conflicting Policies

The Union is making progress on the completion of the Internal Energy Market and the attainment of its climate action objectives. However, as early as the late 2000s, political observers and market participants had started to question the compatibility of the policies implemented to fulfil these goals.

6 Directive 2001/77/EC of the European Parliament and of the Council of 27 September 2001 on the promotion of electricity produced from renewable energy sources in the internal electricity market, http://eur-lex.europa.eu/legal-content/EN/ALL/?uri=CELEX:32001L0077.

7 Directive 2009/28/EC of 23 April 2009 on the promotion of the use of energy from renewable sources and amending and subsequently repealing Directives 2001/77/EC and 2003/30/EC, http://eur-lex.europa.eu/legal-content/EN/ALL/?uri=CELEX:32009L0028.

8 Conclusions of the EU Council of 23/24 October 2014, http://data.consilium.europa.eu/doc/document/ST-169-2014-INIT/en/pdf.

Operational and financial support provided to RES-E producers on a national basis meant not only that the development of RES-E production happened largely outside the energy market, but also that cross-border trading of energy from renewable sources was constrained.

The fact that financial support to RES-E installation was largely provided on the basis of feed-in tariffs meant that RES-E volumes were directly absorbed by the transmission system operator (TSO) – or another entity, depending on the national setting – without being traded on the market. Combined with priority dispatch provisions, this incentivised RES-E producers to generate as much power as they could, without regard for market prices or system needs. By the end of the 2000s, the vast volumes of RES-E produced had led to a significant shift of the merit order and frequent occurrences of negative prices.

National-based support and arrangements for RES-E also incentivised RES-E developers to install facilities in Member States where financial reward was the most generous or grid access conditions were the easiest, rather than investing according to traditional market-based considerations, that is, where RES-E production would be the most efficient and demand would be highest.

The absence of cross-border tradability of RES-E attributes created another layer of deviation from internal energy market objectives: most RES-E support schemes were designed in a way that would *de facto* restrict the consumption of RES-E to the Member State where it had been produced. This impeded cross-border exchanges in a growing share of the power generated in Europe and led to significant peak – negative and positive – price divergence between Member States at certain times.

Finally, the fast development of RES-E production in Europe also had unexpected effects on the EU ETS: RES-E support policies reduced demand for other types of energy and EUAs, thus acting as a 'substitute' for the EU ETS. As a consequence, prices of EUAs fell drastically, from EUR 30/t in 2008/2009 to less than EUR 5/t in 2015. The lack of consideration of the effect of overlapping policies and the rigidity of the EU ETS therefore dampened the carbon price, which failed to provide investment signals for low-carbon technologies in the energy industry and other ETS sectors.

All these elements, together with the economic downturn of the late 2000s, have left the energy markets in Europe in a fairly precarious state. Investments are mainly driven by state support and focused primarily on RES-E development; operators of flexible generation plants are pitching Member States to set up capacity remuneration mechanisms to finance stranded investments in flexible power generation; and despite efforts to revitalise carbon markets with back-loading of EUAs and the Market Stability Reserve, it is unlikely that the EU ETS will be the driver for investment in low-carbon technologies in Europe until the mid-2020s. This presents a fairly grim picture, only about twenty years after development of the European framework for the electricity sector began.

C Challenges and Opportunities for Market Participants in the Run-up to 2020

No actor in the electricity field can ignore the difficulties faced by the sector, or the ways in which conflicting European and national policies have contributed to worsening the situation. Little of the legislative framework is likely to change until the end of the decade. However, a number of regulatory and market-driven initiatives may contribute to creating a brighter future for the electricity sector in anticipation of a recovery in the European economy.

I Developing a Coherent and Comprehensive Market Design

The third energy package was admittedly one of the most ambitious frameworks adopted by the EU in any of the sectors it covers. Directive 2009/72/EC and Regulation (EC) 714/2009 set elaborate rules for the unbundling of system operation from generation/supply activities, and for cross-border trading of energy products. These rules were later to be expanded by more precise regulations on various operational, connection and market aspects of the electricity sector in the form of network codes and binding guidelines. By the end of 2015, as many as ten network codes and binding guidelines were being developed by ENTSO-E on the basis of Framework Guidelines established by ACER.

The Capacity Calculation and Congestion Management (CACM) guideline, the first of these market instruments to be officially adopted and turned into a Regulation,[9] establishes the rules for capacity calculation and allocation as well as congestion management practices in the day-ahead and intraday timeframes; it also paves the way for parallel rules applicable to the forward timeframe in the Forward Capacity

[9] Regulation (EU) 2015/1222 of 24 July 2015 establishing a guideline on capacity allocation and congestion management, http://eur-lex.europa.eu/legal-content/EN/TXT/?uri=uriserv:OJ.L_.2015.197.01.0024.01.ENG.

Allocation (FCA) guideline, which was adopted in September 2016.[10] The third market code in preparation, the Electricity Balancing (EB) guideline, focuses on the balancing timeframe and establishes a framework for a gradual move towards compatibility and harmonisation of balancing arrangements and products across Europe. This guideline deals with a highly complex matter and was not expected to be adopted before the first half of 2017.

The three market guidelines will be the cornerstones of the future design of electricity markets. Based on Regulation (EC) 714/2009, they define the rights and obligations of market participants, TSOs and power exchanges (as Nominated Electricity Market Operators or NEMOs). All parties have put a great deal of effort into making sure these Regulations will provide a sound framework to facilitate the exchange of electricity throughout Europe, facilitate competition between market participants without discrimination and phase out privileges enjoyed by either TSOs or particular market participants.

However, the adoption of the guidelines is only the first step towards ensuring a properly functioning market design: the guidelines indeed only set out the principles for the development of specific methodologies, which so far remain undefined, ranging from the harmonised capacity calculation methodology and the common grid model to the methodology for coordinated re-dispatching and countertrading or fall-back procedures. Implementation of the guidelines, with elaboration of methodologies at the local, regional or European levels, will therefore be key to the success of the third package.

Pre-implementation projects are being run to speed up change, such as the ENTSO-E-led bidding zones review (CACM), the Balancing Pilot Projects and various other initiatives in the balancing timeframe (EB) and the adoption of EU Harmonised Allocation Rules (FCA). While well-intentioned, some of these projects may end up defeating their own ends, as demonstrated in part by the development of the Harmonised Allocation Rules, which led some TSOs to reduce the volumes of forward transmission rights to be allocated in exchange for the higher firmness requirements set in the rules.[11] Continuous attention will therefore be

[10] Regulation (EU) 2016/1719 of 26 September 2016 establishing a guideline on forward capacity allocation, http://eur-lex.europa.eu/legal-content/EN/TXT/?uri=uriserv%3AOJ.L_.2016.259.01.0042.01.ENG (accessed 1 December 2016).

[11] The second version of the EU Harmonised Allocation Rules were submitted by all EU TSOs to their respective regulators for approval in June 2016, and, as of November 2016,

required to ensure full compliance – not only of the guidelines, but also of all methodologies and rules established on their basis – with the fundamental objectives of Regulation (EC) 714/2009, namely 'setting fair rules for cross-border exchanges in electricity, thus enhancing competition within the internal market [and] facilitating the emergence of a well-functioning and transparent wholesale market with a high level of security of supply in electricity'.[12]

II Financing Investment in Capacity Adequacy

Since the beginning of the decade, the significant changes in power-generation patterns mentioned above have called the viability of investments in power generation into question. Many European utilities' decisions regarding investment in combined cycle gas turbine (CCGT) plants in the mid-2000s proved uneconomical as a result of the financial crisis and the plummeting demand for electricity linked to the economic downturn. Regulatory interventions impacting the market also bear their share of responsibility for the declining viability of existing assets and investors' lack of willingness to invest in new capacity – whether generation, demand-response or storage.

The excessive risk and uncertainty for investors, combined with a lack of sufficient forward price signals, has created an environment that deters decisions on new capacity investments, and even promotes divestments and/or mothballing of existing capacities in certain cases. Economic conditions associated with inconsistent regulatory actions tend to dampen price signals in electricity markets, preventing the market from delivering an optimal level of capacity, or at least a level of capacity with which decision-makers feel comfortable. Despite recent adequacy reports showing that Europe is generally in a situation of over-capacity,[13] fears of capacity shortage in the medium to long term

were in the process of being adopted by national regulators. More information at www.entsoe.eu/major-projects/network-code-implementation/Harmonisation-of-Allocation-Rules/Pages/default.aspx.

[12] Article 1 of Regulation (EC) 714/2009, http://eur-lex.europa.eu/legal-content/en/ALL/?uri=CELEX:32009R0714.

[13] According to ENTSO-E's *Scenario Outlook & Adequacy Forecast 2014–2030*, 'generation adequacy is expected to be maintained during the entire forecast period (in Scenario B and Scenario EU20, and in each reference point), however with a considerably decreasing margin level in the 2020s. It must be noted however, that under conservative Scenario A, at the winter reference points at and beyond 2020, the level of adequacy

have led some Member States to design and implement capacity remuneration mechanisms (CRMs).

Like many policy interventions of this type, the possible introduction of CRMs has the characteristic of creating uncertainty in the investment sphere as soon as discussion of the question begins. The way in which Member States handle the topic has an influence on the final outcome of the decision, as shown by the debate in Germany, where a full-blown CRM was rejected by the government but a mix of capacity and grid reserve – with rather similar effects to those of a CRM on the energy market – is in the process of being approved as a result of intense political and industry pressure.

The introduction of CRMs should never serve as an excuse to halt progress on improving market design and market conditions. If enacted, CRMs should be developed in such a way that they naturally phase out when the need for financial support to existing and/or new-build capacity disappears. Sharpening price signals in wholesale markets and improving market participants' ability to respond to such signals, encouraging more liquidity and greater competition, should always be the primary goals of Member States in this sense. The role of forward markets as a tool to highlight market fundamentals far in advance and signal the need for new-build years ahead is key. Forward transmission rights have underpinned the development of cross-border liquidity in the continental wholesale power market since the early years of liberalisation. Much remains to be done to guarantee fully firm access to the maximum available cross-border transmission capacity throughout Europe. Bilateral long-term contracts between market participants – strictly unbundled from any kind of transmission capacity reservation – also have a role to play in the design of current and future electricity markets to secure financing for investments further ahead than the period covered by forward transmission rights.

Whichever market instruments are chosen by investors to guarantee a return on their investment, they should remain available and free from the potential negative impact of ill-conceived policy interventions. This is true of CRMs, which have a damping effect on long-term markets. It is also true for European legislation aimed at securing financial markets and which becomes applicable to energy markets as 'collateral damage', such

becomes negative, underlining the need for further investments compared to what is confirmed today'. More details are available at www.entsoe.eu/Documents/SDC%20documents/SOAF/141031_SOAF%202014-2030_.pdf.

as the Markets in Financial Instruments Directive (MiFID) II and the European Market Infrastructure Regulation. Specific commitments have been made by the European Commission to accommodate the needs of energy market participants with regard to these pieces of legislation that were designed for financial markets, and to avoid redundancies with regard to the Regulation on Energy Market Integrity and Transparency (REMIT). While discussions are ongoing with regard to implementation of MiFID II, some market participants, are concerned about the effect a full-blown application of MiFID II to energy markets would have on competitiveness, energy security, affordability and sustainability.

III Letting the Value of Flexibility Emerge

A third challenge that will arise before the end of the decade will be that of responding to the electricity system's, and hence the market's, increasing need for flexibility. With a continuously growing share of intermittent RES-E in Europe's power-generation portfolio, flexible capacity – whether generation, demand or storage – will increasingly be required to ensure the security of electricity supply.

The different power systems in Europe have different needs and require different types of flexible capacity. Considering the current situation of over-capacity in Europe, system needs are generally satisfied in most EU Member States at the moment. Flexible capacity is thus already valued in the market, but its value remains low, as system needs are easily satisfied by existing installations.

Looking forward, considering both the closure/mothballing of a number of CCGT plants and the expansion of RES-E generation, a recovery in the European economy in the coming years could result in a situation where it is harder for the existing capacity, or where the existing capacity may even be insufficient, to meet system needs. Flexible capacity, that is, capacity that best responds to the requirements of the electricity system with the fewest limitations, will then be rewarded in the market.

However, for flexible capacity to be appropriately valued in the market, prices should reflect the reality of electricity supply and demand. A number of measures will be required in order to let the full value of scarcity emerge, including:

- the adaptation and harmonisation of price caps and floors at a level that does not interfere with market forces in setting the price;

- the guarantee that market participants will not be unduly constrained to bid in the market at high prices in times of scarcity;
- the removal of exit and entry barriers, such as licensing requirements and licensing or supervisory fees or measures preventing free market exit;
- the promotion of market-based congestion management practices by TSOs, without discriminating between national and foreign market participants;
- the swift entry into force of more efficient transmission capacity allocation solutions, such as the issuance of fully firm forward transmission rights at every bidding zone border in Europe, in both directions; the implementation of flow-based capacity calculation in day-ahead markets throughout Europe; and the recalculation of intraday capacity after the clearing of the day-ahead flow-based market coupling;
- greater clarity in the rights and duties of TSOs in times of scarcity, especially in the case of scarcity situations in several markets;
- the eradication of all regulated end-user energy prices, except in cases of fuel poverty.

These measures will increase the efficiency of the market and improve price signals during episodes of scarcity, ensuring that all types of capacity – generation, demand and storage – can be used properly and valued based on a level playing field. They will provide incentives for new entrants, and opportunities for all market participants. Temporary price spikes will be absorbed thanks to better usage of resources across borders and market participants' hedging of their market risks. They will allow the energy market to provide the appropriate signals for the use of and need for each type of product and/or asset.

Scarcity-reflective prices are also necessary indicators for investments in generation, demand-response management and storage. Without price spikes, it is unlikely that the demand-side response will be appropriately valued in the market. Decision-makers should not be scared of scarcity and increased price volatility. Prices should be allowed to reflect the value of scarcity during times of system stress and high demand for power; similarly, when energy is abundant, prices should be allowed to reflect the value of displacing capacity and even become negative – which would provide signals for storage operators/investments – if they are not caused by non-market-based policies. Likewise, volatile energy prices, when not induced by flaws in the market design – for example, lack of transparency, excessive reserve margins or any other distortion to price formation – are a sign that the market is reacting appropriately and quickly to demand and supply signals. Natural volatility of the markets does not lead to greater systemic risks or higher prices for end-consumers, provided that these customers are fully aware of market risks and are able to use the appropriate hedging instruments available, or to outsource these activities.

D Ensuring Market-Based Promotion of Climate-Friendly Generation by 2030 and Beyond

Beyond the measures mentioned above to safeguard the internal energy market and improve its design, European and national decision-makers will need to start rationalising and aligning energy and climate action policies. The integration of these two policy streams under the authority of the same Commissioner at European level is a clear sign that the extent of the problem is understood in the political sphere. However, further reforms will be needed to ensure the viability of a secure, sustainable and affordable energy market in Europe.

I Putting the ETS Back at the Centre of the Climate Action and Tackling Overlapping Policies

As mentioned above, the EU ETS was designed as a European, market-based tool to ensure the reduction of greenhouse gas emissions. Experience over the past decade has shown a number of flaws in the design of the mechanism and exposed its lack of flexibility to adapt to changing market conditions. Appropriately tailored and dimensioned, the EU ETS nonetheless has the potential to deliver decarbonisation in the electricity sector, on a truly European basis and potentially at a much lower cost than other existing policies. In the words of the European Commission, the EU ETS is – or should be – 'Europe's flagship tool for tackling climate change and to place the EU on track towards a low-carbon economy'.[14]

The market for EUAs was oversupplied by 2.1 billion in 2013, and the European Commission expects this surplus to rise to 2.6 billion by 2020. A variety of factors may explain this situation. Some may be considered historical or temporary, such as the overly generous allocation of free allowances to certain industries by Member States, or the declining demand for energy and manufactured goods linked to the economic

[14] European Commission press statement of 15 July 2015, http://europa.eu/rapid/press release_IP-15-5358_en.htm.

recession. Others are truly structural, such as the progressive decoupling of emissions from gross domestic product (GDP) owing to the ongoing move towards a less energy-intensive economy, or the effect of overlapping climate action policies. As a consequence, the price of EUAs has fallen to a level that impedes the EU ETS providing meaningful signals for investment in low-carbon technology.

The Market Stability Reserve, formally approved by the EU Environment Ministers on 18 September 2015,[15] is a first step in the direction of reform of the EU ETS. This measure will enable the absorption of part of the surplus of EUAs in the market starting in 2019, targeting the EUAs back-loaded in 2014 and the unallocated allowances of Phase 3. It will bring flexibility to a mechanism that is otherwise very rigid, without the need to undergo lengthy negotiations in the European Parliament and Council. However, the adopted extraction rate of 12 per cent remains low in comparison to the current oversupply of EUAs, and the mechanism is not likely to bear fruit until the mid-2020s.

Further reform is therefore needed. The European Commission launched discussions in the summer of 2015, but hefty negotiations between the European Parliament and Council were still ongoing as of November 2016. For the EU ETS to fulfil its role, a reform should focus on the following points.

- *A review of the industry benchmarks and activity levels used in determining the pool of EUAs available to the market* – more frequent reviews of benchmarks and activity levels should be introduced, possibly every three years. Such reviews should strike a balance between investor certainty and reflection of the latest technological developments in low-carbon technology in order to bolster change.
- *Consideration of overlapping policies and interactions between the ETS Directive and other relevant energy and climate action legislations* – the overlap between different climate policies directly undermines the EU ETS by reducing demand for energy and EUAs, thus acting as a 'substitute' for the EU ETS. EU institutions should ensure that carbon reductions made through out-of-market policy mechanisms have no effects on the dynamics of the EU ETS. Carbon reductions achieved through out-of-the-market policies should be accounted for and

[15] European Council press statement of 18 September 2015, www.consilium.europa.eu/en/press/press-releases/2015/09/18-greenhouse-gas-emissions-creation-of-market-stability-reserve-approved/.

factored in *ex ante* in a transparent manner in the determination of EUA volumes for Phase 4. Possible review mechanisms could be established to ensure a transparent periodic review of the cap and volumes during Phase 4.

- *The gradual switch towards full auctioning of allowances* – full auctioning is required to ensure efficient price discovery and transition to a low-carbon economy at the lowest cost to society. The proportion of auctioned allowances should increase with the progressive decrease in carbon leakage concerns – which may remain temporarily under specific protection via free allowances, according to transparent assessment rules preferably run by EU institutions in order to avoid national interference.
- *The extension of the EU ETS to more sectors and small installations* – trading of EUAs should be gradually extended to other carbon-intensive sectors, and the EU ETS cap adjusted accordingly. The addition of the heating and transport sectors is an obvious first step, with agriculture possibly included at a later stage. Small installations of ETS sectors, which are not currently covered by the ETS Directive, should also be integrated.

From an international perspective, cap-and-trade systems for greenhouse gas emissions are the only mechanism proven to be compatible on a global scale. Reinforcing the EU ETS would allow delivery of carbon abatement in Europe at the lowest cost, but would also signal to Europe's international partners its willingness to pursue decarbonisation efforts in a manner that remains compatible with policy options adopted elsewhere, for example, in the most progressive US states and Canadian provinces.

II Phasing out Grid Privileges for RES

As mentioned earlier in this chapter, if policy-makers wish to pursue the adventure of market liberalisation and at the same time push the decarbonisation of the electricity sector, a fundamental reflection on the drivers of climate action policies is needed. Market participants, including RES-E installation operators,[16] have called for reform of the privileges

[16] According to the European Wind Energy Association, 'the Energy Union should establish a well-functioning European energy market to facilitate trade, enable fair competition and drive down costs for consumers and taxpayers'. More details of the EWEA's positions regarding RES-E integration in the market can be found at www.ewea.org/policy-issues/energy-union/.

enjoyed by RES-E producers. The diversity of support schemes for renewables currently in place in EU Member States is no longer compatible with the completion of the single electricity market if these schemes conflict with the EU ETS price signal. New EU legislation in this field is needed and the European Commission was expected to publish a review proposal of the Renewable Energy Directive and other instruments such as the third Electricity Directive and the Cross-Border Electricity Regulation at the end of 2016.

Reform should tackle all the privileges enjoyed by RES-E producers. This concerns, first of all, operational and connection privileges: all market participants should be exposed to balancing responsibility, including RES-E installations operating in countries mandating balancing responsibility for new-built facilities but operating under old contracts. Priority dispatch and priority grid access for renewables also need to be abandoned where they still exist. The aim of these measures is for generators to adapt their production to the needs of the system, for flows to follow prices and not be prioritised based on feed-in of specific technologies and for all capacity providers to be subject to the same connection procedures and costs.

Financial privileges enjoyed by RES producers should also be reformed. Many Member States have established support for RES-E producers in the form of feed-in tariffs. These mechanisms, which guarantee fixed financial returns to RES-E installations for each MWh produced, irrespective of market prices or system needs, have effectively segregated RES-E generators from the market. They also constitute trade barriers – implicitly or explicitly, depending on national legislation – which distort the internal market. All this limits liquidity and competition in power markets and, arguably, constitutes quantitative restrictions on cross-border trade.[17] All things being equal, these considerations also apply to direct or indirect support mechanisms for electricity generated from nuclear and fossil fuels.

EU Member States have had too much room for manoeuvre in relation to the implementation of EU legislation on renewable energy. The absence of coordination between national renewable energy support schemes leads to variety and incompatibility, which is certainly not in the spirit of an EU-wide internal market. Cooperation mechanisms

[17] See Advocate General Bot's opinion in the Ålands Vindkraft case (C-573/12), issued on 28 January 2014, http://eur-lex.europa.eu/legal-content/EN/TXT/?uri=CELEX:62012CC0573.

provided for in the Renewables Directive have proved unpopular: only one has been created, between Sweden and Norway. This has resulted in a pronounced fragmentation of the regulatory framework around renewables. Increasing the share of renewable energy in the consumption mix without at the same time ensuring the harmonisation and tradability of renewable attributes has made the integration of large volumes of electricity from renewable sources into the wholesale market increasingly difficult.

A fundamental reform aimed towards the harmonisation of financial support for RES-E producers is also needed in terms of public acceptability of climate action policies. RES-E support costs levied on end-consumer bills have come to comprise more than 50 per cent of retail customers' electricity bills in some Member States. Citizens' energy costs drastically increased at the same time that electricity market prices were going down; the justification for energy market liberalisation was rendered more difficult while it lowered wholesale prices. Moving the financing of schemes to support renewables and other climate action policies from electricity bills to state budgets could be a temporary solution. This would have the benefit of aligning private incentives and socially optimal investment signals and reducing the cost of RES and demand-side response development. It would also limit and factor in the development of 'behind the meter' solutions, and spread the costs of climate policies evenly.

III Ensuring the Financing of RES through the Market

The fundamental question in the post-2020 environment will be how to ensure the coexistence and financing of all types of generation, demand-side response and storage in a liberalised environment. There is a real possibility that it will never be possible for RES-E, at least for certain technologies, to be financed entirely according to competitive market rules. Nascent, immature technologies are likely to require public support for many years to come, preferably through R&D budgets at the EU or national levels. For others, notably on-shore wind power and solar photovoltaic installations, everything possible should be done to ensure their financing through market means, including a stringent application of the European Guidelines for Environment and Energy Aid.

Greater awareness should be raised regarding the transfer effects between the carbon price and support scheme, as a stronger carbon price will be largely passed through the wholesale power price. Provided that meaningful reforms are implemented in the ETS Directive, RES-E installations could obtain a significant part of their remuneration through this price signal, as they would not have to surrender emission allowances for their production.

Direct financial support for RES-E, where still necessary, should take account of market prices and be awarded competitively. Recent reforms in Germany and Great Britain to implement feed-in premium and contract for difference systems, respectively, are heading in the right direction. In both cases, the operator is guaranteed a fixed return – per MWh produced – but has the obligation to commercialise the energy produced on the market. Specific provisions avoid the production of power at times when the system does not need it, that is, when market prices fall below zero. Also, both schemes operate on the basis of public tenders, which should guarantee that the best available projects are awarded support, at the lowest cost possible for society.

Looking further ahead, such public tenders could be organised at the European level. Considering the pan-European RES development target for 2030, this could become a solution acceptable to more Member States than originally expected, provided that appropriate cost-sharing agreements can be drawn up.

In the longer term, and especially for technologies like on-shore wind and solar photovoltaic systems, the need for public support should disappear. At this point, ensuring the cross-border validity of renewable electricity attributes for marketing purposes when imported and exported across borders will be key. In this sense, the market for Guarantees of Origin will play a growing role in supplementing the energy market and the EU ETS in the financing of energy from renewables.

E Conclusion

To yield the desired benefits in terms of competition and sustainability, the energy and climate framework for the years to come should contribute to creating a secure, liquid and well-functioning energy market through the full harmonisation and integration of regulatory measures introduced in recent years. A high priority for decision-makers will be to provide greater coherence between the EU ETS and other EU climate policies in order to limit distortions to the internal energy market to a minimum. Mechanisms to promote generation from renewables must be closely controlled to ensure a level playing field for investment in renewable energy production, to meet the RES-E development targets in a cost-efficient manner and to preserve Europe's internal electricity market and the EU ETS.

The establishment of sustainable energy markets throughout Europe for the overall long-term benefit of the economy and of society must be retained as a key goal of the EU's energy policy and strategy. Well-functioning energy markets have the capacity to provide clear signals to optimise supply and demand, let the value of flexible capacity emerge and enhance security of supply. By allowing market fundamentals to continue to form the backbone of EU energy and climate policy beyond 2020, they will provide a cost-efficient realisation of the EU climate objectives, improve the security of Europe's energy supply and offer an efficient model for others to follow.

7

The North American Experience

FEREIDOON P. SIOSHANSI*

A Introduction

US greenhouse gas (GHG) emissions peaked in 2005 and have been gradually declining since then. This, however, is mostly a reflection of the abundance and unprecedented low cost of shale gas and other contributing factors, not the result of any national policy. A recent proposal from the Environmental Protection Agency (EPA), orders by the Federal Energy Regulatory Commission (FERC) and a number of other federal initiatives are likely to result in gradually declining GHG emissions. The most important development, however, is not a binding federal mandate but rather the unexpected emergence of distributed energy resources (DERs), which are propelled by rapid technological advancements, falling prices and state-level supportive policies.

This chapter is organized into five sections, as follows. Section B provides an overview of US GHG emissions, their sources and trends; Section C an overview of federal initiatives affecting GHG emissions; and Section D an overview of state-level initiatives. Section E looks at California's climate law and its cap-and-trade scheme, the only one in the United States at present, and Section F explains why the rapid rise of DERs, supported by state-level policies, is likely to dwarf all other measures directed at promoting green electricity generation. Section G presents the conclusions.

B US Greenhouse Gas Emissions

Even though on paper – and on the map – the United States and Canada appear as single countries, both are composed of federated states, with significant autonomy among individual states, in the case of the United

* This chapter was concluded in December 2015 and therefore reflects the state of play at the time of writing.

States, and provinces, in the case of Canada. This includes regulations, or lack thereof, when it comes to promotion of green electricity generation and/or trade. This chapter is, for the most part, focused on the United States, with only minor references to Canada.[1]

US GHG emissions peaked in 2005 and have been gradually declining since then, mostly owing to the abundance and unprecedentedly low cost of shale gas, but also due to other contributing factors. According to recent IEA statistics, they are projected to continue declining until 2040.[2]

The United States is currently the second largest GHG emitter worldwide, having been overtaken by China in the mid-2000s. Needless to say, any initiative to reduce global emissions must include the United States if it is to succeed or to be meaningful. Canada's contribution, by comparison, is relatively minor. Yet Canada is among the top ten global emitters, which collectively account for over three-quarters of global emissions. As other contributors to this volume have noted, cross-border electricity trade and international trade in green certificates is an effective means of encouraging countries with opportunities to cut GHG emissions at a lower cost to do so while being compensated by those where abatement costs are comparatively high. In reality, however, this is not necessarily happening, as states and countries such as Denmark or California embark on unilateral efforts to reduce their GHG footprint even when the *relative* costs of doing this may be excessive compared to the costs in other countries, such as India or Indonesia, to name just two.

US states vary considerably in terms of the carbon intensity of their economies, which reflects historical patterns of economic development, the contribution of heavy industry to state gross domestic product (GDP), relative energy prices and energy and environmental policies, or lack thereof.[3] Wyoming stands out as the most carbon-intensive in the

[1] It must, however, be noted that Canadian provinces, like US states, vary from one extreme to another in terms of their environmental and climate perspectives. Ontario, for example, has virtually phased out its entire coal-fired fleet following a political decision to do so. More recently, Alberta has proposed to phase out its coal-fired plants by 2030. Hydro-based Quebec, the Maritime Provinces and British Columbia are far more concerned about climate issues, perhaps because they enjoy very low carbon intensities, than some of the other provinces that are heavily dependent on fossil fuels and exports of the same.

[2] US Energy Information Administration, *Monthly Energy Review*, September 2013, and the *Annual Energy Outlook 2014*.

[3] This applies to Canadian provinces, as noted in note 1.

United States, while the District of Columbia – not a state, but a federal district – appears at the opposite end of the spectrum.[4]

Political pundits mockingly point out that the District of Columbia only produces hot air, and therefore few emissions. A number of populous states, including California and New York, rank low on the carbon-intensity scale, partly because most of their GDP is generated from non-energy-intensive sectors such as finance, services, high-tech, education, medical services and light, high-value-added manufacturing or agriculture.

To gain a better understanding of where the US emissions are coming from, it is helpful to attribute GHG emissions to various sectors of the economy. According to EIA statistics, the power-generation sector is the single biggest contributor, followed by transportation.

Clearly, all sectors of the economy contribute to the total, and all have to clean up, so to speak, if total emissions are to be reduced over time. Yet it is also obvious that it makes sense to focus on power generation and the transport sector, since this is where the opportunities to reduce emissions are greatest.

Within the power sector, coal is currently the single biggest contributor to total generation, despite the fact that natural gas and renewables have been gaining market share in recent years. Over time, the contribution of primary fuels in the US generation mix has declined, indicating a gradual shift to cleaner fuels. Nuclear power, which is virtually carbon-free, has been gradually losing market share, however, mostly due to economic factors.[5]

As is widely known, the recent abundance of shale gas and its low price in North America explains, for the most part, the rapid conversion from coal to natural gas-fired generation in the United States. The Environmental Protection Agency (EPA)'s recent restrictions on emissions from existing and new coal-fired plants are likely to intensify the gradual erosion of coal's share in the power-generation sector.

The US Energy Information Administration (EIA) projects that by the mid-2030s, if not sooner, natural gas will overtake coal as the primary fuel in power generation in the United States.[6] This will have a significant long-term impact on US GHG emissions, since natural gas emits roughly half as much CO_2 per unit of electricity generated as coal does.

[4] 'Review & outlook', *Wall Street Journal*, 4 June 2014.

[5] EIA, *Annual Energy Review*, September 2012. [6] IEA, *Annual Energy Outlook 2014*.

Another major factor contributing to the overall expected drop in US GHG emissions is the fact that demand for electricity in the United States is projected to grow at a rather sluggish rate – below 1 per cent per annum, or even less, depending on whose projections are used. This low demand growth rate, plus state-level mandatory renewable portfolio standards (RPS), further described in Section D, means that overall emissions from the US power sector will be falling for the foreseeable future.

The contribution of coal in the generation mix varies between states, from negligible or nil, as in the case of California, to near total dependence, as in the case of Wyoming.[7] This disparity partly explains why the US Senate was unable to pass a federal cap-and-trade bill, even though the House had passed it with a slim majority, during President Obama's first term in office.

C Federal Initiatives to Reduce US GHG Emissions

In August 2015, the Environmental Protection Agency (EPA) issued the Clean Power Plan (CPP), which requires overall US emissions to be reduced by 32 per cent from their 2005 levels by 2030. With a few exceptions, noted below, there are no binding, mandatory or overarching laws, legislation or regulations *specifically* focused on promoting green electricity generation or trade at the federal level. The last time federal legislation on a nationwide cap-and-trade scheme was attempted, it (barely) passed the House but did not clear the Senate. Political gridlock in Washington makes it nearly impossible for similar legislation to pass the US Congress, certainly before the end of President Obama's current term, and possibly thereafter, depending on the outcome of the 2016 election.

The only exceptions known to the author are the following, which arguably have the potential to reduce US GHG emissions over time even though they are *not* as specific or *effective* as a federally mandated carbon tax and/or a federally binding cap-and-trade scheme, as was proposed but ultimately defeated by Congress during President Obama's first term in office.

- *Clean Air Act* and amendments to the same, which – broadly speaking – have provided the EPA with a mandate to restrict polluting forms of

[7] 'Black, brown & green states of America', *New York Times*, 2 June 2014.

generation, particularly from coal-fired plants, since the mid-1970s, when the original law was passed by the US Congress.[8] The EPA, while highly effective in reducing emissions of nitrogen oxides (NOx), sulphur oxides (SOx), mercury, soot and other toxic pollutants from the power-generation sector, has thus far had limited success in curbing GHG emissions – but this may change if the agency succeeds with its recent proposal, further described below.

- *Federal Energy Regulatory Commission* (FERC) policies encouraging increased and coordinated investment in electricity transmission, which directly and indirectly affects green electricity generation and trade. These policies, including Order 1000, encourage additional investments in transmission in part to facilitate transmission and integration of renewable forms of energy into the interconnected grid, as renewables tend to be plentiful in areas that are far from major load centres. FERC has consistently and forcefully advocated the practice of 'socialising' the costs of additional transmission when determination of cost causation and attribution of benefits become excessively complicated or convoluted across vast interconnected grids such as those prevalent in the United States.[9]
- *EPA* policies and regulations specifically focused on GHG emissions as well as restrictions on emissions of certain pollutants, which encourage green electricity generation.

Particularly noteworthy is EPA's proposal, unveiled in early June 2014, to reduce overall US GHG emissions by 30 per cent from the 2005 level by 2030. If it survives multiple expected legal and procedural challenges by those who oppose it, this proposal will be the most significant attempt at the federal level to address climate change.

Given the political situation in Washington DC, few believe any other major environmental legislation is likely to be passed over the next couple of years, and, depending on the outcome of the 2016 elections, even after that. The politics of coal, like all politics, are very much

[8] The Clean Air Act and its amendments give the EPA the responsibility for regulating emissions that are deemed to be harmful to health and safety – which applies directly to toxic emissions such as arsenic and mercury. The argument is less clear in the case of GHG emissions, which some argue are not harmful. In a seminal decision, the US Supreme Court backed the EPA's regulation of GHG emissions, yet the agency's ability to do so effectively remains to be seen.

[9] Similar practices are in place to various degrees in wind-rich states including Texas and California, where investments in transmission expansion are deemed socially beneficial to the achievement of wider societal goals such as meeting RPS standards.

determined on a state-by-state level and will be strongly dependent on the outcome of the next Congressional elections.[10]

It should be noted that federal production tax credits (PTC) and investment tax credits (ITC), as well as federal loan guarantees, R&D support and a number of other schemes, have also played a significant role in promoting the development of renewables over the past two decades.

D State-Level Initiatives

Developments with respect to environmental regulation in the United States are largely driven by state-level policies, which, in turn, are driven by local politics at state, local or occasionally regional levels. However, since electricity flows across state borders or, in the case of Canada,[11] national ones, it is also regulated by FERC (and its counterpart in Canada).

In this context, it is worth highlighting three broad categories of state-level initiatives that directly or indirectly support, subsidise and/or promote green energy generation:

- state-level renewable electricity generation targets or mandates;
- state-level energy efficiency requirements, appliance efficiency standards and/or building codes;
- distributed generation, notably from customer-side solar photovoltaic (PV) systems.

The best example of the first is mandatory and non-binding RPS currently in effect in twenty-nine states and the District of Columbia. While the target levels and deadlines vary from state to state, by most estimates, these programmes have been responsible for the bulk of the investments in renewable generation in states that require them since they were enacted, mostly in the 1990s and early 2000s.

For example, according to a recent study,[12] from 1999 to 2013, 69 per cent of US wind capacity installations were located in states with RPS

[10] *The Economist*, 'Wealth transfer and politics of coal', 7 June 2014.

[11] What is particularly noteworthy is that most Canadian transmission lines run north–south rather than east–west, mostly for provincial political reasons. This means that much of what happens in Canada comes under FERC's purview.

[12] R. Wiser and M. Bolinger, *2013 Wind Technologies Market Report* (Springfield, VA: US Department of Energy, 2014).

policies.[13] In 2013, the corresponding number was 93 per cent. Similar patterns are seen with solar, biomass, small hydro, geothermal and other forms of renewable generation. This demonstrates the significance of mandatory state-level RPS requirements even though there is no federal requirement for renewables currently in effect in the United States.

Partly as a result of these state-level initiatives, assisted by federal production tax credits and other support schemes, a number of states in America now generate more than 25 per cent of their electricity needs from wind power alone, while the corresponding number for the United States as a whole was 4.1 per cent at the end of 2013.

Examples of energy efficiency standards and measures include a number of programmes implemented at state level – some, such as in California, dating back to the mid-1970s. The details and scope of these programmes also vary, but to the extent that they reduce overall electricity consumption, they result in significant reductions in GHG emissions, since less generation is needed, all else being equal. Based on 2011 State Energy Efficiency Scorecard Rankings developed by the American Council for an Energy-Efficient Economy, states with aggressive energy efficiency programmes tend to reduce the growth rate of energy consumption and peak demand – both significant contributors to GHG emissions – over time.

There is a wide range of variation in spending levels, in how the programmes are funded and in who is responsible for their implementation and enforcement. A number of states, for example, have introduced energy efficiency resource standards (EERS), which work quite similarly to mandatory RPS laws. A number of states have systems of penalties and rewards to further encourage utilities to invest in energy efficiency. In a number of states, for example, utility profits have been 'decoupled' from sales volume – hence removing any disincentive to engage in energy efficiency on the part of investor-owned utilities.

To the extent that energy efficiency measures reduce electricity consumption, they clearly reduce the need for generation, which in turn reduces GHG emissions, depending on the prevailing energy-generation mix. California, which has had one of the oldest, most aggressive and comprehensive energy efficiency programmes in the United States since the mid-1970s, has managed to keep its per capita electricity consumption flat for more than thirty years, in contrast to the United States as a whole.

[13] 2013 Wind Technologies, August 2014.

Table 7.1 *Emissions in pounds of CO_2 per MWH of generation, US average California and PG&E*

CO_2 emissions for delivered electricity in pounds of CO_2 per MWh		
US average	California average	PG&E
1,216	659	445

More generally, it has been noted that differences among states in per capita electricity consumption are correlated with the two main political parties. Democratically leaning states – so-called blue states – tend to have lower per capita consumption rates than Republican-leaning states – so-called red states. Other variables, including customer attitudes towards energy, cultural and political factors, income and education levels and climate, as well as the state's overall political agenda or priorities, may have a bearing on states' per capita electricity consumption. Overall, it is fair to say that energy efficiency policies, building codes, energy efficiency standards for appliances, retail prices, climate, energy intensity and a number of other variables combine to produce dramatic differences in per capita electricity consumption among the states.[14]

As a result of these differences and the prevailing generation mix, states' carbon intensity varies greatly. Table 7.1 shows emissions in pounds of CO_2 per MWhr of generation for the United States, California and the Pacific Gas & Electric Company, based in San Francisco, which is among the biggest of the US utilities. This Table illustrates how persistent policies and politics applied over time combine to produce dramatic results.

As further explained in Section E, Table 7.1 also illustrates that in the case of California, taking further steps to reduce the state's already low carbon footprint may be politically less daunting than it would be for a state like Wyoming, which is nearly 100 per cent dependent on coal and has the highest per capita electricity consumption and the highest carbon intensity of any state in America. By contrast, reducing GHG emissions would be more expensive at the margin in California than in Wyoming – because the latter has plenty of low-hanging fruit that can be had at relatively low cost. This clearly demonstrates the advantages offered by cross-border trade among states and/or countries where vast opportunity and cost disparities exist.

[14] See www.energytrends.org.

As already pointed out, however, many such cross-state or cross-border opportunities are not explored or fully made use of owing to rigidities in local, state or regional regulations and policies. For example, the state of California limits the amount of electricity generated from renewables imported from out of state that may qualify towards a state's RPS requirement. While there are good arguments for such restrictions,[15] they impede cross-border trade when a country such as the United States as a whole would be better off.

Finally, since the mid-1990s, a number of states have adopted net energy metering (NEM) laws, which typically give customers a credit for any self-generation in excess of domestic consumption. Most of the self-generation is currently from solar rooftop PV systems, which have become cost-competitive with grid-supplied electricity in sunny regions with high retail tariffs and supportive NEM laws. Many experts believe that solar and other forms of self-generation, say from fuel cells, will reach grid parity in many parts of the United States that have high retail tariffs as well as in other parts of the world, which means that self-generation will be cheaper than buying electricity from the grid at prevailing retail tariffs.

As further described in Section F, together, these developments – namely the growth of renewables, continued improvements in energy efficiency and customer self-generation – are likely to have a much bigger impact on GHG emissions in the United States, and perhaps other mature OECD economies, than any other possible measures or policies. Electrification of the transport system is perhaps the only other significant development that can have a major impact on GHG emissions, provided that the electricity consumed is generated from clean resources.

E California's Climate Law and Its Cap-and-Trade Scheme

The state of California, the most populous and, perhaps, prosperous in the United States, would be the eighth largest economy in terms of GDP if it were an independent nation. Not only is it an economic powerhouse, but California also stands out as an outlier among the US states when it comes to energy and environmental policies. As already mentioned, California has the lowest per capita electricity consumption in the United States, one of the lowest carbon intensities and the lowest level of

[15] Among these arguments is local job creation, which would not take place if investments in renewables are made outside the California state boundary.

emissions per unit of electricity generated, and spends less on energy relative to the size of its GDP than the other major states in the country.

But what makes California unique is that it is currently the *only* state in America with an aggressive and binding climate change law. This law, passed in 2006, requires aggregate *state-level* emissions to fall to their 1990 level by 2020, and to fall 80 per cent below their 1990 level by 2050.

California intends to achieve this despite a growing population requiring more energy, water, transportation and infrastructure. It requires a multi-faceted series of programmes collectively contributing to reach the intended targets. The major contributors include the programmes and initiatives already mentioned, plus those listed below:

- energy efficiency efforts, building codes and appliance standards;
- 33 per cent *new* renewables by 2020;[16]
- a *cap-and-trade* (C&T) scheme;
- agreements to collaborate with other states or regions;
- low-carbon fuel standards;
- advanced clean cars, including electric vehicles (EVs);
- urban planning, zero net energy buildings;
- solar rooftop PV systems;
- energy storage systems;
- mandatory commercial recycling;
- a high-speed rail system;
- water conservation and efficiency measures;
- a myriad of other programmes.

The discussion below is focused on California's C&T scheme, which is managed by the California Air Resources Board (CARB), an agency whose boss reports directly to the Governor and is in charge of implementing the state's climate bill.

At the heart of the C&T scheme is a plan to gradually reduce the number of tradable allowances over time to 334 million metric tonnes of carbon dioxide equivalents (MMTCO2e) by 2020 – the first major target of the climate bill.[17] This requires a reduction of 3.4 per cent per annum until 2020.

It will *not* be easy. Between 2000 and 2012, state-wide total GHG emissions barely nudged downwards, from 466 to 459 MMTCO2e, despite

[16] Existing renewables such as old hydro and geothermal resources do not count towards the 33 per cent RPS. In effect, by 2020, the renewable portion is likely to exceed 50 per cent of total generation.

[17] *CARB Scoping Plan, 2008.*

Table 7.2 *Reductions to be achieved by California to meet its 2020 target*

Source of emissions	Amount of reduction (MMTCO2e)
Energy	25
Transportation	23
High global warming potential gases	5
Waste	2
C&T scheme	23

efforts on multiple fronts.[18] This goes to show just how difficult, and expensive, it will be for California to meet its 2020 target.

To meet the 2020 target, California has to reduce its business-as-usual 2020 emissions, projected to be 509 MMTCO2e, by 78, to 431. This must be achieved through the reductions listed in Table 7.2.

The C&T scheme, in other words, is expected to contribute roughly 30 per cent of the emission reductions by 2020. The rest will come from the other programmes managed through different schemes, policies, agencies and regulations.

To make matters worse, in the period after 2020, the low-hanging fruit will no longer be available. To meet the 2050 target, California would have to reduce its GHG emissions by 5.2 per cent per annum for thirty years, by roughly 11.4 MMTCO2e per year, despite a growing population.[19]

To put things into perspective, California's current GHG emissions constitute roughly 2 per cent of global emissions. Assuming successful implementation of its climate bill, the state's contribution is expected to be reduced to roughly 1 per cent of global emissions. Consequently, the state's Climate Bill is widely regarded as symbolic or inspirational. It is intended:

- first, to show that it is technically feasible;
- second, to show that it is economically viable – that is, it will not ruin the state's economy, prosperity and comparative advantage;
- third, possibly to inspire other states, regions and countries – most importantly the US federal government and the US Congress – to follow suit.

[18] *CARB's First Update to the Climate Change Scoping Plan: Chapter IV, Accomplishments and Next Steps*, May 2014.

[19] *CARB's May 2014 Scoping Plan.*

Not surprisingly, California has been on a mission – with missionary zeal, one might add – to convince others to join its green, low-carbon energy agenda by adopting similar targets and by tying their programmes to those of California. Its track record to date has not been stellar, with several failed attempts to bring neighbouring states or Canadian provinces into the fold. Although California has found a few sympathisers in Oregon, Washington, British Columbia and some other states and Canadian provinces, its attempts have mostly been met with indifference, reluctance and even hostility, notably from carbon-heavy Alberta and Wyoming and other so-called red states in the western United States. The sole successful case has been Quebec, an odd choice given that it sits on the opposite side of the continent, in a different country, and is blessed with ample supplies of hydroelectric power – making carbon and GHG a virtual moot issue. More than 95 per cent of Quebec's generation is from hydro resources, making it among the cleanest energy mixes in the world.

Auctions held by CARB so far have been deemed a 'success' by most measures, with significant volumes traded and prices hovering around 12 MMTCO2e.[20]

The proposal put forward by the EPA in June 2014, previously mentioned in Section D, offers an opportunity for various state and regional programmes, including the north-east's Regional Greenhouse Gas Initiative, to potentially link their efforts and, possibly, trade emission allowances or credits across state boundaries. Such efforts, however, currently suffer from complex fungibility issues, which need to be resolved over time.

Moreover,[21] in 2015, California's C&T programme will expand beyond electric utilities, refineries and industry to include fuel distributors and suppliers. The second and third compliance periods, covering 2015–2017 and 2018–2020, may pose challenges, and it is possible that current rules governing California's programmes will need fine-tuning or revising. In addition, to provide more certainty to Californian businesses and consumers, CARB plans to set a 2030 emission target and describe its vision for post-2020 carbon reductions.

Under most market scenarios, carbon allowance prices are anticipated to remain slightly above or near the allowance auction minimum reserve price during most of the 2013–20 period. However, the market could

[20] See InterContinental Exchange (ICE).

[21] This discussion is based on a contribution from Andy Van Horn of Van Horn Consulting.

tighten significantly with rapidly rising prices, particularly after 2017 if the Climate Bill's complementary programmes fail to deliver their projected reductions.

As previously mentioned, California has always recognised that, by itself, the state's climate bill – known as Assembly Bill 32 or AB 32 – cannot substantially reduce global or even US carbon emissions. According to Andy Van Horn, an expert on California's C&T market, a major challenge will be for California to reach out again to other western states to develop a viable regional GHG reduction programme that might satisfy the EPA's proposed Clean Power Plan, unveiled in June 2014.

If California hopes to provide a model for other states and countries to emulate, attracting other states and satisfying the EPA will necessarily involve simplifying elements of AB 32's current market design, as well as successfully conducting complex negotiations with neighbouring states. These efforts must be made at the same time as the C&T programme more than doubles in size in 2015, and while other states are also trying to decide how to address the EPA's proposed regulations for new and existing electric power plants.

F Why the Rise of the DERs Matters

DERs refer to a two-sided coin: one side is energy efficiency; the other is distributed self-generation. Taken together – and adding some storage and other peripheral smart devices – this may portend the end of the electric utility industry as we know it. That is what many believe, at least with regard to mature economies of the OECD. And if this prognosis is correct, or even partially correct, it has significant implications for traditional means of electricity generation, GHG emissions and electricity trade – the focus of this section.

To understand why DERs matter, it is instructive to first examine the energy efficiency side of the coin. If consumers collectively and aggressively invest in energy efficiency opportunities, the already modest projections of demand growth in mature economies of the world will be even smaller. This means falling sales volumes for the industry, whose traditional business model is to recover nearly all costs through volumetric tariffs. If sales fall, retail tariffs have to rise – since the industry's costs are mostly fixed – which will encourage even more energy conservation and energy efficiency – a death spiral.

Why would consumers invest in more energy efficiency? The main reasons are as follows:

- Retail tariffs are already high in many countries and are poised to rise even higher, partly as a result of pressures to move towards a cleaner energy mix.
- Energy efficiency, which has always been cheap, is becoming even cheaper with continued improvements in efficiency of appliances, lighting and all manner of other devices.
- Buildings, where most electricity is consumed, are becoming more efficient, in some cases approaching zero net energy (ZNE) levels, which means they can generate nearly as much power as they consume.

For example, light-emitting diodes (LEDs), which are becoming cheaper by the day, have the potential to cut electric lighting load in half. This means that, since lighting constitutes roughly one-quarter of electricity consumption in most industrialised economies, LEDs alone have the potential to reduce one-quarter of the load by half. Street lighting, another major consumer of electricity, can be made 97 per cent more efficient using the latest advances in lighting technology. Examples such as these abound.

In short, consumers, confronted with rising retail tariffs and associated taxes and levies, have strong incentives to conserve energy, and the opportunities to do so abound and are expanding.

These and numerous other factors, too many to mention,[22] have combined to produce flat or even falling consumption patterns in a number of jurisdictions in the United States and elsewhere. Australia, where retail tariffs have nearly doubled in the past five years, has experienced flat or falling aggregate consumption in the past few years.[23] In the United States, where per capita electricity consumption has reached saturation levels, total consumption *can* be managed to be flat or even decline over time with reasonable and feasible application of higher energy efficiency standards and building codes, according to a study by the Institute for Electric Innovation.

The potential implications of the *other* side of the DER coin are even more stunning. With the rapidly falling costs and equally rapid improvements in technology, customer self-generation, also called distributed generation (DG) or distributed renewable generation (DRG), is becoming not only physically feasible but economically compelling.

[22] For examples see F. Sioshansi (ed.), *Energy Efficiency: Towards the End of Demand Growth* (Academic Press, 2013).

[23] See *Clean Energy Australia Report*, 2013.

Consumers in many parts of the world with high retail tariffs, including California, Hawaii, New York, Germany and Australia, are finding solar rooftop PV systems to be cost-effective, given the prevailing tariffs and net energy metering (NEM) laws or feed-in tariffs (FITs). Australia, a country with virtually no solar generation in 2000, now has more than one million roofs fitted with solar panels.[24] In some states, one in five residential customers now have solar PV systems.

A similar trend has been seen in Hawaii, where average residential retail tariffs are around 40 cents/kWh. By contrast, solar-generated power can be leased for around 16 cents/kWhr.

The net result is that a growing number of consumers are becoming *prosumers* – they produce nearly as much as they consume, at least during the sunny parts of the day. The recent rise of solar PV systems in the United States amounts to a stunning 120-fold increase since 2003, from 97 MW to 12 GW by the end of 2013, and rising.[25] The aggregated effect of thousands of solar PV systems in individual homes in places such as California has had the effect of dramatically changing the traditional pattern of generation and consumption.

Add the rapid rise of intermittent renewables and the daily load pattern is projected to change dramatically by 2020 in California. And, as dramatic as these trends are, that is not the end of the DER story. As the cost of energy storage drops, technologies such as batteries are likely to become more compelling and commonplace. A prosumer with an electric vehicle (EV) can become virtually self-sufficient within a self-contained micro-grid. It is not hard to imagine entire neighbourhoods, universities, hospitals, shopping malls and office buildings becoming first ZNE, and then, after adding some storage and/or self-generation, virtually self-sufficient electrical islands, with little need to buy electricity from the grid, relying on it primarily for reliability, load balancing and backup services.

Such scenarios, considered outlandish only a few years ago, are now moving into the mainstream. For further information on the rapid advances in distributed generation, see this author's recent edited book on the subject.[26]

What are the implications of advances in DERs for green electricity and trade? The answer to this question depends critically on assumptions about the speed with which DERs may evolve and their penetration in

[24] Ibid. [25] Environment America Research, *Lighting the Way*.

[26] F. Sioshansi (ed.), *Distributed Generation and Its Implications for the Utility Industry* (Academic Press, 2014).

different parts of the world. Although they vary, projections for the United States suggest that within a decade, solar PV systems are likely to become cost-competitive with grid-supplied electricity in large areas of the country – so-called *grid parity*.

As recently as 2010, rooftop solar PV systems in the United States cost around US$ 6.40 per watt installed. By the first quarter of 2014, the cost had dropped to US$ 2.35. In 2010 a total of 340 MW of solar PV systems were installed in the United States, whereas in the first quarter of 2014 alone, 1,330 MW were installed. Similar trends apply to other regions of the world with high retail tariffs. If such trends persist – or accelerate, as many expect – they may lead to several important outcomes within developed economies:

- Volumetric electricity sales – that is, kWhs purchased from incumbent utilities through traditional networks – are likely to dwindle over time, certainly within mature economies with high and rising retail tariffs and rising levels of DERs.
- Electricity generation from large centralised thermal power plants, especially from coal-fired units, is also likely to dwindle as the need for baseload generation diminishes.
- Electricity generation from peaking and cycling units is likely to remain viable, since such resources will increasingly be needed to complement intermittent renewables plus fluctuations in self-generation.
- Energy storage technologies are likely to become more important in a future that features a large percentage of intermittent renewables and distributed self-generation.
- *Prosumers* will become more active participants in balancing load and generation, including distributed self-generation and storage, perhaps through increased ownership of EVs.
- New and innovative tariffs will emerge to facilitate the transition to a distributed, intermittent future with bidirectional flows and complicated transactions among market participants.
- The roles of the grid operator and distribution network would have to be redefined in the emerging environment.[27]
- Transmission will remain vital and will have to assume a more active role in integrating and delivering intermittent power from remote locations to major load centres while keeping the network stable and reliable.

[27] This is already taking place, at least in populous states of New York and California; refer to *EEnergy Informer*, June and September 2014 issues for further details.

- The grid – the network that connects consumers, producers and their intermediaries – will become vastly more important and central to the overall operation of the complicated transactions that will take place among market participants.

What would these developments portend for cross-border trade and green electricity? This is difficult to predict and even more difficult to generalise. For the United States, the generation sector is expected to become greener over time as existing old coal-fired plants are decommissioned, either because they are inefficient and marginally viable and/or because they will not meet stringent emission requirements imposed by the EPA. Few, if any, new coal-fired plants are expected to be built, also due to new EPA proposed restrictions. Since most new investments are expected to be renewable and gas-fired, overall US GHG emissions can be expected to drop. Whether a national market for green certificate trading will evolve in the United States is a matter of speculation, since the EPA's proposals have only recently been released.

Regarding the DER trends highlighted above, one can speculate that green credit may be attached to energy efficiency and/or green self-generation investments in recognition of their contribution towards lowering GHG emissions, as well as other environmental benefits. Zero net energy buildings and low-emission vehicles, for example, can be encouraged through green certificate schemes. California already offers green certificate credits to buyers of EVs and similar schemes could be adopted elsewhere.[28]

As described in the preceding sections, despite the lack of strong federal initiatives such as a national C&T bill or carbon tax – which would naturally lead to a national green trading regime – many changes are happening as a result of EPA regulations, both those already imposed and those that have been proposed. Moreover, individuals, communities, cities, states and individual companies are making unilateral and voluntary decisions to reduce their carbon footprint. Consider the following examples:

- The city of Palo Alto already operates a carbon-neutral utility system.
- Tesla's massive new battery factory, which is under construction in Nevada, is designed to run on 100 per cent renewable energy.

[28] California's low-emission vehicle scheme is controversial since it gives a rebate to buyers of electric and hybrid vehicles, who tend to be affluent. Opponents of the scheme complain that the poor are subsidising the rich who can afford expensive Tesla cars, for example.

- Apple's new office building in Cupertino, CA, will be a ZNE building.
- Lancaster and Sebastopol, two communities in Southern and Northern California, have already adopted strict mandatory solar ordinances on all new buildings.
- The cities of Boulder, Colorado, Austin, Texas, and a number of others have adopted carbon neutrality targets.
- California's zero net energy (ZNE) requirements are expected to come into effect in 2020 for all new residential construction, and will apply to all new commercial buildings starting in 2030.

There is, undoubtedly, plenty of potential in organised, coordinated, centralised green electricity trade. The reality in the United States, at least up to now, is that much of the change that is taking place comes from smallish, uncoordinated, voluntary and mostly unilateral efforts by individuals, companies, communities or states to do some good.

G Conclusions

To date, the North American experience in encouraging green electricity generation and trade – both in the United States and in Canada – shows a lack of resolve and leadership at the federal level. Yet in both countries, significant progress is being made as individual states, provinces, cities and communities, as well as a growing number of corporations, take concrete steps to promote renewable energy while encouraging investments in energy efficiency. Progress is likely to be mostly bottom-up, rather than top-down.

Looking forward, the electric power sector is undergoing fundamental transformations that will gradually make generation cleaner and more decentralised, reducing GHG emissions from the power sector. Investments to expand and modernise the transmission and distribution network and to make it smart are likely to facilitate the integration of growing amounts of intermittent renewable and distributed energy resources, also leading to a reduction in GHG emissions.

In this context, the debate on cross-border electricity trade should also be revisited in view of the developments highlighted in this chapter. Policy-makers focused on encouraging green electricity trade must remain cognizant of fundamental changes taking place within the electric power space, particularly with regard to DERs.

8

Developing ASEAN Green Electricity in the Context of the ASEAN Community 2025

SUFIAN JUSOH

A Introduction

The Association of Southeast Asian Nations (ASEAN) represents an important region in Asia and the world. It consists of ten nations: Brunei, Cambodia, Indonesia, Lao People's Democratic Republic (PDR), Malaysia, Myanmar, the Philippines, Thailand, Singapore and Viet Nam. ASEAN Member States have a combined market of more than 620 million people and have seen positive economic growth for many years, associated with different levels of economic development. With this vast intra-regional market, which attracts both domestic and foreign direct investors, ASEAN is gearing up towards a higher level of integration now called the ASEAN Community, which was declared at the 27th ASEAN Summit in Kuala Lumpur, Malaysia in November 2015.

Strong economic growth in ASEAN has resulted in high energy demand in the region. According to the fourth ASEAN Energy Outlook, in 2013, ASEAN accounted for about 8.5 per cent of the world population, consumed about 4.5 per cent of the world's primary energy and was accountable for 5.7 per cent of total global energy production. As ASEAN economies are steadily expanding and its main energy indicators are below global averages, the fourth ASEAN Energy Outlook predicts that demand is expected to outstrip supply unless strong measures are taken to manage growth: this creates challenges regarding energy security and sustainable development.[1]

Few ASEAN Member States have so far been able to utilise new technologies to produce green electricity either for domestic consumption or for export.[2] In developing green electricity, ASEAN Member

[1] ASEAN Center for Energy, *Fourth ASEAN Energy Outlook* (Jakarta: ACE, 2015).

[2] Green technology is defined as technology whose use is intended to mitigate or reverse the effects of human activity on the environment; green energy is defined as natural energetic

States face many challenges relating to technology, raw materials, pricing, international trade law and environmental and sustainability concerns.

This chapter examines the issues and challenges facing ASEAN Member States in ensuring that the ASEAN Member States fulfil the energy policy targets as part of the establishment of the ASEAN Community.[3] Section B will examine the energy sources in ASEAN, before we turn to a discussion on green electricity and the ASEAN Community in Section C. This will be followed in Section D by an examination of the private sector-initiated and -funded green electricity projects in ASEAN Member States. Following this, the issues and challenges faced by ASEAN Member States in the implementation of green electricity projects are described in Section E. The chapter ends by drawing some conclusions.

B ASEAN and Energy Sources

Despite continued global economic uncertainty, the ASEAN region is expected to be able to maintain its dynamic growth momentum, underpinned by the strength of domestic demand in the medium term.[4] Growth for ASEAN is predicted to be about 7.1 per cent between 2016 and 2025, significantly more than the 5.4 per cent growth seen between 2004 and 2014. The Organisation for Economic Co-operation and Development (OECD) recently predicted that the real gross domestic product (GDP) growth rate in the Southeast Asian region will average 5.4 per cent per annum between 2014 and 2018, as against 5.5 per cent in 2000–7.[5]

Economic and population growth in ASEAN Member States has led to higher demand for energy and electricity. According to the Asian Development Bank, Southeast Asia, which includes ASEAN Member States, experienced Asia's fastest growth in primary energy supply from 1994 to 2009, at 6.6 per cent per annum, followed by east Asia, at 6.2 per cent per

processes that can be harnessed with little pollution caused, such as anaerobic digestion, geothermal power, wind power, small-scale hydropower, solar energy, biomass power, tidal power, wave power and some forms of nuclear power.

3 ASEAN Secretariat, *ASEAN 2025, Forging Ahead Together* (Jakarta: ASEAN, 2015).

4 Organisation for Economic Co-operation and Development, *Economic Outlook for Southeast Asia, China and India 2014: Beyond the Middle-Income Trap* (Paris: OECD, 2014).

5 Indonesia is projected to be the fastest growing economy within the ASEAN-6, with an average annual growth rate of 6.0 per cent in 2014–18, followed by the Philippines, with a growth rate of 5.8 per cent.

annum; south Asia, at 4.2 per cent per annum; and central and west Asia, at 1.4 per cent per annum.[6]

The two main energy sources utilised by the ASEAN Member States in this period were coal and gas. This is due to the construction of coal-fired power plants in Indonesia, Malaysia, Philippines, Thailand and Viet Nam, which began in the 1990s, and the installation of natural gas-fired power plants in Indonesia, Malaysia, Philippines, Thailand, Viet Nam and Singapore during the same period.

The third ASEAN Energy Outlook forecast that with an assumed GDP growth rate of 5.2 per cent per annum from 2007 to 2030, final energy consumption in ASEAN will grow at an average annual rate of 4.4 per cent, from 375 Mtoe to 1,018 Mtoe, assuming the business-as-usual (BAU) scenario during this period.[7] By contrast, an average growth rate of 3.6 per cent per year, to reach 843 Mtoe in 2030, is predicted under the alternative policy scenario (APS), which analysed the impact of the energy-saving goals and action plans for renewable energy in each of the ASEAN Member States. However, the fourth ASEAN Energy Outlook has revised the figures and predicts that ASEAN will require 1,685 Mtoe of Total Primary Energy Supply (TPES) in 2035, from 619 Mtoe in 2013.[8]

Some ASEAN Member States are rich in various sources of energy, such as oil, natural gas and coal, and have large potential sources of renewable energy, including wind, hydro and geothermal. However, these resources are not evenly distributed in the ASEAN region. Whereas Indonesia has enormous resources of geothermal potential and Lao PDR is the largest producer of hydro power, accounting for 26,500 MW – it is sometimes referred to as 'the Battery for ASEAN' – Brunei and Singapore, by contrast, have limited energy and alternative energy resources.

C Green Electricity and the ASEAN Community

Vulnerability to changes in oil prices and the need to take into account other considerations, such as sustainable development and climate change, have encouraged ASEAN Member States to rethink their

[6] Asian Development Bank, *Primer Energy Statistics in Asia and the Pacific (1990–2009)* (Manila: Asian Development Bank, 2013); Asian Development Bank. *Asian Development Bank, Energy Outlook for Asia and the Pacific* (Manila: Asian Development Bank, 2013).

[7] ASEAN Center for Energy, *Third ASEAN Energy Outlook* (Jakarta: ACE, 2011).

[8] ASEAN Center for Energy, *Fourth ASEAN Energy Outlook*.

Table 8.1 *Objectives of the AEC*

AEC blueprints			
Single market and production base	A highly competitive economic region	A region of equitable economic development	A region fully integrated into the global economy

approach towards alternative sources of energy. Green energy is one of the areas of concern addressed in the ASEAN Community, as reflected in the ASEAN Community Blueprint 2025 *Forging Ahead Together*, where the AEC committed to developing supportive policies towards innovation and a science-based approach to green technology and development.[9]

To accelerate ASEAN's transformation into an economically stable region, at the 12th ASEAN Summit, held in 2007 in the Philippines, ASEAN leaders agreed to the Cebu Declaration on Acceleration of the Establishment of an ASEAN Community by 2015. The Cebu Declaration committed ASEAN Member States 'to hasten the establishment of the ASEAN Economic Community by 2015 and to transform ASEAN into a region with the free movement of goods, services, investment, skilled labour, and the freer flow of capital'.

ASEAN leaders agreed to the ASEAN Economic Community Blueprint (AEC Blueprint) at the 13th ASEAN Summit held in Singapore in 2007. The AEC Blueprint, as summarised in Table 8.1, consists of four key pillars, namely a single market and production base and achieving the statuses of a highly competitive economic region, a region of equitable economic development and a region fully integrated into the global economy.

Achieving the objective of being a highly competitive economic region requires ASEAN to strive for energy cooperation among its Member States. The Member States agreed to energy cooperation to obtain a

[9] ASEAN Community Blueprint 2025 *Forging Ahead Together*. The ASEAN Declaration 1967 states that among the aims and objectives of ASEAN is the acceleration of economic growth, social progress and cultural development in the region in order to strengthen the foundations for a prosperous and peaceful community of Southeast Asian nations. ASEAN Member States originally agreed to realise the formation of AEC in January 2015, but the date was later postponed to 31 December 2015.

secure and reliable supply of energy, including biofuel, which is crucial to support and sustain economic and industrial activities.[10]

In terms of green electricity, ASEAN's objective is to achieve renewable energy development through connections to the grid. The AEC Blueprint plans the implementation of an ASEAN Power Grid (APG) project involving fourteen electricity interconnection projects, which will enable the optimisation of the region's energy resources for greater security of supply.

The project provides opportunities for private sector involvement through investment, including financing and technology transfer.[11] To accelerate investment and private sector participation in the APG, the AEC Blueprint recognises the need for greater investment through innovative financing schemes to attract greater private sector involvement, and for removal or relaxation of the obstacles to cross-border investment in financing regional infrastructure projects.

In addition to its AEC Blueprint, ASEAN has also adopted the ASEAN Socio-cultural (ASCC) Blueprint. This Blueprint outlines the plan for ASEAN and its Member States to pursue, among others, the ideal of a more environmentally friendly community by facilitating access to applied science and technology, ensuring environmental sustainability, addressing global environmental issues, promoting environmentally sound technology and responding to climate change and addressing its impacts.

The Bali Declaration on ASEAN Community in a Global Community of Nations, or 'Bali Concord III', reaffirms the ASCC's aims and principles with regard to sustainable development, the environment and climate change, with an emphasis on balancing economic growth, social development and environmental sustainability for the attainment of the Millennium Development Goals (MDGs). In addition, the Initiative for ASEAN Integration (IAI) Strategic Framework and the IAI Work Plan 2 (2009–15) provided plans for utilising the development and distribution of energy resources and technologies in order to narrow the development gap among ASEAN Member States.

[10] ASEAN Secretariat, *The ASEAN Economic Community Blueprint* (Jakarta: ASEAN Secretariat, 2008), para. 53. Biofuel is mentioned as an aspect of energy security and infrastructure, which includes other regional collaborations in the Trans-ASEAN Gas Pipeline (TAGP) and the ASEAN Power Grid (APG) Projects.

[11] Ibid., para. 54.

To enhance ASEAN energy security, accessibility and sustainability, its Member States agreed to the formulation of the ASEAN Plan of Action for Energy Cooperation (APAEC) 2010–15. APAEC's components included the APG, the Trans-ASEAN Gas Pipeline, coal and clean coal technology, renewable energy, energy efficiency and conservation, regional energy policy and planning and civilian nuclear energy.

Under APAEC, power supplied through the APG was produced through a mix of oil, gas, coal, hydro, geothermal and other renewable energy sources. It provided access to energy for roughly 66 per cent of ASEAN citizens. The underlying goal of this project was to encourage interconnection projects, initially on a cross-border bilateral basis, then gradually to expand to a subregional basis, and finally to a totally integrated Southeast Asian power grid.[12]

APAEC also predicts that renewable energy (including geothermal, solar, wind, tidal and wave energy) should have grown at a rate of 9.1 per cent among ASEAN Member States by 2030. The target established in APAEC 2004–9 to increase renewable energy by 10 per cent was met, and APAEC 2010–15 aimed not only to increase renewable energy by 15 per cent, but also to reduce the impact of energy usage in the ASEAN region. Hydropower and biofuels were the focus for renewable energy.[13]

The AEC and ASCC Blueprints and APAEC address the vulnerability of ASEAN, as a region reliant on fossil fuel, and the need to move towards alternative fuels to generate electricity by strengthening the development of renewable energy sources such as biofuels, biomass and solar.

Energy issues are also an important element of the ASEAN Vision 2025, which states that the AEC 2025 is envisioned to create a more dynamic and resilient ASEAN, capable of responding and adjusting to emerging challenges through robust national and regional mechanisms that address, among others, energy security issues.[14] The AEC Blueprint 2025 envisions green development by developing a sustainable growth agenda that promotes the use of clean energy and related technologies, including renewable energy through green technology, and also enhances sustainable consumption and production and includes it in national development plans.

[12] ASEAN Center for Energy, *ASEAN Plan of Action for Energy Cooperation 2010–2015* (Jakarta: ASEAN Center for Energy, 2010).

[13] Ibid. [14] ASEAN Secretariat, *ASEAN 2025, Forging Ahead.*

Table 8.2 *Generation of green electricity in seven ASEAN Member States (2012)*

ASEAN Member State	No. of projects	MW generated
Indonesia	14	896
Lao PDR	18	4,586
Malaysia	5	157
Myanmar	1	575
Philippines	16	1,053
Thailand	23	492
Viet Nam	21	635

In particular, the AEC Blueprint 2025 identifies several strategic measures to be used, including fostering policies supportive of renewable energy and setting collective targets accordingly; developing a framework to support the deployment and utilisation of efficient and low-carbon technologies, and calling for international support to ensure ASEAN access to mechanisms that foster low-carbon technologies more affordably; and promoting the use of biofuels for transportation.

D Development of Green Electricity in Selected ASEAN Member States

Most of the ASEAN Member States have now developed policies and commenced green electricity projects, either at the national level, by connecting generated electricity to the grid, or through stand-alone rural electrification projects. In the whole of east Asia, in 2012, financial closure was reached in renewable energy projects expected to generate 1,533 MW of power with private participation, with total project costs of US$ 2,876 million.[15]

Table 8.2 shows the generation of green electricity in most ASEAN Member States in 2012.

Of the ASEAN Member States, the most active was Lao PDR, with 4,347 MW in pipeline projects, and 239 MW or US$ 501 million in financed projects. In 2012 in Lao PDR, 239 MW of renewable energy projects with private participation reached financial closure, with total

[15] World Bank, *Private Participation in Renewable Energy Database*, http://ppi-re.worldbank.org/data (accessed 24 September 2014).

project costs of US$ 501 million. For pipeline projects, the total potential installed capacity reached 4,347 MW. The most active sponsor of renewable energy projects was Sinohydro of China, which was responsible for 1,156 MW in pipeline and 173 MW, or US$ 430 million, in financed projects. Large hydro was the predominant technology, accounting for 4,339 MW in pipeline and 239 MW, or US$ 501 million, in financed projects. Most of the projects were foreign-owned, with only one project owned by Electricite de Laos, the Lao national electricity company.

Lao PDR exports electricity from many of its hydropower plants. These exports and those of metals, together with tourism, helped Lao PDR to achieve robust economic growth of 7.5 per cent in 2010. The country maintained an average GDP growth rate of 8.1 per cent annually between 2005 and 2010, underpinned by substantial investment in mining and hydropower. This upward economic trend was linked to increases in primary energy demand, which showed an annual growth rate of 7.3 per cent during the same period.

Development of power-generating capacity is fundamental to the economic and social development of the country. The government plans to effectively utilise its hydropower both for export and to meet domestic demand. To this end it has promoted the addition of hydropower capacity intended for export, since the export of electricity to neighbouring countries is a major economic activity.

Two hydropower plants, Nam Theun 2 (1,070 MW) and Nam Ngum 2 (615 MW), started running at full capacity in 2010 and 2011, respectively, and boosted hydropower output significantly.[16] Lao PDR also gives priority to providing all the country's households with access to electricity. The electrification rate needs to be improved, as non-commercial energy sources such as firewood and charcoal are still used to meet most of the primary energy demand (62.4 per cent in 2010). The country has targeted an increase in the percentage of households with access to electricity from approximately 67 per cent in 2010 to 80 per cent in 2015 and 90 per cent in 2020. Of particular importance is the facilitation of electrification in rural areas.

Lao PDR aims to diversify into energy resources other than hydropower. The first coal-fired power plant in Hongsa district, composed of 625MW generators, was completed at the end of 2015 and was to provide power to Thailand and Lao PDR. This will lead to a major change in the

[16] Asian Development Bank, *Asian Development Outlook 2012* (Manila: Asian Development Bank, 2012).

country's energy mix. The potential of new and renewable energy sources such as solar and biofuels is also being examined for use in off-grid power generation in remote areas.

In the BAU scenario, Lao PDR's final energy demand is projected to increase from 2.4 Mtoe in 2010 to 6.0 Mtoe in 2035, with an annual growth rate of 3.8 per cent. Lao PDR is no longer fully dependent on hydropower for power generation given the 2015 completion of the Hongsa Powerplant, where coal is also used to generate electricity. In the APS, new and renewable energy will be added to the electricity-generation mix towards 2035. This reflects Lao PDR's renewable energy policy, which aims to increase the share of renewable energy to meet 30 per cent of energy demand by 2025.

Indonesia is a member of the G20 and is one of the fastest growing economies within ASEAN: its GDP is expected to grow from US$ 274.4 billion (constant 2000 $) in 2010 to US$ 1,045.2 billion in 2035 at an annual rate of 5.5 per cent.[17] In addition to economic growth, Indonesia is expected to experience population growth from 239.9 million in 2010 to 285.8 million in 2035, at an annual rate of 0.7 per cent, which will add to the electricity demand in the country.

Under the BAU scenario, Indonesia's final energy demand is projected to increase from 156.4 Mtoe to 305.2 Mtoe at an annual rate of 2.7 per cent. All types of energy – excluding biomass, which will register negative growth – will grow steadily, with electricity, oil and gas showing growth rates above 4.5 per cent up to 2035. Demand for non-commercial biomass, which accounted for about 70 per cent of energy demand in 2010, will continue at a constant level of around 45.0 Mtoe to meet the fuel needs of rural areas.

Demand for electricity will increase at a rapid rate of 6.2 per cent per year (2010–35), from 7.6 Mtoe in 2010 to 34.1 Mtoe in 2035. The energy mix will be dominated by oil, which will meet 30.1 per cent of primary energy demand in 2035, followed by coal (28.8 per cent); natural gas (20.7 per cent); others, including biomass, geothermal and new energy sources (19.5 per cent); and hydro (0.8 per cent).

Diversification of energy sources away from oil to coal is projected to progress. Meanwhile, the implications of such diversification for both the local and the global environment should be noted. The outlook assumes

[17] Asian Development Bank, *Primer Energy Statistics in Asia and the Pacific (1990–2009)* (Manila: Asian Development Bank, 1990); Asian Development Bank, *Energy Outlook for Asia and the Pacific* (Manila: Asian Development Bank, 2013).

the deployment of new and renewable energy sources, including geothermal and wind, particularly in the APS, to reflect domestic availability of resources. Meanwhile, the introduction of these technologies will have to be supported by policies and measures. For instance, new and renewable energy sources will require measures to cope with the rise in electricity tariffs and to resolve issues arising from resettlement of local residents living near the sites. Concerted efforts are necessary across the relevant ministries, as well as by local governments, to make progress on the deployment of such energy sources. This will involve recognition of the long-term benefits for the Indonesian economy of achieving green growth and of utilising domestic fossil fuels efficiently.

Despite being a large country with abundant biomass resources, in 2012 Indonesia only had 99 MW of renewable energy projects with private participation that reached financial closure, with total project costs of US$ 288 million. There were pipeline projects where the total potential installed capacity reached 796 MW. The most active sponsor in renewable energy was the Itochu Corporation of Japan, with 340 MW in pipeline projects. Geothermal was the most active technology, with a total of 610 MW in pipeline projects. Other projects involved solar, small hydro, large hydro and wind, but there were no biofuel- or biomass-based projects. There was a mix of investors from Indonesia, Japan, India, Australia and South Korea.

In 2012 in Malaysia, only 13 MW of renewable energy projects with private participation reached financial closure, with total project costs of US$ 36 million. For pipeline projects, the total potential installed capacity reached 144 MW. Solar was the most active technology, totalling 112 MW in pipeline projects. Other projects involved a mix of energy sources including bio-energy, solar and geothermal. With the exception of one company from the United Kingdom, the investors were Malaysian companies. Faced with a rise in domestic energy demand and maturing hydrocarbon reserves, however, Malaysia has attempted to safeguard depleting reserves by restricting the production of oil and gas, rejuvenating mature fields and intensifying exploration activities.

To ensure its energy supply, Malaysia has pursued energy diversification and improvement of its energy efficiency. The principal policy implemented for energy diversification was the 1981 Four-Fuel Policy, which aimed to reduce dependence on oil and to optimise a fuel mix of oil, gas, hydro and coal for use in electricity generation. Since the incorporation of renewable energy in 2001, this has been known as the Five-Fuel Policy.

The utilisation of renewable energy is underpinned by policy implementation. As a result of the 2010 National Renewable Energy Policy and Action Plan, Malaysia enacted the Renewable Energy Act 2011 and the Sustainable Energy Development Authority Act 2011 to set up the framework for a new feed-in tariff mechanism. For the transport sector, the 2006 National Biofuel Policy promotes the production and consumption of biofuel, and ensures that biodiesel is of acceptable quality, by establishing an industry standard.

In terms of the electricity-generation mix, natural gas is likely to retain the dominant share over the outlook period in both cases. Meeting the objective of creating a more balanced generation mix will result in reduced dependence on natural gas and the expansion of hydro and new and renewable energy up to 2035. Hydroelectric generation is projected to grow strongly during the outlook period as a result of the development of large hydro projects in Sarawak as part of the Sarawak Corridor of Renewable Energy initiative, which plans to develop a total of 20,000 MW over twenty years.

The share of new and renewable energy, mostly from biomass and solar sources, will also increase from 1.0 per cent in 2010 to 2.7 per cent under the BAU scenario. APS will account for a much larger share – 6.0 per cent – in 2035, supported by the implementation of the feed-in tariff mechanisms.

Malaysia will continuously seek to ensure long-term energy security to sustain its economic activity and meet growing energy demand. This would call for improved initiatives on the supply side to maintain oil and gas production – for example, through enhanced oil recovery and exploration of marginal fields – and the development of viable new and renewable energy resources, such as solar, wind and biofuel. Malaysia is endowed with natural resources which have the potential to be exploited for the purposes of generating greener electricity.[18]

In the Philippines, in 2012, 538 MW of renewable energy projects with private participation reached financial closure, with total project costs of US$ 1,008 million. The most active sponsor in renewable energy was the Korea Water Resources Corporation, with 246 MW, or US$ 441 million

[18] For discussion of issues relating to green electricity generation in Malaysia, see S. M. Shafie, T. M. I. Mahlia, H. H. Masjuki and A. Ahmad-Yazid, 'A review of electricity generation based on biomass residue in Malaysia', *Renewable and Sustainable Energy Reviews* 16 (2012), 5879–889.

in financed projects. Small hydro was the most active technology, totalling 120 MW in pipeline and 246 MW or US$ 441 million in financed projects.

In Thailand, there were twenty-three projects, amounting to 492 MW. In 2012, 152 MW of renewable energy projects with private participation reached financial closure in the country, with total project costs of US$ 294 million, and projects in the pipeline with installed capacity of 340 MW. Solar was the predominant technology, totalling 184 MW in pipeline and 142 MW or US$ 274 million in financed projects. Most of the project sponsors were Thai companies, with a few projects being undertaken by Japanese and Vietnamese companies.

In Vietnam in 2012, 121 MW of renewable energy projects with private participation reached financial closure, with total project costs of US$ 169 million. Pipeline projects reached a total installed capacity of 515 MW. The most active sponsor in renewable energy was Enfinity of Belgium, with 125 MW in pipeline projects. Wind was the predominant technology, totalling 323 MW in pipeline projects. Some other projects consisting of small hydro and large hydro were undertaken, mostly by Vietnamese companies.

E Issues and Challenges in the Development of Green Electricity in ASEAN

The above examples show that most major projects undertaken in the seven ASEAN Member States discussed above are based on geothermal, hydro and solar energy, with fewer projects involving bio-energy. Most of the projects outlined above serve country- and regional-level electricity demand.

However, one of the issues facing ASEAN Member States is the electrification of rural areas, especially in the less developed Member States such as Cambodia, Laos PDR, Myanmar and Viet Nam. It is estimated that almost 200 million people in the ASEAN region still lack access to modern and reliable energy services.

I Project development issues

Based on my experience as a consultant to governments in the ASEAN Member States, rural electrification normally requires between 3 and 5 MW. Small hydro is the norm, with biomass and bio-energy projects slowly taking shape on the drawing board. Rural electrification through small hydro faces fewer challenges, as hydro does not require any processing of raw materials before being used as fuel. Rural electrification based on biomass and bio-energy requires in-situ processing and electricity generation facilities.

Typical bio-energy electrification requires:

a. Raw material from bio-energy sources. Such material has to be obtained from third-party suppliers or developed by the project developers themselves. For example, any project using palm waste will have to secure the material from palm processing mills. This exposes project proponents to further obstacles. First, palm waste is also used in palm processing mills' boiler systems, leaving only about half of the waste for sale, thus further reducing the availability of these finite resources. Second, the waste for sale is subject to further competing demands, not only from rural electrification projects but also from other users of heating and boiler systems, locally and abroad. Third, the waste itself has become a commodity and is thus subject to supply-and-demand and price fluctuations. Further study is needed to determine the impact of supply and demand of the waste-cum-commodity on the viability of rural electrification. Thus, the best option is for the project proponents to be in control of the raw material, either by working with the 'owner' of the raw material or by growing biomass plants such as acacia or lucaena themselves.
b. Development of two electricity-generating facilities. One is for in-house electricity usage, especially to process the biomass into a more efficient fuel. Biomass has to be dried to reduce the moisture content to a certain percentage, and then further processed either by palletising or by pulverising. Power generation for the public can only be initiated after the completion of the in-house power-generating facility. The two power-generating systems will add to the cost of setting up biomass-powered rural electrification.
c. Supporting infrastructure development such as local grid, roads and other facilities will add further costs to be borne by the project proponent.

Project proponents will also need to set the price of the electricity units supplied to the rural system. They may be able to set their own rates subject to the approval of the regulators if the electricity is supplied directly to the local population. In most cases, proponents prefer to sell the electricity to the grid operated by the national electricity company and to be compensated or paid for the power supplied. The second option saves the proponents from having to deal with electricity distribution, which may include charging and collecting payment from end-users, who may not be able to afford to pay the bill.

It is evident that setting up rural electrification schemes involves many steps, which may also carry costs. Such costs include payment of royalties if the project requires use of third-party technology; purchase of hardware and software such as burners and boilers and a computer system, which may be subject to import tariffs; facilities from banks and financiers and financing costs; and international and domestic transportation.

Thus, development of a green electricity project, such as a biomass-based scheme, will face the following issues:

a. High generation costs. There are several reasons why costs may be high, such as the need to provide a dual generating system and to develop support infrastructure. High generation costs may also arise from the need for project proponents to develop in-house technologies or purchase technologies from third parties.
b. Land. This is a potential cost if a proponent wishes to be in control of its own biomass supply.
c. Trade facilitation and customs procedures. An import tariff is often charged on the hardware and software needed for the project's implementation. In some circumstances, project proponents may wish to import raw biomass materials from other countries. Biomass is classified as an agricultural product, or as agricultural waste, and will be subject to phyosanitary regulations.
d. Transportation.
e. Funding and finance charges.

Experience has shown that high generation costs and low incomes, as well as adverse topographical conditions and high transportation costs, are obstacles to the electrification of remote areas in some ASEAN Member States.

II Trade Issues

Discussions on development of green energy projects also bring out trade issues. Project proponents currently face high tariffs on hardware imported for use in green electricity projects. For example, a steam boiler, which may also be used for other purposes, is generally classified as a boiler (Harmonized Commodity Description and Coding Systems (HS) Code 84021910), and this applies across the board irrespective of whether it is used in a green electricity project or in another industry. One has either to pay the import tariff or lodge an appeal against the tariff on the grounds that the boiler is to be used in green electricity development. Moreover, the HS Code does not make provision for any new code for a specially designed biomass-powered boiler.

Thus, it is suggested that the negotiations on environmental goods and services currently taking place at the World Trade Organization (WTO) should take into account the inventions and innovations of new products either with a dual purpose or specially made for green electricity generation. In addition, parties negotiating preferential trade agreements may want to consider the tariff concessions in the trade-in-goods chapter, or to discuss them specifically in the environmental chapter.

In addition, the commoditisation of waste or biomass results in competing demands from export markets and may thus cause a shortage of feedstock for domestic green electricity. A domestic green electricity generator would stand to gain from any restrictions on the export of raw material, as this could secure the supply of feedstock and help keep prices down due to reduced demand from export markets. The issue is whether any of the ASEAN Member States, which are all members of the WTO, are able to restrict exports of raw biomass materials to be used in green electricity production.

The decision of the WTO Appellate Body in *China – Measures Related to the Exportation of Rare Earths, Tungsten, and Molybdenum* would be of interest to any ASEAN Member State when considering a demand to restrict export of raw biomass material to other countries.[19] Early analysis of the decision of the Appellate Body in this case shows that a WTO Member may impose an export restriction if it meets the requirement of Article XX (g) of the General Agreement on Tariffs and Trade (GATT) on 'measures related to the conservation of natural resources'. However,

[19] WT/DS431/AB/R, WT/DS432/AB/R, WT/DS433/AB/R, 7 August 2014. On 23 June 2009, the United States requested consultations with China with respect to the latter's restrictions on the export from China of various forms of raw materials. The United States considered that China, through these measures as well as any amendments, replacements, related measures and implementing measures, was in violation of Articles VIII, X and XI of the GATT 1994 and Paragraphs 5.1, 5.2, 8.2 and 11.3 of Part I of the Protocol on the Accession of the People's Republic of China ('Accession Protocol'), as well as China's obligations under the provisions of paragraph 1.2 of Part I of the Accession Protocol (which incorporates commitments in paragraphs 83, 84, 162 and 165 of the Report of the Working Party on the Accession of China). The United States considered that the measures also appear to nullify or impair the benefits accruing to the United States directly or indirectly.

based on the Appellate Body's decision, export restrictions may be imposed under Article XX (g) if the same measure is imposed on domestic industries as well.

III Environmental Issues

Development of one's own biomass or firewood plantation is the best option for enabling one to control the supply of the feedstock. However, as with climate change, air pollution and water, the link between energy security and land-use change is complex. For example, development of firewood plantations may lead to deforestation.

One conservative estimate suggests that 15 per cent of land-use changes are caused by clearing forests for fuel wood and energy crop plantations.[20] Forest clearing will cause increased release of carbon. Indonesia, Malaysia and Thailand are the largest producers of palm oil in the world, and palm waste is now a major feedstock for bio-energy-based projects. As a result of the twin pressures of fuel wood collection and biofuel production, in Southeast Asia as a whole, deforestation has been five times higher than the global average, and ten times the average for the rest of Asia.

This situation contributes to the paradox of wishing to contribute to the production of sustainable green electricity on the one hand, and not wishing to cause damage to the environment on the other. Some scholars and global institutions concerned with bioenergy recognise the environmental dilemmas that the large-scale production of palm oil can present. These include encroachment into protected areas, affecting water systems, displacing food production and harbouring unsustainable land-use practices that may not only cancel out GHG emissions for decades, but also lead to widespread ecological despoliation.[21]

F Conclusions

The above discussion shows that ASEAN Member States face many challenges in ensuring that green electricity trade in the region meets the objective of the ASEAN Community 2015. The focus of ASEAN Member States is still on major hydro and solar sources, which are

[20] B. Sovacool, 'Environmental issues, climate change, and energy security in developing Asia', *Asian Development Bank Working Paper Series* No. 399 (June 2014).

[21] Ibid.

abundant in the region. However, there is a need to start looking into different energy mixes, such as bioenergy and geothermal, as a source of green electricity. ASEAN Member States require a continuous electricity supply to meet the demands of development and industrialisation if they are to attain the objective of making the ASEAN region highly competitive in economic terms, as envisaged in the AEC Blueprint.

Nevertheless, ASEAN Member States have to take into account that, in many cases, the move towards green electricity requires a new policy approach to tackle issues incidental to the energy transformation. Such issues include ensuring sustainable development, meeting the targets of the MDGs and the Sustainable Development Goals (SDG) and ensuring that green electricity in itself does not become the bane of climate change policies.

ASEAN Member States need to make a concerted effort to tackle pertinent issues in rural electrification, technology development and innovation in the field, raw material supply and trade facilitation. To ensure that green electricity becomes a contributor to a sustainable and green ASEAN Community, ASEAN Member States must take a holistic approach to tackling the issues and challenges raised in this chapter.

Bibliography

ASEAN Center for Energy. (2010). *ASEAN Plan of Action for Energy Cooperation 2010–2015*. Jakarta: ASEAN Center for Energy.

(2011). *Third ASEAN Energy Outlook*. Jakarta: ACE.

(2015). *Fourth ASEAN Energy Outlook*. Jakarta: ACE.

ASEAN Secretariat. (2008). *The ASEAN Economic Community Blueprint*. Jakarta: ASEAN Secretariat.

(2015). *ASEAN 2025, Forging Ahead Together*. Jakarta: ASEAN.

Asian Development Bank. (2012). *Asian Development Outlook 2012*. Manila: Asian Development Bank.

(2013a). *Asian Development Bank, Energy Outlook for Asia and the Pacific*. Manila: Asian Development Bank.

(2013b). *Primer Energy Statistics in Asia and the Pacific (1990–2009)*. Manila: Asian Development Bank.

Organisation for Economic Co-operation and Development. (2014). *Economic Outlook for Southeast Asia, China and India 2014: Beyond the Middle-Income Trap*. Paris: OECD.

Shafie, S. M., Mahlia, T. M. I, Masjuki, H. H. and Ahmad-Yazid, A. (2012). A review of electricity generation based on biomass residue in Malaysia. *Renewable and Sustainable Energy Reviews*, **16**, 5879–889.

Sovacool, B. (2014). 'Environmental issues, climate changes, and energy security in developing Asia', *Asian Development Bank Working Paper Series* No. 399, June.

World Bank. *Private Participation in Renewable Energy Database*, http://ppi-re.worldbank.org/data (accessed 24 September 2014).

9

The African Experience

CALLIXTE KAMBANDA

A Introduction

The regional electricity trade in Africa started in the 1950s, with cross-border electricity trading in Southern Africa between the Democratic Republic of Congo (DRC) and Zambia through a bilateral agreement and an interconnector of 1,700 km with a 1 × 220 kV line. In 1955, interconnection between Uganda and Kenya over a 132 kV line was commissioned. In the 1960s Zambia was connected to Zimbabwe through a 2 × 330 kV line and in 1975 Mozambique was connected to South Africa through a 533 kV high-voltage, direct current (HVDC) line of 1,400 km. Other bilateral agreements were signed on electricity trade in other parts of the continent in the 1970s and 1980s. A trilateral agreement was signed in February 1984 between three countries – Rwanda, Democratic Republic of Congo (DRC) and Burundi – to create a tri-national company called SINELAC, which was mandated to operate and manage a common electrical power plant, the Rusizi II, with a capacity of about 45 MW (commissioned in 1989).

The creation of the regional power pools (RPPs) started in 1985, with the signing of an inter-governmental memorandum of understanding (MoU) between the Southern African Development Community (SADC) countries to create the Southern African Power Pool (SAPP). This is now the most advanced power pool on the continent. SAPP introduced the short-term energy market (STEM) in April 2001, which operates daily and hourly electricity trade contracts. This fuelled the development of a competitive energy market in the form of the day-ahead market (DAM) in 2003 (in which short-term contracts are made anonymously through the power pool and guarantees are required).[1]

1 SAPP Annual Report 2013, www.sapp.co.zw/docs/22867%20Annual%20Report%20New.pdf, 10/12/2014 (accessed 9 December 2014).

In 1999, the Economic Commission of West African Countries (ECOWAS) began the creation of the West African Power Pool (WAPP), which was finally established by the ECOWAS heads of state and government in December 2001. WAPP is the only African power pool to have been established by a decision of heads of state and government. The other African power pools were established based on a MoU signed by the ministers in charge of energy. The WAPP Master Plan study of 2011 foresaw a number of projects expected to help achieve the desired energy trade (which includes 10 per cent renewable energy) by 2020. The power trade in WAPP is currently under bilateral or trilateral agreements and the energy trade has not yet started.

Following the success of SAPP and WAPP's establishment, the Union of African Power Utilities (UPDEA) assisted countries in Central and Eastern Africa in establishing their respective power pools, namely the Central African Power Pool (CAPP) in 2003 and the Eastern Africa Power Pool (EAPP) in 2005. EAPP and CAPP are still in the nascent stage, but they have completed their medium- and long-term master plans: 2013–38 for EAPP and 2005–25 for CAPP. In 1989 the Arab Maghreb Union (comprising five member states: Algeria, Libya, Mauritania, Morocco, and Tunisia) set up the *Comité Maghrébin de l'Electricité* (COMELEC), which has a very light structure in comparison to other regional power pools. It is worth noting that the future electricity trade in Africa will be dominated by the African Regional Power Pools (ARPP).

B Current Status of the ARPP

I COMELEC

The electricity trade in COMELEC countries (except Mauritania) is operated through a regional electricity network in the Maghreb region (Morocco, Algeria, Tunisia and Libya). The key interconnections in the region include: (i) 400 kV connections between Spain, Morocco, Algeria and Tunisia and (ii) 220 kV connections between Algeria–Tunisia–Libya and Egypt. As mentioned above, COMELEC has a very light structure, with the Secretary-General being the only technical staff member coordinating the work of the various technical commissions, each of which is headed by a member utility.

An assessment conducted by the Infrastructure Consortium for Africa in 2011 reported that although the countries in this region are well interconnected, except for power exports from Spain to Morocco,

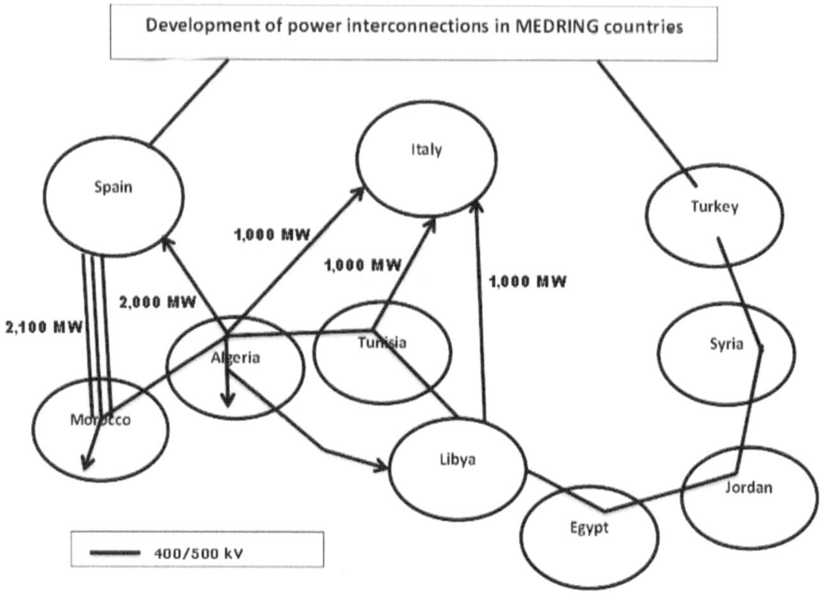

Figure 9.1 Interconnections among MEDRING countries
Source: Adapted from: ICA, Africa Regional Power Status Report, 2011.

regional power trade between Algeria and Morocco and Algeria and Tunisia amounted on average to only 5–16 per cent of interconnection capacities.[2] The total installed capacity of COMELEC countries is 27,347 MW. The electrification rate in the region is over 90 per cent, except in Mauritania (where the rate was 39 per cent in 2009). In 2002–4 a study was conducted prior to the linkage of five countries under the Egypt–Libya–Tunisia–Algeria–Morocco (ELTAM) project with a 500/400 kV transmission line. Implementation of the interconnection was planned to take place from 2010 to 2015. Another important project in the region was the MEDRING (Mediterranean Ring). The feasibility study for this project was conducted from 2001 to 2003 with the aim of developing an interconnection loop for the countries around the Mediterranean basin, including the Maghreb countries, the Middle East and Southern Europe, as illustrated below in Figure 9.1.

[2] ICA, *Africa Regional Power Status Report*, 2011, www.icafrica.org/fileadmin/documents/Knowledge/Energy/ICA_RegionnalPowerPools_Report.pdf (accessed 9 December 2014).

Today, Libya is connected to Tunisia and Egypt through a 220 kV connection. The interconnection between Morocco, Algeria and Tunisia operates in synchrony with the European grid (Union for the Coordination of the Transmission of Electricity; UCTE). Synchronisation between the Libyan system and Europe has yet to be achieved. It should be noted that these projects have been negatively impacted by the political instability in the region.

The legal and regulatory framework in COMELEC has been mainly driven by a study on Algeria, Morocco and Tunisia's gradual integration into the European Union (EU) electricity markets. This was launched with the signing of the Protocol Agreement by the three Maghreb countries and the EC (European Commission) in 2003. The objective was to achieve harmonisation of the signatory countries' legal and regulatory frameworks to create a market for electricity and enable them to meet European standards in order to integrate with the EU market.

The electricity market in COMELEC has yet to move from being driven solely by national policies towards regional policies that allow the development of a regional market, including regional master plans, development of the market rules and harmonisation of the grid code.

II SAPP

SAPP is a specialised SADC institution. Since its creation in 1985, SAPP has evolved significantly, and it is now the only power pool in Africa that operates a real electricity market. SAPP publishes a monthly market performance report, which indicates the participation level in the market, market liquidity, traded volumes (particularly in the DAM), transmission constraints and market clearing prices. For example, in June 2014, total traded volume in SAPP, DAM and post-DAM (PDAM) was 24,774.20 MWh, of which 23,204.20 MWh (94 per cent) was traded in DAM. According to SAPP's market performance report for June 2014, the transmission constraints meant that '7,541.00 MWh that was matched in DAM during the month of June 2014 could not be traded resulting in the SAPP losing USD 15,082 in potential revenues that could have gone a long way in supporting the market trading platform'.[3]

[3] Southern African Power Pool, *Market Performance Report*, June 2014, www.sapp.co.zw/docs/DAM%20Market%20Monthly%20Performance%20Report%20-%20June%202014.pdf (accessed 9 December 2014).

More than 70 per cent of the total installed capacity in the region uses coal to generate electricity and only 17 per cent of electricity is generated by hydropower plants.[4] With the South African Renewable Energy Independent Power Producer Procurement Program (REIPPPP), the share of renewable energy technologies in the SAPP energy mix will change. The first round of REIPPPP achieved financial closure in October 2012. Wind and solar technology dominated, with nine projects, totalling 417.1 MW, for solar PV and seven projects, totalling 562.5 MW, for wind. The REIPPPP target is 3,725 MW of power generated from renewable energy for 2010–30.[5] The SAPP list of priority regional projects comprises fourteen generation plants (four for coal, one for gas and nine for hydro), amounting to a total of 13,015 MW, at an estimated cost of US$ 21.6 billion up to 2018, in addition to US$ 5.6 billion for six major regional interconnections. A clear shift in the future of the energy mix is evident. In the priority list of projects, hydropower plants (80 per cent) predominate, whereas coal (13.8 per cent) and natural gas (6.2 per cent) account for much lower percentages.

In terms of a regulatory framework in the region, in July 2002 SADC established a formal association of electricity regulators, the Regional Electricity Regulatory Association (RERA). RERA's mission is to facilitate harmonisation of regulatory policies, legislation and standards and to be a platform for effective cooperation among energy regulators within the SADC region.

III West African Power Pool (WAPP)

WAPP was created in December 2001. It has a very strong organisational structure, with some of the policy decisions being taken by heads of state and government of the member countries. In 2010, power generation capacity in WAPP reached 46,049 GWh. Hydropower generation is mainly dominated by Nigeria (43.4 per cent) and Ghana (40.9 per cent). In 2010, the total installed generation capacity in WAPP was 13,927 MW – predominantly thermal capacity (70 per cent) and hydropower (30 per cent).[6] The WAPP Master Plan completed in October 2011 included an economic analysis, which supported intensive development of hydroelectricity in WAPP. There was also considerable interest

[4] ICA, *Africa Regional Power Status Report*, 2011.
[5] www.ipprenewables.co.za/#index.php.
[6] ICA, *Africa Regional Power Status Report*, 2011.

in developing renewable energy, with the aim of achieving 10 per cent of the total installed capacity in the region by 2020. This would reduce certain areas' dependency on either hydroelectricity or thermal (gas) resources. The strong commitment to renewable energy in the WAPP region was illustrated by the establishment of the ECOWAS Centre for Renewable Energy and Energy Efficiency (ECREEE) in July 2010.

Based on Nigeria's experience in creating the National Bulk Energy Trading Company, the WAPP Master Plan study recommended establishing an 'aggregator unit/entity', which could gather the small demands of individual countries into a bulk demand, expedite negotiations of bankable power purchase agreements (PPAs) for electricity supply at the regional level and secure appropriate payment security instruments from multilateral guarantee agencies.

As is the case for the rest of the power pools on the continent, with the exception of SAPP, electricity trade in WAPP takes place only through bilateral agreements. Some transmission lines cross many countries, such as the transmission line linking the Manantali hydropower plant to three countries: Senegal, Mauritania and Mali. WAPP has not yet started operating an electricity market. WAPP has, however, established the necessary institutions, for example, the regional coordination centre. The power pool is in the process of building up the required physical infrastructure to enable it to trade electricity in a market structure.

In 2008 ECOWAS established the Regional Electricity Regulatory Authority (ERERA), whose main responsibilities are the regulation of cross-border electricity connections and trading among ECOWAS member states. This includes: (i) the establishment of transparent tariff-setting methodology for regional power pooling; (ii) adopting technical regulations; (iii) monitoring of regional market operations; (iv) resolving disputes among regional market participants; (v) contributing to the development of regional energy policy and (vi) assisting in building capacity of national regulatory bodies.

IV Central African Power Pool (CAPP)

The CAPP was created in 2003 and acts as a specialised ECCAS (Economic Community of Central African States) institution, comprising ten member countries: Angola, Burundi, Cameroon, Chad, Congo, Gabon, Equatorial Guinea, Central African Republic (CAR), DRC and Sao Tomé. CAPP's mandate is to implement ECCAS energy policy. Estimated total energy consumption by CAPP member countries in 2009 was 307 GWh, whereby three countries were responsible for 83 per cent of total consumption (Angola 24 per cent, Cameroon 27 per cent and DRC 32 per cent).[7] CAPP's installed capacity in 2009 was 6,250 MW, with hydropower representing 79 per cent (4,730 MW) and thermal power 19 per cent (1,405 MW). The same ratio as for the consumption figures above applies to the generation capacity. Three countries again have the lion's share – 83 per cent of the total installed capacity – namely DRC, Angola and Cameroon.

One of the challenges faced by almost all the power pools except for SAPP is a lack of up-to-date data. For example, when ICA conducted its assessment of power pools in 2011, the most recent data available were from 2009 for CAPP and from 2010 for WAPP and SAPP. A study conducted for CAPP in December 2009, *Etude sur l'Interconnection des Réseaux Electriques des pays membres de la Communauté Economique des Etats de l'Afrique Centrale* (*Study on Electrical Interconnection Network for ECCAS Countries*), noted several weaknesses, such as a lack of clear legal provisions for independent power producers in most of the ECCAS countries and a lack of harmonised standards, including for technical operators, among others.[8]

Electricity trade in CAPP is conducted mainly through bilateral agreements, with the exception of the trilateral agreement between Rwanda, DRC and Burundi on a common power plant at the border between Rwanda and DRC, the Rusizi II hydropower plant (45 MW). With the planned mega-hydropower plant, Grand Inga (up to 42,000 MW), the region could supply the whole continent with very cheap renewable energy. But a lot of preparatory work will need to be done, including the establishment of a clear and adequate legal and regulatory framework allowing private sector participation in DRC's energy sector. Public funding will not be sufficient for the Grand Inga project. Studies conducted on the development of the Grand Inga suggested doing it in several phases, with capacities varying from 4,800 MW up to about 7,000 MW per phase. Possible scenarios for transmission lines have been suggested depending on demand growth in WAPP, SAPP and EAPP. Implementation would require linking all the African regional power pools.

[7] Pool Energétique de l'Afrique Centrale (PEAC), *Présentation générale du secteur électrique en Afrique Centrale*, Europe Aid/126679/C/SER/CG; WYG International, 1 May 2011.

[8] EIRE ECCAS, *Etude sur l'Interconnection des Réseaux Electriques des pays membres de la Communauté Economique des Etats de l'Afrique Centrale – Rapport de l'Etude de Faisabilité*, December 2009.

V *Eastern African Power Pool (CAPP)*

Following its establishment in 2005, EAPP progressed very quickly. The EAPP System Master Plan and Grid Code Study and the study on the Regional Market Operation Center (RMOC), both released in 2011, anticipated that the DAM in the region would be fully operational by 2017 – assuming that EAPP promptly mobilised all the required resources, namely US$ 670 million per annum for the interconnectors from 2013 to 2016. In addition, all power utilities in the region were planning to invest US$ 14,000 million for their own generation, transmission and distribution projects. The biggest challenge will be filling the huge financing gap, because less than US$ 1,000 million will come from the main donors for the region (the World Bank, Arab Bank for Economic Development in Africa and African Development Bank).

The EAPP region has a huge hydro potential of up to 32,319 MW, according to its Master Plan Study, and 50 per cent of the potential sites are in Ethiopia. The study, which made projections for the period 2013–38, concluded that if all the proposed generation and transmission projects were to be implemented as planned in the basic scenario (first level of coordination), there would be net benefits for the region of US$ 25,194 million, deriving from total investments of US$ 172 million. The annual net benefits were estimated at US$ 969 million. In the optimum scenario, the regional net benefits over the study period would increase to US$ 32,451 million (an increase of 28 per cent) and the annual net benefits to US$ 1,248 million. One of the assumptions of the optimum scenario was that all interconnection projects being studied at that time would be implemented at the earliest date (2015–16). This applied to the following interconnections: Tanzania–Uganda, Tanzania–Kenya, Uganda–Kenya, Uganda–Rwanda; lines for Rusumo hydropower plant; Ethiopia–Kenya, Ethiopia–Sudan and Sudan–Egypt.

C Current Challenges of Regional Electricity Trade in Africa

The 2011–40 Program for Infrastructure Development in Africa (PIDA) Energy Outlook foresees an increase in energy demand from 590 TWh to 3,100 TWh and growth in the installed capacity from 120 GW to 700 GW (considering only units above 50 MW). To meet the future demand, total projected energy investment (capital expenditure) by 2040 of US$ 43 billion per annum is required, of which US$ 5.4 billion per annum will be needed for investment in regional transmission lines. Regional integration would save about US$ 33 billion per annum.[9]

I Lack of Physical Infrastructure

Although all power pools are working hard to promote electricity trade, the level of cross-border electricity traded on the continent within the power pools ranges from only 0.2 per cent (in CAPP) to 7.5 per cent (in SAPP).[10] The reasons for this low level of electricity trade are numerous, but one of them is the lack of interconnectors with associated transmission stations and switchyards. The electricity trade will reach a more satisfactory level when trade starts to take place not only within the power pools but also between them. Currently only COMELEC countries are interconnected with EAPP through the Sudan–Ethiopia interconnection but, as already mentioned, the trade is based only on bilateral agreements. Nonetheless, the near future looks promising if some of the key interconnectors are completed. These include the Zambia–Tanzania–Kenya (ZTK) interconnector, which will connect EAPP to SAPP, and the ongoing effort to interconnect Ethiopia and Kenya, which will strengthen the electricity exchange in EAPP. Other major interconnection projects are ongoing in East African countries to link the electricity grids of Kenya, Uganda, Tanzania, Rwanda, Burundi and Eastern DRC through the Nile Equatorial Lakes Subsidiary Action Program project.

II Poor Planning and Lack of Maintenance

Since the 1980s, Africa's infrastructure sector has faced challenges owing to poor planning, which has resulted in electricity supply shortages, particularly in the 2000s. In 2008 approximately thirty countries were facing an electricity supply deficit, imposing a heavy burden on economic growth and productivity. A number of countries are still either facing an electricity supply deficit or are obliged to use expensive diesel generators. The deficit has been exacerbated by the sustained demand for electricity, which stood at 4.5 per cent per annum from 2001 to 2005, while the

9 African Development Bank, *PIDA*, 2011, www.au-pida.org/documents/ENERGIE/EOE-Energy-Outlook.pdf (accessed 9 December 2014).

10 ICA, *Report on Regional Power Status in African Power Pools*, 2011, www.icafrica.org/fileadmin/documents/Knowledge/Energy/ICA_Regional PowerPools_Report.pdf (accessed 9 December 2014).

generation capacity expanded only by about 1.2 per cent per annum.[11] Owing to a low level of maintenance, most countries cannot operate their total installed capacity, because some generators or the whole power station are out of service.

III Poor Performance of the Power Utilities

The interconnection of electricity grids requires a certain degree of harmonisation of standards as well as a good level of performance of the interconnected systems. In a power pool, where the number of member utilities is usually around ten or more, it is very difficult to operate an interconnected system satisfactorily if one or more utilities are not maintaining their electrical network at the desired level of stability and to the required standards. In the worst case, there is a risk of total blackout.

A World Bank study in 2010 found that the inefficiencies of African power utilities were costing the region about US$ 2.7 billion per year, or 0.8 per cent of the region's GDP. The biggest share of inefficiencies consists of distribution losses and revenue under-collection. The same study estimated that average distribution losses were 23.3 per cent, while the average collection ratio was 88.4 per cent. It was also noted that utilities with a low level of efficiency were having difficulties in meeting the demand for electricity and that these areas had the lowest electrification rate.[12]

D Opportunities for the Regional Electricity Trade in Africa

I Strong Political Will, Economies of Scale, Energy Security and Energy Mix

The rapid growth of energy demand, associated with a growing middle class in Africa, has pushed individual countries and RECs to act strategically and to consider regional planning and integration as a good solution to the problems raised. Thus, the African Union Commission (AUC), NEPAD Agency (NPCA) and the African Development Bank have developed a continental and consensual programme – PIDA – which provides a vision for the development of African infrastructure

[11] World Bank, *Africa's Infrastructure – A Time for Transformation*, 2010. http://siteresources.worldbank.org/INTAFRICA/Resources/aicd_overview_english_no-embargo.pdf.

[12] Ibid.

(2011–40).[13] Looking carefully at the results of the study, one can deduce that implementing regional power projects would have a positive impact on: (i) kWh cost, due to economies of scale (implementation of big projects serving many countries); (ii) energy mix (countries with dominant hydro potential supplying those with dominant thermal (gas and coal) potential) and (iii) increased access to modern energy services, which in turn will trigger increased access to clean water and an improved health-care system. PIDA projections show that sustained African GDP growth, coupled with social progress and increased access to electricity, will boost energy demand in Africa by an average 5.7 percent annually up to 2040. Furthermore, the realistic integration scenario would save US$ 860 billion over the 2014–40 period, equivalent to 17 percent of the cost of electricity, representing an annual gain of US$ 33 billion. If African countries and regional specialised institutions prepare well enough by setting up an adequate legal and regulatory framework and a generally conducive environment for investments, there would be a tremendous opportunity for investors to fill this huge gap in the sector.

II Huge Potential for Renewable Energy Sources

Africa has a huge hydropower potential, of which only about 7 per cent is being exploited. The potentials of the DRC and Ethiopia alone amount to approximately 60,000 GW. In most African countries, particularly in sub-Saharan Africa, daily solar radiation is between 4 kWh and 6 kWh per square metre, and the Great Rift Valley has largely untapped geothermal resources, estimated at 9,000 MW.[14] But to exploit these resources for the benefit of Africa's people there is a need to address the challenge of capital intensity of the renewable energy (RE) investments and, to some extent, to remove the subsidies on environmentally unfriendly fossil fuels that make RE generation uncompetitive in the short term. Hence, since the benefits of RE are generally medium- to long-term, governments should establish incentive mechanisms that will allow the private sector to invest in RE projects. These mechanisms could include feed-in tariffs, tax incentives and setting targets for RE share in

[13] www.afdb.org/fileadmin/uploads/afdb/Documents/Generic-Documents/PIDA%20brief%20Energy.pdf.

[14] UNEP, *Financing Renewable Energy in Developing Countries,* 2012, www.unepfi.org/fileadmin/documents/Financing_Renewable_Energy_in_subSaharan_Africa.pdf.

the total energy mix, among others. According to a study conducted by UNEP in 2012, only eight countries in sub-Saharan Africa have set renewable energy targets for their economies.[15]

The additional advantage of RE is that it increases the energy mix and the subsequent energy security. For example, in Eastern Africa, where generation is predominately based on hydro, an electricity deficit resulted from the severe drought that affected the region in the 1990s. Therefore, relying on one energy resource reduces energy security. That is why if countries in a power pool have a good energy mix and there is a good power pool electricity trade in the region, this provides an excellent opportunity for the countries in the region to optimise the use of their energy resources according to availability and costs. This in turn increases the cost efficiency and the security of supply.

E Conclusion

The electricity trade in Africa up to now has been based mainly on bilateral agreements. With the exception of the SAPP, the African power pools are still in their developmental stages, building up their institutional capacity and their skills mix. The SAPP is the most advanced power pool, currently operating DAM and PDAM.

The main challenges faced by regional electricity trade in Africa include lack of physical infrastructure (due in part to lack of access to capital), poor planning and lack of maintenance and poor performance of the power utilities.

Nonetheless, considering the strong demonstrated political will to overcome the challenges, combined with sustained economic growth and Africa's abundance of energy resources, there is a window of opportunity for investors, provided the countries strengthen their efforts towards setting up a conducive environment for investments.

The role of RE technologies in support of African development is undisputable, considering the countries' huge hydro, solar and geothermal potentials. Thus, for a sustained energy security and energy mix it is imperative for the countries to have targets for the share of RE in their energy mix, combined with additional incentives for attracting RE investments.

[15] Ibid.

Bibliography

African Development Bank. (2011). *Program for Infrastructure Development in Africa (PIDA)*, www.afdb.org/fileadmin/uploads/afdb/Documents/Generic-Documents/PIDA%20brief%20Energy.pdf (accessed 9 December 2014).

Central Africa Power Pool (PEAC). (2011). *Présentation générale du secteur électrique en Afrique Centrale.* Nottingham: WYG International/Europe Aid/126679/C/SER/CG.

Eastern Africa Power Pool (EAPP). (2012). *Presentation at IGAD Investments Conference*, www.icafrica.org/fileadmin/documents/ICA_sponsored_events/IGAD_HoA_Conf_2012/EA%20Power%20pool_IGAD_Conf_March%202012.pdf (accessed 10 December 2014).

Economic Community of Central African States (ECCAS). (2009). *Etude sur l'Interconnection des Réseaux Electriques des pays membres de la Communauté Economique des Etats de l'Afrique Centrale – Rapport de l'Etude de Faisabilité.*

Infrastructure Consortium for Africa (ICA). (2011). *Regional Power Status in Africa Power Pools*, www.icafrica.org/fileadmin/documents/Knowledge/Energy/ICA_RegionalPowerPools_Report.pdf (accessed 9 December 2014).

Kambanda, C. (2013). *Power Trade in Africa and the Role of Power Pools*, www.afdb.org/en/blogs/integrating-africa/post/power-trade-in-africa-and-the-role-of-power-pools-12101/ (accessed 9 December 2014).

Southern Africa Power Pool. (2014). *Annual reports 2013*, www.sapp.co.zw/docs/22867%20Annual%20Report%20New.pdf (accessed 9 December 2014).

Southern Africa Power Pool. (2009). *Annual reports 2008*, www.sapp.co.zw/documents/SAPP%20annual%20report.pdf (accessed 9 December 2014).

United Nations Environment Programme (UNEP) Finance Initiative. (2012). *Financing Renewable Energy in Developing Countries.* Geneva: UNEP, www.unepfi.org/fileadmin/documents/Financing_Renewable_Energy_in_subSaharan_Africa.pdf (accessed 9 December 2014).

Western Africa Power Pool. (2013). *Activity Report of the Secretary General*, www.ecowapp.org/?page_id=10 (accessed 9 December 2014).

World Bank Group. (2010). *Africa's Infrastructure – A Time for Transformation.* Washington, DC: World Bank.

10

Latin American Experiments in Promoting Green Electricity

CAROLINA M. LEMBO AND BELISA E. ELEOTERIO

A Introduction

Latin America leads the world in the exploration of renewable energy, such as hydroelectricity and efficient biofuels. In 2011, installed electric power capacity in the region was estimated at about 270 GW, most of which was already provided through renewable energy sources.[1]

Major developments of renewable energy are underway in Latin America. Solar, wind, geothermal, ocean, small-scale hydropower and advanced bio-energy are now ready to play major roles, alongside hydropower, in meeting the continent's energy needs.[2]

This chapter aims to explore Latin American experiments in promoting green energy. It is divided into three main parts. First, it seeks to elucidate how energy trade and regional integration can foster the wider use of renewable energy. Second, the relationship between energy and sustainable development is highlighted. Finally, some initiatives being implemented in Latin America are described.

B Electricity Integration Enables Expansion of Renewables

The reality of the energy sector has changed greatly in recent decades, with the underlying theme being renewable energy. Including renewable energy in a country's energy mix helps to reduce the cost of electricity for the country as a whole. Since it offers more options for access to electricity, the incorporation of renewable energy technologies into the existing electric power systems can guarantee energy security, reduce import

[1] W. Vergara, C. Alatorre and L. Alves, 'Rethinking our energy future: a White Paper on renewable energy for the 3GFLAC Regional Forum', *Inter-American Development Bank Discussion Paper* No. IDB-DP-292 (2013), p. 5.

[2] IADB, *Energía renovable*, www.iadb.org/es/temas/energia/se4allamericas/energia-renovable. 17688.html (accessed 6 December 2016).

dependency, improve local environment and health statuses and reduce levels of greenhouse gas (GHG) emissions.[3]

Renewable energy is becoming increasingly cost-competitive where resources are available. If supported by strong enabling policies at the public level and substantial investment from the private sector, renewable energy could supply a much larger share of the world's energy by 2030.[4]

The complementarity between the production of renewable energies and the increased security of energy supply is widely acknowledged.[5] Security increases as a nation reduces the risks associated with vulnerability to supply disruption, and economic instability may be reduced by making the nation's economic infrastructure less dependent on fuel price volatility.[6] This rationale could easily be applied at a regional level, where an integrated grid of electricity with complementary renewable sources could guarantee improvement of the energy supply and continuity of service in distribution networks.

The Brazilian Interconnected National System (SIN) is a good example of how the above-mentioned complementarity can work at the national level. The system for electricity production and transmission in Brazil is highly interconnected, supplying 98 per cent of the Brazilian market.[7] To guarantee continuity of supply, the National System Operator (ONS) has to coordinate, supervise and monitor the operation of the entire Brazilian energy mix, within which hydroelectric sources dominate.

The expansion of installed capacity planned for 2022 will be based predominantly on renewable sources such as hydroelectric, wind and biomass-based power plants. These sources are expected to supply

[3] IPCC, *Renewable Energy Sources and Climate Change Mitigation: Special Report of the Intergovernmental Panel on Climate Change* (Cambridge: Cambridge University Press, 2012), p. 20; www.ipcc.ch/pdf/special-reports/srren/SRREN_Full_Report.pdf (accessed 6 December 2016).

[4] Sustainable Energy for All, *Renewable Energy*, www.se4all.org/about-us_our-ambition_renewable-energy (accessed 6 December 2016).

[5] S. P. A. Brown and H. G. Huntington, 'Energy security and climate change protection: complementarity or tradeoff?', *Energy Policy*, 36(9) (2008), 3510–13.

[6] M. A. Abdullah, A. P. Agalgaonkar and K. M. Muttaqi, 'Assessment of energy supply and continuity of service in distribution network with renewable distributed generation', *Applied Energy*, 113(C) (2014), 1015–26.

[7] Portal Brasil, *Sistema Interligado Nacional atende 98% do Mercado brasileiro*, www.brasil.gov.br/infraestrutura/2011/12/sistema-interligado-nacional-atende-98-do-mercado-brasileiro (accessed 6 December 2016).

approximately 41 per cent of the Brazilian energy mix, which will bring environmental benefits for the country.[8]

However, the integration of some generation sources (such as wind and solar energy) into energy transmission systems may cause energy instability in a given geographical area. Increasing cross-border trade in the grid can play a major role in helping overcome this challenge. Enabling countries to gain access to a more diversified energy portfolio may help to the extent that that they can increase or decrease their output more easily and quickly so that the risk of intermittence is reduced and it is possible to keep the overall system balanced at all times.[9]

C Energy Trade and Integration as a Pathway to Achieving Sustainable Development

Economic development has been strongly associated with increasing energy use and increased carbon dioxide (CO_2) emissions. The implementation of the concept of sustainable development is closely linked to the rational and sustainable use of energy sources. As noted by the United Nations (UN) Secretary-General's Advisory Group on Energy and Climate Change, 'energy is at the heart of most critical economic, environmental and developmental issues facing the world today. Clean, efficient, affordable and reliable energy services are indispensable for global prosperity.'[10]

Agenda 21, formulated in the context of the United Nations Conference on Environment and Development (1992), stresses the importance of fostering the creation of programmes that promote energy efficiency and the development of technologies that allow renewable sources to be used more widely:

> 9.11. The basic and ultimate objective of this programme area is to reduce adverse effects on the atmosphere from the energy sector by promoting policies or programmes, as appropriate, to increase the contribution of

[8] Brasil, Ministério de Minas e Energia, *Plano Decenal de Expansão de Energia 2022*, p. 384, www.epe.gov.br/Estudos/Documents/PDE2022.pdf (accessed 6 December 2016).

[9] H. Bahar and J. Sauvage, 'Cross-border trade in electricity and the development of renewable-based electric power: lessons from Europe', *OECD Trade and Environment Paper* No. 2013/02 (2013), p. 9, www.oecd.org/officialdocuments/publicdisplaydocumentpdf/?cote=COM/TAD/ENV/JWPTE(2012)20/FINAL&docLanguage=En (accessed 6 December 2016).

[10] United Nations, *Energy for a Sustainable Future* (New York: United Nations, 2010), p. 7, www.un.org/millenniumgoals/pdf/AGECCsummaryreport[1].pdf (accessed 6 December 2016).

> environmentally sound and cost-effective energy systems, particularly new and renewable ones, through less polluting and more efficient energy production, transmission, distribution and use. This objective should reflect the need for equity, adequate energy supplies and increasing energy consumption in developing countries, and should take into consideration the situations of countries that are highly dependent on income generated from the production, processing and export, and/or consumption of fossil fuels and associated energy-intensive products and/or the use of fossil fuels for which countries have serious difficulties in switching to alternatives, and the situations of countries highly vulnerable to adverse effects of climate change.[11]

The Johannesburg Declaration on Sustainable Development, issued ten years after the establishment of Agenda 21 (2002), proposed, as a means to change unsustainable patterns of consumption and production, 'the increased use of renewable energy resources, more efficient use of energy, greater reliance on advanced energy technologies, including advanced and cleaner fossil fuel technologies, and the sustainable use of traditional energy resources, which could meet the growing need for energy services in the longer term to achieve sustainable development'.[12]

In 2005, another important development milestone was reached at the World Summit, where important themes were debated and consolidated, including the importance of promoting clean energy to combat climate change, as well as to achieve sustainable development and the eradication of poverty:

> 50. We face serious and multiple challenges in tackling climate change, promoting clean energy, meeting energy needs and achieving sustainable development, and we will act with resolve and urgency in this regard.[13]

Similarly, the document adopted at the Conference on Sustainable Development in 2012, called *The Future We Want*, recognises the role of energy in sustainable development and the eradication of poverty. It

[11] United Nations, *Agenda 21*, United Nations Conference on Environment and Development, Rio de Janeiro, Brazil, 3–14 June 1992, http://sustainabledevelopment.un.org/content/documents/Agenda21.pdf (accessed 6 December 2016).

[12] United Nations, *Report of the World Summit on Sustainable Development*, Johannesburg, South Africa, 26 August–4 September 2002, *Plan of Implementation of the World Summit on Sustainable Development*, Section III.20(d), www.unmillenniumproject.org/documents/131302_wssd_report_reissued.pdf (accessed 6 December 2016).

[13] United Nations, 2005 World Summit Outcome, Resolution adopted by the General Assembly, 16 September, New York, United States, Section II.50, www.un.org/womenwatch/ods/A-RES-60-1-E.pdf (accessed 6 December 2016).

also emphasises the challenges of ensuring access to energy.[14] Still on the energy issue, the document supports the UN Secretary-General's initiative to promote Sustainable Energy for All, which focuses on access to energy, energy efficiency and renewable energy sources.[15]

I The Three Pillars of Sustainable Development's Relationship with Energy

During the United Nations Conference on Environment and Development held in Rio de Janeiro in 1992, a new concept of development was coined – a sustainable one, based on three inseparable pillars: economic, social and environmental. The particularities of each of the pillars will be addressed below.[16]

1 Economic Pillar: A Secure Supply of Energy and Competitiveness

The relationship between the energy sector and economic development is evident, as a country can only develop its services and its industry sectors when there is electricity. It follows that a nation requires not only a supply of energy that is secure, but also one that is constant, to ensure the competitiveness of its private sector.

One way to ensure a secure supply of energy is the creation of public policies for the promotion of energy from renewable sources. This is because 'dependence on energy imports, whether of fossil fuels or the technology needed for implementation of renewable energy, represents a potential source of insecurity in the supply of energy for both developing and industrialized countries'.[17] Thus, one should seek an energy mix that leads to complementarity between energy sources and a more independent and diverse energy portfolio. This would allow developing countries to reduce their dependence on fossil fuels and meet their energy needs in

[14] United Nations, *The Future We Want*, Report of the United Nations Conference on Sustainable Development, 20–22 June 2012, Rio de Janeiro, paras. 125–26, www.uncsd2012.org/content/documents/814UNCSD%20REPORT%20final%20revs.pdf (accessed 6 December 2016).

[15] Ibid., para. 129.

[16] C. M. Lembo, *Energia e o Sistema Multilateral de Comércio: perante o Paradigma do Desenvolvimento Sustentável* (São Paulo: Atlas, 2015).

[17] IPCC, *Renewable Energy Sources and Climate Change Mitigation: Special Report of the Intergovernmental Panel on Climate Change* (2012), p. 717, http://srren.ipcc-wg3.de/report/IPCC_SRREN_Full_Report.pdf (accessed 6 December 2016).

the long term through sharing regional resources and offering incentives for the provision of affordable energy.

2 Environment Pillar: Environmental Conservation

The environment pillar, according to the Brundtland Report, is related to the idea of limitations imposed on the environment's ability to meet present and future needs through the state of technology and social organisation.[18] It is thus imperative to change patterns of consumption and production that jeopardise the conservation of the environment, as in the case of energy sources. Therefore, it is necessary to increase energy efficiency and the use of renewable sources in order to reduce emissions of GHGs that cause climate change.

Strategies that enable the necessary changes to be made to the current patterns of consumption and energy production should include the integration of renewable sources into coherent energy systems. These should be influenced by efficiency and energy-saving measures in order to ensure: (i) energy savings on the demand side; (ii) greater efficiency in energy production; and (iii) replacement of fossil fuels by various sources of renewable energy. Consequently, plans for the implementation of large-scale renewable energy schemes must include strategies for the renewable sources' integration into coherent energy systems.[19]

Especially with regard to Latin American countries, and in addition to optimisation of the use of energy sources, the development of coordinated electricity markets and strengthening of environmental policies and regulations are fundamental to foster sustainable and environmentally favourable energy resources for power generation.

3 Social Pillar: Overcoming Energy Poverty and Achieving Sustainability

Access to modern, clean and environmentally sound energy is a key condition for overall improvements in the quality of life and for sustainable development worldwide. Environmental policies and poverty mitigation are two sides of the same coin.

[18] United Nations, *Report of the World Commission on Environment and Development: Our Common Future* (1987), www.un-documents.net/our-common-future.pdf (accessed 6 December 2016).

[19] H. Lund, 'Renewable energy strategies for sustainable development', *Energy*, 32(6) (2007), 912.

Agbemabiese maintains that access to modern, clean and environmentally sound energy is a key precondition for overall improvements in the quality of life and sustainable development all over the world.[20] It fulfils the basic human needs for nutrition, warmth and light and is an engine for economic growth. Indeed, energy and human development are inextricably linked.[21]

Access to energy should be considered a pathway leading to social and economic development in low-income communities, helping to reduce poverty and increase family income. One way to ensure the viability of programmes designed to facilitate access to electricity for the poor is the establishment of special tariff mechanisms that are socially oriented, economically sustainable and enforced by a legal framework.

In order to accomplish this complex paradigm shift from a carbon-intensive energy mix to a more sustainable one that makes greater use of renewable sources, a combination of command and control mechanisms, and even market ones, is necessary to assist the public and private sectors in making this transition. The ultimate aims of these mechanisms should be to transform or to deploy new energy matrices to reduce GHG emissions, and to increase sustainability and socioeconomic benefits.

Brazil adopted this approach when its 'Electricity for All' programme was launched in 2003.[22] The programme was initiated in response to the finding that more than two million people in Brazil's rural areas were living without electricity in 2000. The people without access to electricity had the lowest rates of human development in Brazil, according to the Brazilian government.[23] The indicator known as the Human Development Index (HDI) evaluates not only the economic development of the population of a country, but also the quality of life of its people.[24]

The objectives of the Electricity for All programme were to fight energy poverty and promote socioeconomic development. The viability

20 L. Agbemabiese, 'A framework for sustainable energy development beyond the grid: meeting the needs of rural and remote populations', *Bulletin of Science, Technology & Society*, 29(2) (2009), 151–8.

21 Ibid.

22 Brazil, *Programa Nacional de Universalização do Acesso e Uso de Energia Elétrica – Luz para Todos*, Law 10.762/2003 and Decree 4.873/2003.

23 Brazil, Ministério de Minas e Energia. *Programa Luz para Todos* (2003), http://luzparatodos.mme.gov.br/luzparatodos/Asp/o_programa.asp (accessed 6 December 2016).

24 C. M. Lembo, *La relación de los pilares del desarrollo sostenible con la universalización del acesso a la energia: la experiencia brasileña* (Congresso Internacional sobre Accesso Universal a Servicios Públicos de Energía, Peru, 2013).

of the programme was ensured through the establishment of a social tariff. Thus, people who had a per capita family income of half the minimum wage or less would receive discount rates between 10 per cent and 65 per cent, according to consumption.

The Brazilian experience with the Electricity for All programme is recognised not only as an opportunity for South–South cooperation, where the Brazilian government can share its experience with other countries seeking to bring about universal access to electricity for their populations, but also as a model for the Sustainable Energy for All initiative, as mentioned above. This initiative was launched by the United Nations in 2011, with the aim of promoting access to energy, energy efficiency and renewable energies as a means to eradicate poverty and move towards sustainable development and global prosperity.

D Latin American Integration Experiments in Green Electricity

Latin America is rich in energy resources. However, the asymmetry in individual countries' capacities prevents the achievement of energy independence.[25] As previously discussed, electricity cooperation and integration increases suppliers' and consumers' access to modern electricity services and products. Moreover, it provides greater security and reliability to the power sector, as well as fostering improvements in technical operation.[26]

Ensuring long-term energy security remains one of the most vital steps towards electricity trade in the region, and there is also a need to address the lack of strong institutions capable of enforcing mutually beneficial energy policies. This implies not only defining development plans with common objectives but also developing adequate regulatory approaches to effectively address energy issues.

The creation of regulatory institutions is key to national and regional coordination. They are necessary not only to ensure that operational and

[25] F. E. Gonzáles and S. Weintraub, 'Energy use and climate change: western hemispheres partnership', *Partnership for the Americas Commission, Background Document* BD-01 (May 2008), p. 3, www.brookings.edu/~/media/Projects/latin%20america/energy_climate_gonzalez.PDF (accessed 6 December 2016).

[26] F. J. Burgos, *Regional Electricity Cooperation and Integration in the Americas: Potential Environmental, Social and Economic Benefits* (Washington, DC: Department of Sustainable Development of the Organization of American States, 2007), p. 8, www.oas.org/DSD/Documents/RegionalElectricityCooperationandIntegration.pdf (accessed 6 December 2016).

market rules and environmental regulations are not breached, but also to achieve social, economic and environmental goals.[27]

The Inter-American Development Bank (IADB) plays an important role in the region, supporting efforts by Latin American countries to reduce poverty and inequality and promoting integration to foster competitiveness and energy efficiency with a view to enabling economic growth and sustainable development.[28]

Improving energy efficiency in Latin America is closely related to diversification of the energy matrix by sustainably exploiting renewable sources. Programme financing is part of the IDB's commitment to developing mechanisms to support the long-term funding of renewable energy and clean technology projects in the region, which stimulate innovation, job creation and green economic growth.

Analysed below are some initiatives that foster electricity trade and integration among Latin American countries.

I *Continental Integration Initiatives*

In the second half of the twentieth century, Latin American nations started to adopt mechanisms to facilitate trade integration. Initially broader in scope than they are now, they constituted forums to stimulate dialogue among countries in the region. These projects ultimately brought about sub-regional initiatives with various purposes, including energy integration.

1 Regional Energy Integration Commission (CIER)

CIER is a non-governmental organisation (NGO) created in 1964, which comprises several companies and bodies of its Member States' and Member Associates' electrical energy sectors.[29]

The main objective of CIER is to promote the integration of regional electricity sectors through activities aimed at:

(i) increasing the efficiency and management of energy companies;
(ii) promoting cooperation and the exchange of information, experiences and documents on technical, economic and legal matters;

[27] Ibid., p. 19.

[28] IADB, 'Renewable energy', www.iadb.org/en/topics/energy/renewable-energy,19008.html (accessed 6 December 2016).

[29] The CIER Members are Argentina, Bolivia, Brazil, Chile, Colombia, Ecuador, Paraguay, Peru, Uruguay, Venezuela and CECACIER (Regional Committee for Central America and the Caribbean), as well as more than 260 companies.

(iii) training professionals;
(iv) conducting studies to improve the quality of service provided by the companies;
(v) developing projects with a regional focus;
(vi) compiling the general specifications and technical standards applicable to each Member;
(vii) promoting harmonisation of statistics in the region, to be gathered in a database;
(viii) promoting the rational use of energy, with an emphasis on the effects on the environment;
(ix) promoting integration and energy trade between the continent's electricity markets.[30]

CIER is conducting several studies with the aim of providing alternative solutions for the electric power sector, thereby contributing to the resolution of issues that may impact the energy supply of the region. To meet its objectives, it supports the creation of projects in each of its areas of activity, including energy trade.

The *Study of the Electricity Trade among the Andean Markets, Central America, and Mercosur* (CIER 15 Project) examined the feasibility of creating electricity interconnection projects that align with the domestic policies of each country in order to ensure maximum benefits to the final users. The results of CIER 15 show that there is a potential for deepened energy integration in the region, with more efficient use of resources, including renewable ones; better quality of service delivery; more efficient rates; and a more competitive environment.[31]

2 Latin American Energy Organization (OLADE)

OLADE was created in response to the need to design energy policies for Latin American and Caribbean countries, especially after the international energy crisis of the early 1970s. The organisation seeks to support the independent development of energy resources and capabilities by: promoting a rational and effective policy for source exploitation; processing and marketing energy resources; coordinating and promoting

[30] CIER, *Proyecto CIER 15 Fase II*, Informe Final (2011), http://sg.cier.org.uy/Publicaciones/Revista.nsf/4b042312b0364a4c03257140006a24cb/.db83e9241649d7138325738e00435475?OpenDocument (accessed 6 December 2016).

[31] C. A. Cavalcanti, C. M. Lembo and V. H. Thorstensen, *The Regulation of the International Energy Trade: Fuels and Electricity* (São Paulo: FIESP, 2013), pp. 291–3.

direct negotiations between Member States;[32] facilitating cooperation; promoting means to ensure the free movement and the use of different means of transportation of energy resources; and going beyond the search for common energy policies as a means of integration.

OLADE's contribution to energy integration in Latin America is demonstrated through a variety of activities aimed to strengthen the institutional capacities of the region's energy ministries (e.g. through training), the industry's knowledge of government (e.g. through regional events and expert networks) in technical studies and projects and the compilation and systematisation of statistical data on the region's energy sector. OLADE also monitors various countries' energy-related legislation, in an effort to facilitate the sharing of knowledge and information on these countries' regulations.

3 South American Energy Council

The South American Energy Council, set up to coordinate the regional energy policy of the Union of South American Nations (UNASUR),[33] proposes strategies that will meet the objectives required for the achievement of South American energy integration through:

(i) the promotion of energy security of the region;
(ii) the promotion of regional energy trade;
(iii) the strengthening of regional energy infrastructure;
(iv) the creation of complementary mechanisms between the national state-owned enterprises producing energy from hydrocarbons and other types of energy;
(v) aid for trade and technology transfer, as well as training of personnel;
(vi) the encouragement of regional energy development in order to provide a rational and sustainable consumption model that preserves natural resources and the environment;
(vii) the promotion of industrialisation and development of the energy sector and its regional complementation;

[32] The OLADE members are: Argentina, Barbados, Belize, Bolivia, Brazil, Chile, Colombia, Costa Rica, Cuba, Ecuador, El Salvador, Granada, Guatemala, Guyana, Haiti, Honduras, Jamaica, Mexico, Nicaragua, Panama, Paraguay, Peru, Dominican Republic, Suriname, Trinidad and Tobago, Uruguay and Venezuela.

[33] The UNASUR members are: Argentina, Bolivia, Brazil, Chile, Colombia, Ecuador, Guyana, Paraguay, Peru, Suriname, Uruguay and Venezuela.

(viii) the promotion of harmonisation of regulatory and commercial aspects related to energy integration;
(ix) the incorporation of regional integration factors in national energy planning;
(x) the promotion of efficient use of energy and exchange of successful experiences in this area;
(xi) the promotion of the development of renewable and alternative energy sources (e.g. biofuels, wind, solar, nuclear, ocean, geothermal and hydro);
(xii) the encouragement of partnerships between the public and private sectors;
(xiii) the maintenance of bilateral, regional and sub-regional agreements, as well as the negotiation of future agreements;
(xiv) the promotion of a balance between energy-producing countries and consumers of energy;
(xv) the promotion of convergence of national energy policies, taking into account the current legal framework in each country.[34]

The Action Plan for Regional Energy Integration, adopted in 2010, establishes, among other objectives, an increase in the share of renewable energy sources and alternative energy sources in the South American region coming from the development of their own resources.

II Regional Electrical Integration

1 Mesoamerica Energy Integration Project

In 2005, the six Central American countries (Guatemala, El Salvador, Honduras, Nicaragua, Costa Rica and Panama), plus Belize, the Dominican Republic, Mexico and Colombia, adopted the Mesoamerica Energy Integration Program, which consolidated several of the earlier electricity integration initiatives, in order to support the region's energy market integration and increase electricity generation.

The project is intended to substantially improve energy security, reduce the costs of electricity supply and strengthen the electricity sector in order to increase the competitiveness of the Mesoamerican region. It

[34] OLADE, *Unasur: un espacio que consolida la integración energética* (2012), www.olade.org/sites/default/files/publicaciones/UNASUR%20-%20Un%20espacio%20que%20-completo.pdf (accessed 6 December 2016).

was also designed to enhance the production of energy from renewable sources and to consolidate a Regional Electricity Market (MER).

Two main programme components are envisaged: (i) infrastructure development (electricity interconnection projects) and (ii) formulation of a legal and institutional framework for the functioning of MER. Among the initiatives under the programme, we highlight the following.

Central American Electrical Interconnection System (SIEPAC) The SIEPAC project, established through the Framework Treaty for the Electricity Market in Central America in 1997, aims to help neighbouring countries to better distribute and commercialise their energy resources in order to reduce waste, protect against price volatility, lower transmission losses and increase overall energy security.[35] The project contributes to the development of a more competitive energy base for the region and provides an opportunity for the integration of renewable energies. Central American countries now have access to a more secure transmission system that offers additional advantages such as reducing the average cost of generation. This cost reduction is also applicable to renewable energies, which will allow better integration of these technologies in the matrix, increase the transmission capacity between countries and enable the interconnected countries' energy demand to be met if a deficit situation arises.[36]

The first phase of the electrical interconnection system was concluded at the beginning of October 2014, when the final 130-km Parrita-Palmar Norte segment in Costa Rica fully integrated Panama into the broader regional grid.[37] Now SIEPAC countries are discussing the execution of a second phase, which aims to encourage the implementation of regional energy projects, especially with respect to renewable sources, that would reduce the burning of fossil fuels and the consequent GHG emissions.

Besides the SIEPAC Project, at least 279 small renewable energy projects have been implemented in Central America with international cooperation and support. These projects concern solar, micro and small

[35] The SIEPAC countries are: Guatemala, El Salvador, Honduras, Nicaragua, Costa Rica and Panama.

[36] SIEPAC, www.proyectomesoamerica.org (accessed 6 December 2016).

[37] Proyeto Mesoamerica, *Costa Rica completa línea de transmisión eléctrica SIEPAC*, San José, Costa Rica (16 October 2014), www.proyectomesoamerica.org/joomla/index.php?option=com_content&view=article&id=731&Itemid=85 (accessed 6 December 2016).

hydro, wind, bioenergy, biogas generation from agro-industrial processes and installation of efficient stoves, among others.[38]

The adoption of these policies has resulted in an improvement in installed capacity use and a reduction in fuel consumption, benefiting the participants and enabling the growth of the electricity market in the region. Nonetheless, there are still concerns relating to regulatory issues and electricity trade.

Regional Electricity Market (MER) MER consolidation is crucial for achieving the objectives of electrical interconnection set out by SIEPAC. Thus, a legal and institutional framework was established to regulate the regional electricity market and to further its efficiency and competitiveness. Providing a solid regulatory framework for energy exchange between countries increases the attractiveness of investment in power-generation projects at the regional level, reduces costs, encourages competition and fosters the development of regional power plants.[39]

The market's operation is based on the rules of the Framework Agreement for the Electricity Market in Central America, adopted in 1996 to ensure the free movement or flow of electricity through their territories for the countries themselves or for third countries in the region, which are subject only to the conditions set out in the treaty framework, its protocols and its regulations.

Mexico–Guatemala Electric Interconnection In recent years, Mexico has set ambitious goals for expanding its access to renewable energy and reducing its dependence on fossil fuels.[40] To this end, it has adopted various legal mechanisms, regulatory frameworks and financial instruments to address energy issues.

[38] 'Al menos 279 proyectos de energia renovable se ejecutan en Centroamérica', *El Diario* (El Salvador, 26 June 2014), www.eldiario.es/economia/proyectos-energia-renovable-ejecutan-Centroamerica_0_275072520.html (accessed 6 December 2016).

[39] Proyecto Mesoamerica, *Mercado Eléctrico Regional (MER)*, www.proyectomesoamerica.org/joomla/index.php?option=com_content&view=article&id=439&Itemid=229 (accessed 6 December 2016).

[40] Mexico is also part of NAFTA, a regional integration initiative that comprises the United States and Canada and which has its own regulatory framework for energy. Nonetheless, this regional initiative is not addressed in this chapter because energy transactions between Mexico and the United States are very limited owing to short cross-grid interconnection, market harmonization and grid compatibility issues.

The country's main objectives are to increase installed capacity and the generation of electricity from renewable sources, boost public and private investment in generating power and in building and expanding the interconnection infrastructure, increase the share of biofuels in the national energy mix and facilitate the development of renewable energy technology.

The Mexico–Guatemala interconnection project meets the objectives set both in the Mesoamerican initiative and in the Mexican renewable energy development plan, and aims to facilitate the Central American Electricity Market's connection with the Mexican Electricity System.

Panama–Colombia Electric Interconnection Building an electrical interconnection infrastructure between Panama and Colombia is also considered a strategic step towards the consolidation of complete regional integration in Central America. The electrical interconnection project between the two countries is part of a regional strategy linking Mexico to Colombia through a 2,500-km transmission line.

This project embodies the Andean Community and Central America's integration with an organised energy market through SIEPAC. The accomplishment of this interconnection will allow access to new sources, optimise energy use, enhance the management of energy surplus and enable the development of biofuels, as well as having environmental benefits.

2 Andean Electrical Interconnection System (SINEA)

SINEA was created in 2011 with the objective of working actively towards the development of a Community framework that enables integration in the participating countries' energy markets. Through the Declaration of Santiago, members agreed that the Andean Community (CAN) Working Groups on planning and regulatory harmonisation would participate in the SINEA to assist them by providing the technical and legal support required for the process of electrical integration in the region. The SINEA also follows the principles of: (i) legal and contractual stability; (ii) freedom of transit; (iii) free access to the remaining capacity of electricity transmission lines; (iv) non-discrimination; (v) competitive pricing and (vi) sustainable development.

In April 2014, the SINEA Council of Energy Ministers signed the Declaration of Lima establishing a roadmap for progressive electrical integration among its members – Peru, Ecuador, Chile, Colombia and Bolivia. The main objective is to provide assurance for exchanges and transactions of electricity, which shall be conducted within a regulatory framework that provides legal certainty, efficiency and equity for participating countries.[41]

3 MERCOSUR

The market framework defined by MERCOSUR for the integration and cooperation of its electricity sector includes not only financial regulation, but also energy efficiency, production of renewable energy, environmental protection and legal harmonisation.[42] In 1998, the group decided to strengthen its efforts towards electricity integration,[43] especially regarding transparency of operations and information, energy supply and a competitive market. Recently, it also advocated more efficient coordination of the regulatory rules and technical activities in support of energy trade.

Member States also signed a framework agreement on regional energy complementation that,[44] although not binding, is intended to contribute to the advancement of regional energy integration in terms of trade, interconnection systems, fostering renewable sources and cooperation on resource exploration among others.

Noteworthy within MERCOSUR is the Action Plan for Biofuels,[45] which aims to encourage the production and consumption of biofuels, technical cooperation, regulatory coherence, transparency and capacity building.

III Binational Integration Projects

1 Itaipu

The Itaipu Hydroelectric Power Plant between Brazil and Paraguay is one of the largest hydroelectric plants generating clean and renewable energy. With twenty generating units and 14,000 MW of installed capacity, it

[41] SINEA, *Declaración de Lima para la Interconexión e Integración Eléctrica Andina* (25 April 2014), www19.iadb.org/intal/intalcdi/PE/CM%202014/14275.pdf (accessed 6 December 2016).

[42] MERCOSUR Action Plan to Year 2000, 5 August 1995, MERCOSUR/CMC/DEC No. 9/95.

[43] MERCOSUR Memorandum of Understanding on Electricity Exchange and Electrical Integration, 23 July 1998, MERCOSUR/CMC/DEC No. 10/98.

[44] MERCOSUR's members are: Argentina, Brazil, Paraguay, Uruguay and Venezuela.

[45] MERCOSUR Memorandum of Understanding to Establish a Special Working Group on Biofuels, 15 December 2006, MERCOSUR/CMC/DEC No. 36/06.

provides approximately 17 per cent of the energy consumed in Brazil and 75 per cent of that consumed in Paraguay. The production record was set in 2013, when the dam generated 98,630,035 MWh.[46]

The plant also plays an important role in carrying out activities and social projects as a way to promote sustainable development and improved quality of life in its vicinity.

2 Yacyretá

The Yacyretá hydroelectric plant between Argentina and Paraguay has twenty generating units, an installed capacity of 3,200 MW and an average annual gross generation of 20,000 GWh.[47] The dam generates approximately 15 per cent of the energy consumed in Argentina and provides significant financial resources to Paraguay,[48] which sells its surplus to Argentina because Paraguay has lower consumption levels and still lacks a high-voltage transmission line to enable electricity transportation.[49]

The Treaty of Yacyretá, signed by the governments of Argentina and Paraguay on 3 December 1973, promotes the generation of low-cost electricity, improving navigability, developing fisheries and providing water for irrigation.

3 Salto Grande

The Salto Grande dam, an initiative of Argentina and Uruguay, was the first binational hydraulic complex in Latin America. The project began in 1938, and the dam was already in operation in 1979. The plant is equipped with fourteen generators, representing an installed capacity of 1,890 MW and annual production of 7,812 GWh. The plant is responsible for the generation of nearly half of the electricity consumed in Uruguay, while it supplies about 7 per cent of Argentina's demand.[50]

[46] Itaipu, www.itaipu.gov.br/energia/geracao (accessed 6 December 2016).

[47] Entidad Binacional Yacyretá, *Características Técnicas*, www.yacyreta.org.ar/index.php/12-paginas/127-2-caracteristicas-tecnicas. (accessed 6 December 2016).

[48] There is no data related to Paraguay's consumption. World Bank, *Yaciretá Hydroelectric Project*, www.worldbank.org/projects/P005937/yacyreta-hydroelectric-project?lang=en (accessed 6 December 2016).

[49] 'Argentina y Paraguai se preparan para rever tratado de la represa Yaciretá', *MercoPress* (Montevideo, 18 March 2014), http://es.mercopress.com/2014/03/18/argentina-y-paraguay-se-preparan-para-rever-tratado-de-la-represa-yacyreta (accessed 6 December 2016).

[50] Salto Grande, *Generation*, www.saltogrande.org/generacion.php (accessed 6 December 2016).

4 Wind Integration between Brazil and Uruguay

In 2012, representatives of the governments of Brazil and Uruguay decided to create a new model for their bilateral relationship, which was intended to translate into an action plan for sustainable development and the integration of Brazil and Uruguay.

The issue of energy integration was included as both governments had expressed the need for cooperation with a view to consolidating the proposed integration of electro-energy systems to be incorporated in a bilateral treaty on the subject, including aspects of operation, marketing, regulation and system expansion. In this context, the presidents recognised the importance of the construction of a 500-kV transmission line between San Carlos (Uruguay) and Candiota (Brazil), to be completed in 2013, creating new possibilities for structuring the exchange of electricity for the benefit of energy security for the two countries.[51]

As regards the development of basic industry linked to the development of wind power, progress in the implementation of two wind farms by the power companies Eletrobras (Brazil) and UTE (Uruguay) should be highlighted, as well as efforts towards industrial development. To this end, both countries have stated their willingness to make every effort to comply with the financing conditions set by the Brazilian Development Bank (BNDES) in order to promote the abovementioned industrial development.[52]

E Concluding Remarks

Current projects on electricity cooperation and energy integration in the Americas are focused on the development of regional grid interconnections that are supported by bilateral and regional agreements. Long-term integration efforts could include investment in enabling

[51] Itamaraty, *Comunicado Conjunto Presidencial: Novo Paradigma para a Relação Brasil-Uruguai*, 31 July 2012, www.itamaraty.gov.br/sala-de-imprensa/notas-a-imprensa/comunicado-conjunto-presidencial-novo-paradigma-para-a-relacao-brasil-uruguai (accessed 6 December 2016).

[52] The BNDES, established in 1952, is the main financing agent for development in Brazil, playing a fundamental role in stimulating the expansion of industry and infrastructure in the country. Brazil's growing global importance, coupled with the interdependence of the globalised world economies, has compelled the BNDES to broaden its horizons and intensify its participation outside Brazil.

infrastructure; modification of institutional and governance frameworks; paying attention to social aspects, markets and planning; and capacity building in anticipation of green energy growth.

Latin American countries are developing various mechanisms to foster sustainable development. Nonetheless, the challenge common to all energy integration projects established in Latin America is in establishing a common regulatory framework, which appears to be a major obstacle to regional integration. It is evident that new rules are needed in order to provide coherence to the international energy trade.

Bibliography

Abdullah, M. A., Agalgaonkar, A. P. and Muttaqi, K. M. (2014). Assessment of energy supply and continuity of service in distribution network with renewable distributed generation. *Applied Energy*, **113**, 1015–26.

Agbemabiese, L. (2009) A framework for sustainable energy development beyond the grid: meeting the needs of rural and remote populations. *Bulletin of Science, Technology & Society*, **29**(2), 151–8.

Bahar, H. and Sauvage, J. (2013) Cross-border trade in electricity and the development of renewables-based electric power: lessons from Europe. *OECD Trade and Environment Paper* No. 2013/02, www.oecd.org/officialdocuments/publicdisplaydocumentpdf/?cote=COM/TAD/ENV/JWPTE(2012)20/FINAL&docLanguage=En (accessed 6 December 2016).

Brasil, Ministério de Minas e Energia, *Plano Decenal de Expansão de Energia 2022*, 384, www.epe.gov.br/Estudos/Documents/PDE2022.pdf (accessed 6 December 2016).

Brazil, *Programa Nacional de Universalização do Acesso e Uso de Energia Elétrica – Luz para Todos*, Law 10.762/2003 and Decree 4.873/2003.

Brazil, *Sistema Interligado Nacional atende 98% do Mercado brasileiro*, www.brasil.gov.br/infraestrutura/2011/12/sistema-interligado-nacional-atende-98-do-mercado-brasileiro (accessed 6 December 2016).

Brown, S. P. A. and Huntington, H. G. (2008). Energy security and climate change protection: complementarity or tradeoff? *Energy Policy*, **36**, 3510–13.

Burgos, F. J. (2007). *Regional Electricity Cooperation and Integration in the Americas: Potential Environmental, Social and Economic Benefits*, OEA, August, www.oas.org/DSD/Documents/RegionalElectricityCooperationandIntegration.pdf (accessed 6 December 2016).

Cavalcanti, C. A., Lembo, C. M. and Thorstensen, V. H. (2013). *The Regulation of the International Energy Trade: Fuels and Electricity*. São Paulo: FIESP.

CIER. (2011). *Proyecto CIER 15 Fase II*, Informe Final, www.cacier.com.ar/Institucional/Proyectos/Documentos/Informe_Final-CIER15.pdf (accessed 6 December 2016).

El Diario, 'Al menos 279 proyectos de energia renovable se ejecutan en Centroamérica' (El Salvador, 26 de Junio de 2014), www.eldiario.es/economia/proyectos-energia-renovable-ejecutan-Centroamerica_0_275072520.html (accessed 6 December 2016).

Entidad Binacional Yacyretá. (n.d.). *Características Técnicas*, www.yacyreta.org.ar/index.php/12-paginas/127-2-caracteristicas-tecnicas (accessed 6 December 2016).

Gischler, C. and Janson, N. (2011). Perspectives for distributed generation with renewable energy in Latin America and the Caribbean: analysis of case studies for Jamaica, Barbados, Mexico and Chile. *IDB Discussion Paper* No. IDB-DP-208.

Gonzáles, F. E. and Weintraub, S. (2008). *Energy Use and Climate Change: Western Hemispheres Partnership*. Partnership for the Americas Commission, Background Document BD-01, May, www.brookings.edu/wp-content/uploads/2016/07/energy_climate_gonzalez.pdf (accessed 6 December 2016).

IADB. *Energía renovable*, www.iadb.org/es/temas/energia/se4allamericas/energia-renovable,17688.html (accessed 6 December 2016).

IEA. *International Energy Agency World Outlook 2011*.

IPCC. *Special Report on Renewable Energy Sources and Climate Change Mitigation*, http://srren.ipcc-wg3.de/report/IPCC_SRREN_Full_Report.pdf (accessed 6 December 2016).

Itaipu. *Geração*. www.itaipu.gov.br/energia/geracao (accessed 6 December 2016).

Itamaraty. (2012). *Comunicado Conjunto Presidencial: Novo Paradigma para a Relação Brasil-Uruguai*, 31 July, www.itamaraty.gov.br/sala-de-imprensa/notas-a-imprensa/comunicado-conjunto-presidencial-novo-paradigma-para-a-relacao-brasil-uruguai (accessed 6 December 2016).

Lembo, C. M. (2013). *La relación de los pilares del desarrollo sostenible con la universalización del acesso a la energia: la experiencia brasileña*. Congresso Internacional sobre Accesso Universal a Servicios Públicos de Energía, Lima, Peru.

(2015). *Energia e o sistema multilateral de comércio: perante o paradigma do desenvolvimento sustentável*. São Paulo: Atlas.

Lund, H. (2007). Renewable energy strategies for sustainable development. *Energy*, **32**(6), 912–19.

MercoPress, 'Argentina y Paraguai se preparan para rever tratado de la represa Yaciretá' (Montevideo, 18 March 2014), http://es.mercopress.com/2014/03/18/argentina-y-paraguay-se-preparan-para-rever-tratado-de-la-represa-yacyreta (accessed 6 December 2016).

MERCOSUR Action Plan to Year 2000, 5 August 1995, MERCOSUR/CMC/DEC No. 9/95.

MERCOSUR Memorandum of Understanding on Electricity Exchange and Electrical Integration, 23 July 1998, MERCOSUR/CMC/DEC No. 10/98.
MERCOSUR Memorandum of Understanding to Establish a Special Working Group on Biofuels, 15 December 2006, MERCOSUR/CMC/DEC No. 36/06.
Ölz, S., Sims, R. and Kirchner, N. (2007). Contribution of renewables to energy security. *IEA Information Paper*, April.
Proyecto Mesoamerica. *Mercado Eléctrico Regional (MER)*, www.proyectomesoamerica.org/joomla/index.php?option=com_content&view=article&id=439&Itemid=229 (accessed 6 December 2016).
(2014). Costa Rica completa línea de transmisión eléctrica SIEPAC, San José, Costa Rica, 16 October, www.proyectomesoamerica.org/joomla/index.php?option=com_content&view=article&id=731&Itemid=85 (accessed 6 December 2016).
Salto Grande. *Generation*, www.saltogrande.org/generacion.php (accessed 6 December 2016).
SIEPAC. www.proyectomesoamerica.org/ (accessed 6 December 2016).
SINEA. (2014). *Declaración de Lima para la Interconexión e Integración Eléctrica Andina*, 25 April, www19.iadb.org/intal/intalcdi/PE/CM%202014/14275.pdf (accessed 6 December 2016).
Sustainable Energy for All. *Renewable Energy*, www.se4all.org/about-us_our-ambition_renewable-energy (accessed 6 December 2016).
Olade. (2012). *Unasur: un espacio que consolida la integración energética*, http://biblioteca.olade.org/opac-tmpl/Documentos/old0235.pdf (accessed 6 December 2016).
United Nations. (1987). *Report of the World Commission on Environment and Development: Our Common Future*, www.un-documents.net/our-common-future.pdf (accessed 6 December 2016).
(1992). Agenda 21, United Nations Conference on Environment and Development, Rio de Janeiro, Brazil, 3–14 June, http://sustainabledevelopment.un.org/content/documents/Agenda21.pdf (accessed 6 December 2016).
(2002). Report of the World Summit on Sustainable Development, Johannesburg, South Africa, 26 August–4 September, Plan of Implementation of the World Summit on Sustainable Development, Section III.20(d), www.unmillenniumproject.org/documents/131302_wssd_report_reissued.pdf (accessed 6 December 2016).
(2005). 2005 World Summit Outcome, Resolution adopted by the General Assembly, 16 September 2005, New York, United States, Section II.50, www.un.org/womenwatch/ods/A-RES-60-1-E.pdf (accessed 6 December 2016).
(2010). *Energy for a Sustainable Future.* www.un.org/wcm/webdav/site/climatechange/shared/Documents/AGECC%20summary%20report%5B1%5D.pdf (accessed 6 December 2016).

(2012). *The Future We Want*, Report of the United Nations Conference on Sustainable Development, 20–22 June, Rio de Janeiro, www.uncsd2012.org/content/documents/814UNCSD%20REPORT%20final%20revs.pdf (accessed 6 December 2016).

Vergara, W., Alatorre, C. and Alves, L. (2013). Rethinking our energy future: a White Paper on renewable energy for the 3GFLAC Regional Forum. *Inter-American Development Bank Discussion Paper* No. IDB-DP-292.

World Bank. *Yaciretá Hydroelectric Project*, www.worldbank.org/projects/P005937/yacyreta-hydroelectric-project?lang=en (accessed 6 December 2016).

PART III

Interconnectivity Issues

11

Interconnections in Energy Transportation: Implications for International Trade Law

YULIA SELIVANOVA

A Introduction

The issue of regulating interconnections in energy transportation has gained prominence due to the rapid development of renewable energy sources and cross-border electricity trade, prompted by climate change policies and renewable energy targets.

Owing to the above developments, large amounts of renewable electricity must be able to connect to electricity networks in the future. Timely grid access arrangements are vital to delivering new generation projects. Regulation of access, on reasonable terms, to transport and distribution networks is thus crucial for the development of clean energy trade. Moreover, long-term investments in energy infrastructure are necessary for such successful integration of clean energy in present and future energy systems. Most renewable energy resources are location-specific, which means that the electricity generated using such resources may need to be transported over long distances. This emphasizes the importance of rules for cross-border transportation of clean energy, especially through fixed infrastructure. Moreover, the ability to sell electricity at peak hours in different time zones is key for the efficiency of clean energy systems. This requires new infrastructure interconnections. Adequate interconnections between regions could take advantage of the total available power and transfer it to the areas in which it is needed, increasing the potential for exploiting renewable energy.[1]

[1] Y. Selivanova, *Clean Energy and Access to Infrastructure: Implications for the Global Trade System*, E15Initiative (Geneva: International Centre for Trade and Sustainable Development (ICTSD) and World Economic Forum, 2015), http://e15initiative.org/wp-content/uploads/2015/01/E15_Clean-Energy_Selivanova_FINAL.pdf (accessed 29 November 2016).

Therefore, integrating more renewable sources into the current system will require reinforcement of the transmission and interconnections infrastructure.[2] The focus of this chapter is on how international trade rules could better accommodate energy trade – including renewable electricity trade – through fixed infrastructure. The chapter will examine issues of fixed infrastructure and respective interconnections from an angle that is pertinent both to electricity and to gas trade, as these two key energy products are not only difficult and costly to transport and store, but also depend on fixed infrastructure for their transportation.

Indeed, transportation of these products is typically carried out through grids and pipelines, and such a method of transportation has low flexibility in terms of capacity. The special characteristics of hydrocarbons and electricity and the fact that a significant part of trade in them takes place through fixed infrastructure lead to significant challenges. The link between energy trade and fixed infrastructure brings two issues to the fore, neither of which is addressed in a specific way by WTO rules:

- the framework for investment in highly capital-intensive infrastructure projects;
- the conditions for access to these networks – that is, access to transportation services whose capacities are not only limited, but are also usually almost entirely filled on a long-term basis.[3]

This chapter analyses the aspects of energy trade via fixed infrastructure that need to be taken into account by policy-makers. While WTO rules are fully applicable to trade in energy materials and products, including natural gas and electricity, they are not well designed for dealing with all aspects of such trade. One such aspect is transportation of energy materials and products via fixed infrastructure. This chapter examines the implications of relevant rules of the General Agreement on Tariffs and Trade (GATT) and the General Agreement on Trade in Services (GATS) for the transportation of energy products via fixed infrastructure, namely the rules on transit in Article V GATT and GATS disciplines related to energy services.

[2] S. Chatzivasileiadis, D. Ernst and G. Andersson, 'The Global Grid', *Renewable Energy*, 57 (2013), 372–83.

[3] A. Mernier, 'Setting the rules of energy trade' in *Fundamentals of the Global Oil and Gas Industry* (London: Petroleum Economist, 2008).

B Characteristics of Energy Transportation and Implications for Trade Rules

Two interrelated characteristics are associated with energy trade via fixed infrastructure. First, energy transport infrastructure possesses the characteristics of a natural monopoly. Second, infrastructure construction projects are highly capital-intensive investments. These characteristics render it challenging to regulate trade in energy products through existing multilateral trade rules.

As noted above, the technical structure of the energy sector leads to the prevalence of natural monopolies in some of its segments, with high fixed costs and technical constraints.[4] The implication of a natural monopoly is that the activity concerned cannot be carried out efficiently in a competitive context and needs to be carried out by a single company. There are elements of natural monopoly in the transportation of natural gas and electricity through fixed infrastructure, as substantial costs are involved in building alternative energy networks.[5] Moreover, from the environmental point of view it is not desirable to construct additional networks when the capacity of the existing network could be expanded instead.[6]

[4] C. Melly, 'Electric power and gas market reform and international trade in services', note 2 in *Energy and Environmental Services: Negotiating Objectives and Development Priorities* (Geneva: United Nations Conference on Trade and Development, 2003) (UNCTAD/DITC/TNCD/2003/3). A natural monopoly is a situation in which production of a good or service by one producer ensures the least costly supply. The public policy dilemma in the case of a natural monopoly arises because society does not wish to suffer from the potential monopoly pricing if natural monopoly characteristics require the existence of a single firm. Subhes C. Bhattacharyya, *Energy Economics: Concept, Issues, Markets and Governance* (London: Springer, 2011).

[5] Y. Selivanova, The WTO and energy: WTO Rules and Agreements of relevance to the energy sector', International Centre for Trade and Sustainable Development, Issue Paper No. 1, 2007. However, supplies of oil and coal are not affected by transportation bottlenecks. The physical characteristics of coal and oil allow them to be traded internationally on the basis of market principles. Their price may be influenced by quality and different levels of taxation and subsidies. However, transportation and distribution constraints do not create the same problems of access that they do in the electricity and gas industries. See P. D. Cameron, *Competition in Energy Markets. Law and Regulation in the European Union* (Oxford: Oxford University Press, 2007), p. 21.

[6] Waelde and Gunst note that even this may change as technology and perceptions evolve. For the transport of natural gas, for example, competing pipelines have been built. T. W. Waelde and A. J. Gunst, 'International energy trade and access to energy networks' in *Energy and Environmental Services: Negotiating Objectives and Development Priorities* (Geneva: United Nations Conference on Trade and Development, 2003) (UNCTAD/DITC/TNCD/2003/3), p. 126.

Regulation of the natural monopolies is needed to minimise unproductive monopoly rent and to protect consumers. In addition, in order to foster competition and greater diversity of suppliers, regulation is needed to ensure that other energy suppliers have access to the fixed infrastructure.[7]

Grid-bound energy is difficult to store. Capacity therefore needs to be available at the right time. For most manufactured goods (transported by rail, road, air or sea), capacity constraints can be dealt with more easily. In energy transportation, the volumes to be transported are programmed on a long-term basis, because the use of pipelines is limited to the transport of specific energy products and significant investments are involved.[8]

It may often be impossible to provide access to transportation to a new user without taking transport capacity away from an existing user, owing to the lack of free capacity. This is because it is unusual to invest in pipeline projects with substantial extra capacity.[9] Although other means of transportation also have their capacity limitations, and they cannot accept unlimited additional shipments, there is usually a certain amount of spare capacity in infrastructures such as motorways and railways.[10] Another difference between these transportation modes and pipelines is that trucks, ships, planes and trains can move to the

[7] Ibid., p. 127.

[8] Construction of transportation pipelines and transmission grids requires substantial investments. Such transportation facilities involve significant economies of scale in the construction phase and, once constructed, have limited capacity for transporting specific energy products. Energy transport infrastructure projects not only necessitate large capital investments, but the sunk costs in this segment are also very high. Therefore the companies that invest in infrastructure projects typically seek exclusive rights over a certain period to recoup their investments, often using long-term contracts with take-or-pay clauses. Cameron comments that the exclusive rights granted often include the ownership or operation of a transmission grid over a specified period of time, as well as an exclusive right to import gas or electricity. See Cameron, *Competition in Energy Markets*, p. 21.

[9] Pipeline investors plan for volumes backed by long-term supply contracts. Energy volumes sold on the basis of such contracts guarantee repayment of loans and interest. It is unusual for financial institutions to invest in pipeline projects with significant extra capacity.

[10] This is because there is a more favourable relationship between fixed and variable costs. If the use of aeroplanes, ships and trucks has a higher proportion of variable costs than a pipeline, transportation service providers can more easily afford to have – at times idle – spare capacities available than can the operator of a pipeline. Such spare capacities also make commercial sense because they allow expansion of supply in response to temporarily increased demand.

location of increased demand at a particular time and on a temporary basis,[11] whereas fixed energy infrastructure cannot be moved.

To sum up, energy trade's dependence on fixed infrastructure implies that traditional trade disciplines might not be sufficient to ensure free trade in energy. Additional measures have to be taken to guarantee the availability of fixed infrastructure and non-discriminatory access to transportation pipelines and networks, distribution systems and storage facilities.[12]

C Liberalisation of the Energy Sector: Implications for Interconnections and Access to Fixed Infrastructure

In many countries, the energy sector has historically been dominated by vertically integrated state-owned companies. Such companies typically occupied monopoly positions and undertook the production, transportation and distribution of energy products as well as all energy services along the energy supply chain.[13] The main characteristics of the regulatory system included exclusive rights to construct and operate networks, vertical integration, guaranteed remuneration on the basis of historical costs and central planning.[14] There was little competition in the sector.[15]

Network-bound industries were perceived as strategic assets of the state and it was considered more efficient for a single vertically integrated company to perform all the functions along the production chain. However, as politically motivated decision-making pervaded the sector, performance was unsatisfactory.[16]

[11] L. Ehring and Y. Selivanova, 'Energy transit' in Y. Selivanova (ed.), *Regulation of Energy in International Trade Law: WTO, NAFTA and Energy Charter* (Alphen aan den Rijn: Kluwer Law International, 2011), pp. 50–1.

[12] Selivanova, 'The WTO and energy'.

[13] C. Gamberale, 'Energy services' in WTO Secretariat, *Guide to the GATS: An Overview of Issues for Further Liberalization of Trade in Services* (Alphen aan den Rijn: Kluwer Law International, 2011), p. 259.

[14] Cameron, *Competition in Energy Markets*, pp. 6–7.

[15] WTO, Council for Trade in Services, Committee on Specific Commitments, Communication from the United States, S/CSC/W/27. In the past, almost all countries had one vertically integrated state-owned power company, which produced, transported and distributed electricity to end-consumers. In the 1990s, several countries started restructuring their power sectors and introducing competition.

[16] Bhattacharyya, *Energy Economics*.

The liberalisation of the energy sector began with the dismantling of the monopoly and electricity production's opening up to competition.[17] As a result of reform around the world it is clear that while transmission may remain a natural monopoly, competition may develop in the generation and distribution sectors. The ownership linkages between these two types of activities can lead to competition distortion because the owner of a transportation system which also produces energy has an incentive to prevent other producers from gaining access to the fixed infrastructure.[18] Thus, regulation of access rights to the transportation networks became necessary.[19]

D Disciplines on Energy Services and Access to Transportation Networks

Given energy trade's dependence of on fixed infrastructure and the natural monopoly character of some fixed infrastructure installations, market access commitments alone might be insufficient to ensure liberalised energy trade. Indeed, trade in energy is often obstructed by difficulties in gaining access to transportation infrastructure. A foreign supplier may be granted market access and establish a power and gas trading and marketing company in a foreign country, but then encounter obstacles when trying to access the transmission network.[20]

Typical barriers include excessive access fees and discriminatory allocation of network capacities by the transmission system operator. These problems are aggravated by non-transparent terms and conditions for calculating access fees and usage conditions.[21] Regulation of access, on reasonable terms, to storage, transport and distribution networks is therefore necessary for the liberalisation of cross-border energy trade. The problem with regulation of third-party access to transportation networks is that such networks are mostly controlled not by governments, which are bound by obligations under international or regional treaties, but by powerful private companies. These companies often claim lack of capacity and charge transportation fees exceeding the cost of

[17] S. Thomas and D. Hall, *GATS and Electricity and Water Sector* (University of Greenwich: Public Services International Research Unit, 2006), p. 17.

[18] D. Parker and S. Van Siclen, 'OECD perspectives on competition policy and regulatory reform in the electricity sector', *OGEL* 1(3) (2003), 4.

[19] Gamberale, 'Energy services', 265.

[20] Melly, 'Electric power and gas market reform and international trade in services', 176.

[21] Ibid., p. 169.

services rendered.[22] Competitive safeguards and an independent regulator controlling the behaviour of market participants are therefore necessary to ensure access to the transportation infrastructure.[23]

Regulation is also crucial to encourage investment and achieve public policy objectives.[24] Private profitability and efficiency have to be considered together with such issues as security of supply, public service obligations and environmental, social and developmental goals. In order to build energy transportation networks and contribute to meeting these objectives, private investments are necessary. However, investors are generally unwilling to invest unless they are certain to get a return on their investments. Private suppliers therefore prefer to serve profitable market segments. Liberalised, unregulated energy markets will not always meet social, developmental and environmental goals – therefore the role of the regulator is important.

WTO rules and commitments ensure market access and non-discrimination, and tackle quantitative measures. However, these rules do not address the anti-competitive practices that are widespread among incumbents controlling different segments upstream and downstream along the energy value chain.

I Pro-Competitive Principles in GATS

WTO agreements do not contain competition rules, although the GATS has some provisions that deal with the conduct of private companies such as monopolies and exclusive service suppliers. Restrictive business practices by incumbent operators are subject to Article VIII and Article IX GATS.

Article VIII GATS is especially pertinent to gas transportation and distribution services. It requires Members to ensure that the incumbent natural monopolist in the transportation and distribution market does not act in a manner inconsistent with the most-favoured-nation (MFN) principle and the Member's specific commitments. However, few commitments have been made by WTO Members in this respect.[25] In

[22] T. W. Waelde and A. J. Gunst, 'International energy trade and access to energy networks', p. 123.

[23] Selivanova, 'The WTO and energy'. [24] Ibid.

[25] M. Cossy, 'Energy services' in Y. Selivanova (ed.), *Regulation of Energy in International Trade Law: WTO, NAFTA and Energy Charter* (Alphen aan den Rijn: Kluwer Law International, 2011), p. 159.

addition, if such a monopoly or exclusive supplier competes in the supply of a service outside the scope of its monopoly rights, the Member concerned has to ensure that the incumbent monopoly does not abuse its position with respect to the Member's specific commitments.[26] Finally, monopoly suppliers acting outside the scope of their monopoly in a sector where a Member has undertaken GATS commitments must respect those commitments.

Arguably, the above GATS disciplines do not address the main anticompetitive practices in energy transport networks, which adversely affect access to and use of energy transport facilities by third parties. According to Article IX, Members recognise that 'certain business practices' other than those addressed under Article VIII 'may restrain competition', and agree to enter into consultation with a view to eliminating these practices. This provision is limited to an obligation to consult and to provide information, and has never been used so far.[27]

In the past, some WTO Members proposed the introduction of pro-competitive safeguards in relation to energy services in WTO, similar to those applicable to telecommunications services. It was underlined that the energy sector had similarities with the telecommunications sector, both being highly regulated markets characterised by large incumbent suppliers and undergoing de-regulation.[28] Market access and national treatment commitments have been supplemented in the telecommunications sector by additional commitments on regulatory principles to ensure pro-competitive conditions in supply of services. These competition disciplines have been set out in the Reference Paper to the GATS Agreement on Basic Telecommunications Services.

In 2001, the United States and Norway proposed to devise a Reference Paper for energy services, modelled on the Reference Paper on telecoms, and rules for cross-border energy trade. The purpose of these additional disciplines would be to ensure transparency in the formulation and implementation of rules, as well as non-discriminatory third-party access to and interconnection with energy infrastructure; non-discriminatory, objective and timely procedures for the transportation and transmission of energy; an independent regulator; and requirements preventing anti-competitive practices. One delegation reportedly proposed considering the effectiveness of a framework for domestic regulation, which would

[26] Selivanova, 'The WTO and energy'. [27] Cossy, 'Energy services', 159.

[28] WTO, Council for Trade in Services – Special Session – Energy Services – Information Note by the Secretariat, JOB(05)/204.

contribute to the creation of a competitive environment, including non-discriminatory, fair and transparent use of networks.[29] However, the negotiations based on these proposals did not get very far.[30]

Although the idea of developing a Reference Paper in the energy sector is very acute, the substantial differences between telecommunications and energy need to be taken into account. Consequently, competition disciplines in the energy sector may need to be more complex than those for telecommunications. Although requirements regarding transparency, independence of regulators and even third-party access are similar in the gas and electricity industries, the energy sector requires more sophisticated regulation. This is mainly because of the possible impact on the environment, as well as issues related to energy efficiency and security of supply, including non-interruption of energy flows.[31]

Additional commitments, either in an Annex to GATS on Energy Services or a Reference Paper, could be used to address competition issues and third-party access to fixed infrastructure, similarly to the additional disciplines for the telecommunications sector contained in the Reference Paper. However, while telecom companies supplying services are unquestionably 'service suppliers' within the meaning of the GATS, the situation in the energy sector is more complex, with the consequence that additional commitments on energy services may not benefit all firms seeking access to transportation facilities.[32] Indeed, in the energy sector, companies that need access to transportation networks are mainly producers and traders of energy products, and not providers of energy services in a classical sense. It has been argued that, as the scope of the GATS is limited to services and services suppliers, companies transporting their own energy goods should in principle fall outside the Agreement and, hence, would not benefit from the above-mentioned

[29] Ibid.

[30] M. Cossy provides the following account of the latest discussion of the Reference Paper on energy services, which took place in the Council for Trade in Services in February of 2010: 'The issue of a Reference Paper for energy services briefly resurfaced in the Council for Trade in Services, in February 2010. Referring to possible additional disciplines, several delegations cautioned against using the telecom Reference Paper as a model because, in their view, there were important differences between telecom and energy networks, as well as among different energy industries, which meant that disciplines in the telecom Reference Paper were not necessarily transposable to the energy sector. Nevertheless, one delegation supported the idea of developing specific disciplines for energy services and was ready to consider the extent to which the telecom Reference Paper could be relevant in that context.' Cossy, 'Energy services', 168.

[31] Selivanova, 'The WTO and energy'.

[32] Cossy, 'Energy services', 159.

competition disciplines.[33] This problem of interpretation may arise because GATS does not provide definitions of services and service suppliers.

The opposite interpretation, however, is also possible – supported by the WTO jurisprudence. The WTO Appellate Body has made it clear on several occasions that application of GATT and GATS is not mutually exclusive. While generation of electricity and production of natural gas are production processes related to a commodity, transportation and transmission are service activities. Companies that are producers of energy commodities can also act as service suppliers when transporting energy products through fixed infrastructure. In this case, vertical integration of the sector does not preclude the application of service disciplines. For example, in *EC – Bananas*, the Appellate Body dealt with vertical integration. Although that dispute concerned an entirely different product – bananas – this judgment has a direct bearing on the trade in energy commodities.[34]

In that case, one of the issues was whether the GATS and the GATT 1994 are mutually exclusive. The Appellate Body concluded that measures that involved a service relating to a particular good or a service supplied in conjunction with a particular good could be scrutinised under both the GATT 1994 and the GATS.[35]

33 Ibid.

34 According to the Panel in that case, the EC banana import licensing regime fell under the scope of the GATS. Article I:1 of the GATS provides that '[t]his Agreement applies to measures by Members affecting trade in services'. In the view of the Appellate Body, the use of the term 'affecting' reflects the drafters' intent to give a broad reach to the GATS. The ordinary meaning of the word 'affecting' implies a measure that has 'an effect on', which indicates a broad scope of application. Furthermore, Article I:3(b) of the GATS provides that '"services" includes any service in any sector except services supplied in the exercise of governmental authority', and that the '"supply of a service" includes the production, distribution, marketing, sale and delivery of a service'. WTO, Report of the Appellate Body, *European Communities – Regime for the Importation, Sale and Distribution of Bananas*, WT/DS27/AB/R, 9 September 1997, paras. 220–1.

35 The Appellate Body noted that, 'given the respective scope of application of the two agreements, they may or may not overlap, depending on the nature of the measures at issue. Certain measures could be found to fall exclusively within the scope of the GATT 1994, when they affect trade in goods as goods. Certain measures could be found to fall exclusively within the scope of the GATS, when they affect the supply of services as services. There is yet a third category of measures that could be found to fall within the scope of both the GATT 1994 and the GATS. These are measures that involve a service relating to a particular good or a service supplied in conjunction with a particular good. In all such cases in this third category, the measure in question could be scrutinized under both the GATT 1994 and the GATS.' Ibid.

Although the operators were engaged in certain activities that did not fall within the definition of 'distributive trade services' in the Central Product Classification, the Appellate Body found 'there is no question that they are also engaged in other activities involving the wholesale distribution of bananas that are within that definition'. The Appellate Body noted that although the operators were vertically integrated with producers, ripeners and retailers,

> even if a company is vertically-integrated, and even if it performs other functions related to the production, importation, distribution and processing of a product, to the extent that it is also engaged in providing 'wholesale trade services' and is therefore affected in that capacity by a particular measure of a Member in its supply of those 'wholesale trade services', that company is a service supplier within the scope of the GATS.[36]

The situation is similar in the transportation of energy commodities. Vertically integrated companies are often engaged in transportation of a commodity through fixed networks: such transportation constitutes a service. However, such companies also perform other functions related to production, importation, exportation, distribution and processing. In the absence of additional disciplines that tackle these issues, such companies would be affected by governmental measures with regard to transportation services as well as obstacles impeding their access to the transportation network created by the incumbent controlling the network.

In the WTO context, there is already a precedent for including such additional obligations with respect to energy transportation. For instance, upon their accession to the WTO, Ukraine and Montenegro undertook an obligation to ensure non-discriminatory treatment in access to and use of pipeline networks – within the technical capacities of these networks – with regard to the origin, destination or ownership of the product transported. This obligation is included in their GATS schedule. Considering that the commitment mentions the products transported, it can be inferred that the owners of energy materials and products are eligible to benefit from this commitment.

In conclusion, the telecommunications and energy sectors are not strictly comparable. Therefore, although the Reference Paper for Basic Telecommunications Services could provide a basis for consideration,

[36] WTO, Report of the Appellate Body, *Canada – Automotive Industry* WT/DS139/AB/R, WT/DS142/AB/R, 31 May 2000, para. 163.

other issues should also be addressed if additional energy services disciplines are negotiated in the WTO. The following issues must be tackled in order to secure an effective pro-competitive framework for energy trade, including third-party access to fixed infrastructure:

(1) providing non-discriminatory, third-party access to and interconnection with energy networks and grids, and other essential infrastructure, whether dominated by government entities or by privately owned companies;
(2) creating an independent regulator separate from and not accountable to any supplier of energy services;
(3) ensuring non-discriminatory, objective and timely procedures for the transportation and transmission of energy;
(4) maintaining appropriate measures for preventing certain anticompetitive practices in the sector;[37]
(5) ensuring transparency in the formulation and implementation of rules, regulations and technical standards;
(6) a requirement for the provision of non-discriminatory and timely information on data relevant for transportation and transmission of energy, such as prices and transmission capacity;
(7) security of supply and non-interruption of energy transportation;
(8) expeditious dispute settlement (as the interruption of energy transportation services can have drastic consequences for consumers).

1 Third-Party Access to Transportation Networks and Other Essential Facilities

Third-party access (hereinafter referred to as TPA) is an essential condition for creating competitive cross-border energy markets and trade.[38] TPA entitles third parties, including the network owner's competitors in production and supply of energy, to use fixed infrastructures. Such use typically requires payment of a reasonable fee and takes place on practical technical terms – either with respect to free capacity only, or on the basis of sharing existing capacity. TPA thus implies the existence of a legal right with enforceable procedures that give practical effect to such

[37] WTO, Council for Trade in Services, Committee on Specific Commitments, Communication from the United States, S/CSC/W/27.

[38] T. Waelde, *Access to Energy Networks: A Precondition for Cross-Border and Energy Services Trade* (Dundee: Centre for Energy, Petroleum and Mineral Law and Policy, 2000).

rights. The right of access to pipelines and grids is one of the most important competition safeguards in network-bound industries.[39]

Even in liberalised markets, 'power and gas markets are confronted with the existence of a single transportation network that must somehow be shared by several competing firms'.[40] Opening the essential infrastructure to all suppliers on a non-discriminatory basis requires the development and enforcement of rules on interconnection and on access to and use of network facilities. Such rules need to address the terms, conditions and fees for connection to existing networks. Furthermore, it is crucial that all policies and procedures for using network facilities are transparent and reasonable.

The WTO telecoms Reference Paper contains disciplines on interconnection, which require ensuring an interconnection with a major supplier at any technically feasible point in the network. For energy services, such access to essential facilities should also be provided under non-discriminatory terms, conditions and rates that are no less favourable than those provided for the supplier's own like services, for like services of non-affiliated service suppliers or for its subsidiaries or other affiliates. Moreover, such terms, conditions and cost-oriented rates must be transparent and reasonable, having regard to economic feasibility. Taking into consideration the physical limitations of energy, access to essential facilities should be provided in a timely fashion. Finally, the access must be sufficiently unbundled so that the supplier would not be obliged to pay for network components or facilities that it does not require for the service to be provided.

Importantly, in telecoms, interconnections must also be provided, upon request, at points in addition to the network termination points offered to the majority of users, subject to charges that reflect the cost of construction of the necessary additional facilities. Taking into consideration the limited capacity of energy networks, the question arises whether the additional disciplines should be devised to provide a similar obligation to expand the capacity, which may include the

[39] Two types of access regime exist: (1) negotiated access (where the majority of binding access terms and conditions are set by the parties) and (2) regulated access (established externally by an authority). Such authorities may be independent regulators, courts, ministries and administrative bodies, competition authorities or industry associations. Considering an integrated transmission system operator's economic incentives to prevent fair access to a network, negotiated access is much less efficient than regulated access. Waelde and Gunst, 'International energy trade and access to energy networks', 129–30.

[40] Melly, 'Electric power and gas market reform and international trade in services', 168.

construction of new grids and/or pipelines. This issue is linked to regulation of rights of way and investment.

The set of pro-competitive principles for energy services must restrict a major supplier's ability to refuse access to all essential facilities.[41] TPA is related to the concept of essential facilities[42] developed in competition law in the United States and the European Union.

According to the US courts' jurisprudence, the following aspects have to be considered under the essential facilities doctrine when analysing the liability of the monopolist: (1) control of an essential facility; (2) inability of a competitor practically or reasonably to duplicate the essential facility; (3) denial of the use of the facility to a competitor and (4) the feasibility of providing the facility.[43] The facilities that are considered essential and that are owned by a company that has a dominant or monopolistic position must be open to access by competitors that could not otherwise enter the market.[44] If competitors could construct the facilities themselves, such access would not be deemed essential.

[41] Different types of infrastructure are associated with different cost structures. For instance, it might be easier for new market entrants to construct oil storage facilities than a gas pipeline or an electricity transmission grid.

[42] The essential facilities doctrine refers to a type of anti-competitive behaviour in which a firm with market power uses a 'bottleneck' in a market to deny competitors entry into the market. It is closely related to a claim for refusal to deal. As it was put by the US courts: 'The essential facilities doctrine imposes liability when one firm, which controls an essential facility, denies a second firm reasonable access to a product or service that the second firm must obtain in order to compete with the first', or '[w]here facilities cannot practicably be duplicated by would-be competitors, those in possession of them must allow them to be shared on fair terms. It is illegal restraint of trade to foreclose the scarce facility'. *Hecht* v. *Pro-Football, Inc.*, 570 F.2d 982, 992 (D.C. Cir. 1997). In the *Terminal Railroad* case, a group of railroads controlling all railway bridges and switching yards into and out of St. Louis prevented competing railway companies from offering transportation to and through that destination. The Supreme Court held this to be an illegal restraint of trade. The court considered the railroad junction to be a bottleneck facility, and access to that facility was essential for the competing companies in order to preserve their ability to compete. Non-members of the acquiring consortium had to be allowed the use of the railroad junction. See *United States* v. *Terminal Railroad Association*, 224 U.S. 383 (1912). See also *B&I Line PLC* v. *Sealink Harbours Ltd. and Sealink Stena Ltd.*, Case IV/34.174, (1992); *Hugin* v. *Commission*, Case 22/78 (1979) ECR 1989; *Volvo* v. *Venk*, Case 238/87 (1988) ECR 6211; *Radio Telefis Eiraenn and Independent Television Publications Limited* v. *Commission (Magill)*, joined Cases 241/91 P and 242/91 P (1995) ECR I-743; *Bronner* v. *Mediaprint Zeitungs- und Zeitschriftenverlag, GmbH & Co.*, Case 7/97 ECR I-7791.

[43] *Associated Press* v. *United States*, 326 U.S. 1 (1945), *Lorain Journal Co.* v. *United States*, 342 U.S. 143, 146–49 (1951), *Otter Tail Power Co.* v. *United States*, 410 U.S. 366, 377–79 (1973), *Aspen Skiing Co.* v. *Aspen Highlands Skiing Corp.*, 427 U.S. 585 (1985).

[44] Waelde and Gunst, 'International energy trade and access to energy networks', p. 127.

The essential facilities doctrine led to inclusion of the definition of 'essential facilities' in the Reference Paper on Basic Telecommunications. According to the Reference Paper, essential facilities means facilities of a public telecommunications transport network or service that: (1) are exclusively or predominantly provided by a single or a limited number of suppliers and (2) cannot feasibly be economically or technically substituted in order to provide a service.

The definition of essential facilities in the telecoms Reference Paper could be the basis for discussion of similar additional disciplines on energy services. In the energy sector, however, in order to guarantee a competitive environment, suppliers have to be allowed access to essential energy infrastructure other than the electric power transmission network and the natural gas pipelines.[45] For example, access to gas and oil storage facilities and liquefied natural gas terminals might qualify under the essential facilities doctrine.

It has been suggested that the essential facilities concept to be inserted in a possible Reference Paper on energy services could contain an illustrative list of what is covered by such definition. The character of essential facilities could then be determined on a case-by-case basis by using as criteria: (1) the aspects of exclusive or predominant provision by a single supplier or a limited number of suppliers and (2) the unfeasibility of technical or economical substitution.[46] The facility will be considered essential if it possesses monopoly characteristics and would be essential for the respective market participants' operations. There must be no alternative to the use of the facility in question. It would not suffice for the market participant simply to be inconvenienced and to bear some losses should the access be refused.[47]

The third-party access regulation needs to deal with the following issues: (1) lack of capacity; (2) tariffs and (3) terms and procedures.

Where capacity is lacking, the regulatory options are pro-rata sharing of limited capacity between commitments to existing users and new entrants, or access to available capacity only. Several options are available

[45] P. C. Evans, *Liberalizing Global Trade in Services* (Washington, DC: AEI Press, 2002), 45.

[46] P. Poretti and R. Rios-Herran, 'A reference paper on energy services: the best way forward?', *OGEL*, 4(4) (2006) (online).

[47] P. C. Evans, 'Strengthening WTO Member commitments in energy services: problems and prospects' in A. Mattoo and P. Sauvé (eds.), *Domestic Regulation and Service Trade Liberalization* (Washington DC: World Bank and Oxford University Press, 2003), p. 183.

in the latter situation: first come, first served; merit order; renewable or source-specific priority; or auctioning. Furthermore, if demand often exceeds the available capacity, an obligation to expand capacity may be imposed. Congestion management procedures are also important. Existing operators could deliberately fill available capacity and interconnectors in order to have a reason to refuse access to their competitors. While anti-competitive use of prior capacity reservation is dealt with by the application of a use-it-or-lose-it principle, unused capacity is made available to short-term markets.[48]

The pricing of interconnections can adversely affect competition in the market. Access to essential facilities should be ensured on the basis of tariffs that are cost-reflective and that do not involve cross-subsidies or discriminate between competing suppliers. Countries' views differ, however, as to the appropriate methodologies for determining tariffs for access to essential facilities. Tariffs could be based on incremental/marginal costs, which are typically low, or on not yet depreciated capital costs, or, in the case of fully depreciated facilities, they could be calculated taking into account the original investment and risk.[49]

Terms of access should be reasonable, should not favour incumbents and must not be discriminatory. Moreover, they should be standardised and publicly available. Procedures should not be cumbersome, nor should they involve high transaction costs. Both tariffs and terms should be approved and publicly available. Lack of transparency could constitute an anti-competitive barrier in itself, as discussed in the following section.

2 Transparency and Availability of the Information

For the competitive markets to work properly, equal, non-discriminatory access to information is key. In the energy sector, information must be available on two levels: the regulatory level and the market level. Regulatory transparency is related to the formulation, promulgation and implementation of rules, regulations and technical standards. Market transparency is related to the business operations of market participants, such as data on the capacity of networks, as well as tariffs and conditions of access.

The general rules of GATS regarding transparency and domestic regulation arguably do not provide sufficient coverage of issues that are crucial to the power and gas industries (e.g. concrete measures

[48] Waelde and Gunst, 'International energy trade and access to energy', 145. [49] Ibid.

implemented by energy regulatory authorities regarding access to transportation networks).

The Reference Paper on Basic Telecommunications addresses the transparency issue. The interconnection procedures applicable to a major supplier must be made publicly available.[50] There is a requirement for transparency of interconnection arrangements.[51] It is considered to be an anti-competitive practice not to make 'available to other services suppliers on a timely basis technical information about essential facilities and commercially relevant information which are necessary for them to provide services'.[52]

Furthermore, where a licence is necessary, the following information has to be made publicly available: (1) the licensing criteria and the period of time normally required to reach a decision concerning an application for a licence and (2) the terms and conditions of individual licences. The applicant has a right to know the reasons behind denial of a licence.[53]

According to the telecoms Reference Paper, any procedures for the allocation and use of scarce resources, including frequencies, numbers and rights of way, must be carried out in an objective, timely, transparent and non-discriminatory manner. The current state of allocated frequency bands must be made publicly available, but detailed identification of frequencies allocated for specific government uses is not required.

Rules ensuring transparency of regulation and licensing procedures similar to those that exist in telecoms should be devised for the energy sector. Fair and non-discriminatory standards should apply to ensure an efficient and fair system for siting, permitting and constructing new transportation networks and other essential infrastructure.[54]

In addition to regulatory transparency, governments should ensure market transparency, which is essential for efficient functioning of the energy markets. For example, withholding, delaying or charging for access to market information could be detrimental to competitive market conditions. The information on spare transmission capacities and congestion charges needs to be available to all energy market participants on a regular, non-discriminatory basis.[55]

[50] Para. 2.3 of the Reference Paper on Basic Telecommunications.

[51] Para. 2.4 of the Reference Paper on Basic Telecommunications.

[52] Para. 1.2 of the Reference Paper on Basic Telecommunications.

[53] Para. 4 of the Reference Paper on Basic Telecommunications.

[54] Evans, 'Strengthening WTO Member commitments in energy services', 183.

[55] Ibid.

3 Pro-Competitive Safeguards

The energy sector has been associated with market concentration, which led to the accumulation of market power. Vertical integration is associated with anti-competitive behaviour. Indeed, the company controlling the network may cross-subsidise a costly in-house supply through network charges.

The most effective way to avoid anti-competitive behaviour by vertically integrated companies is unbundling.[56] This method eliminates the incentives and the ability to conduct anti-competitive practices. The second-best option is to exercise regulatory control over the behaviour of incumbents with a dominant market position.

The telecoms Reference Paper provides for competitive safeguards and prevention of anti-competitive practices. The WTO Members that adopted the Reference Paper are obliged to take appropriate measures to prevent suppliers who, alone or together, constitute a major supplier from engaging in or continuing anti-competitive practices. In particular, the anti-competitive practices referred to in the Reference Paper include:

- engaging in anti-competitive cross-subsidisation;
- using information obtained from competitors with anti-competitive results;
- not making available to other service suppliers, on a timely basis, technical information about essential facilities and commercially relevant information which is necessary for them to provide services.[57]

Similar provisions would be pertinent to the energy industry.

Cross-subsidisation is a serious problem in the energy industries, especially when it concerns activities that involve the incumbent competing with other market participants by using the activity where it has the monopoly position. This problem is especially acute in countries that have not undertaken broad reforms in their energy sector. Without vertical unbundling, which is the most efficient competitive safeguard, the major supplier may cross-subsidise loss-making activities using the benefit gained in the segment in which it has an exclusive right or dominant position.[58]

[56] Ibid. [57] Articles 1.1 and 1.2 of the Reference Paper on Basic Telecommunications.
[58] Poretti and Rios-Herran, 'A reference paper on energy services'.

4 Independent Market Regulator

In practice, TPA would be ineffective without regulation and monitoring by an independent body.[59] An independent regulator should be established to set tariffs for the use of fixed infrastructure transmission and distribution networks, to protect market participants against discrimination and to identify obstacles to competition.[60] Market monitoring can be useful to reduce anti-competitive behaviour and to identify changes that should be made in market rules.[61]

The telecoms Reference Paper contains a principle that the regulatory body should be separate from, and not accountable to, any supplier of basic telecommunications services. The decisions of and the procedure followed by regulators must be impartial with respect to all market participants.[62] Similar provisions should be established for the energy sector.

The decisions of the regulatory authorities should be neutral and transparent. The regulator should have well-defined obligations of transparency, especially regarding its decision-making processes.[63] Moreover, the powers and the objectives of the regulator must be clearly defined, and activities aimed at meeting these objectives should be monitored. The combination of transparency of objectives, powers, processes, decisions and information gives the public clear performance criteria with which to evaluate the extent to which the regulator is fulfilling its role.[64]

[59] Waelde and Gunst, 'International energy trade and access to energy networks', 131.

[60] 'Nexant: interconnected electric networks and competitive electricity markets', *OGEL*, 1 (3) (2003), 6. Monitoring of the following issues could typically be assigned to the regulators: management and allocation of interconnection capacity; mechanisms to deal with congestion within the national system; publication of appropriate information; effective unbundling of accounts to avoid cross-subsidies and the unbundling compliance programme; connecting new producers; conditions of access to storage and to other ancillary services; and monitoring the security of supply. In addition, regulators may be responsible for issuing authorisations and licences; for the organization, monitoring and control of the tendering procedure for generation; for dispute settlement; and for deciding on exemptions to normal access rules for new investments: No author, 'The role of regulatory authorities', *OGEL*, 2(3) (2004), 3.

[61] The system operator rather than the regulator may monitor the market. D. Parker and S. Van Siclen, 'OECD perspectives on competition policy and regulatory reform in the electricity sector', *OGEL*, 1(3) (2003), 11.

[62] Para. 5 of the Reference Paper on Basic Telecommunications.

[63] Parker and Van Siclen, 'OECD perspectives on competition policy and regulatory reform in the electricity sector', p. 12.

[64] Ibid.

The regulator has to be independent not only from regulated companies but also from government authorities, including ministries.[65] Only such full independence will assure potential and actual market participants that the rules are applied in a non-discriminatory, stable and transparent manner.

E Implication of GATT Transit Rules for Interconnection to Fixed Infrastructure

The second set of WTO rules that are relevant to interconnections of fixed infrastructure in energy trade are contained in the disciplines on transit. In particular, GATT transit rules are relevant for access to the energy transportation networks used when the energy flow crosses country borders more than once, that is, when the energy traffic is in transit.

Several energy transit disputes over the past decade supply evidence that issues related to energy transit are among the most important problems of the multilateral trading system with respect to energy trade.[66] The applicable multilateral trade rules in Article V GATT set a general framework that governs energy transit.[67] These rules are not as complete with respect to energy transit via fixed infrastructure such as pipelines and grids as they should be if they are to address all problems pertinent to cross-border energy trade.

[65] Both the IEA and the OECD call for a separation between political and regulatory functions. See, for instance, *Regulatory Institutions in Liberalized Electricity Markets* (Paris: IEA, 2001) and *Power and Gas Regulation – Issues and International Experience*, draft working paper (Washington DC: World Bank, 2001).

[66] Most notably those between Russia and Ukraine in January 2006 and 2009. Y. Selivanova, 'Managing the patchwork: challenges for multilateral agreements in trade and investment' in A. Goldthau and J., M. Witte (eds.), *Global Energy Governance: The New Rules of the Game* (Washington, DC: Brookings Institution Press, 2010), pp. 49–72.

[67] GATT Article V para. 2 establishes 'freedom of transit through the territory of each contracting party, via the routes most convenient for international transit, for traffic in transit to or from the territory of other contracting parties'. No distinctions are allowed 'based on the flag of vessels, the place of origin, departure, entry, exit or destination, or on any circumstances relating to the ownership of goods, of vessels or of other means of transport'. Although contracting parties 'may require that traffic in transit through its territory be entered at the proper custom house', such traffic cannot be subject to any unnecessary delays or restrictions. Furthermore, such traffic shall be exempt from customs duties and from all transit duties or other charges imposed in respect of transit, 'except charges for transportation or those commensurate with administrative expenses entailed by transit or with the cost of services rendered'. Transit charges as well as regulations must be reasonable. GATT para. 5 requires that members accord MFN treatment with respect to all charges, regulations and formalities in connection with transit.

For the purposes of the present discussion, the implications of disciplines contained in Article V GATT for interconnections and access to fixed infrastructure used for energy transportation will be analysed. There are, however, other aspects of transit regulation by Article V GATT that are beyond the scope of the present analysis.[68]

The GATT transit provisions set out the non-discrimination requirement regarding the transit of energy products and materials. This obligation applies equally to those products transported via fixed infrastructure. Doubts have sometimes been raised as to whether Article V of the GATT 1994 applies to energy products and materials, especially gas and electricity, and to transportation methods, notably pipelines and grids. These doubts, however, are unjustified, as will be explained below.

I Coverage of Energy Products and Fixed Infrastructure by GATT Transit Provisions

According to the definition of 'traffic in transit', energy products such as oil and natural gas fall within the scope of Article V GATT when they are transported through the territory of a WTO Member. Indeed, these products are covered by GATT Article V because, in the sense of GATT 1994, they are 'goods'. Based on the GATT text and its object and purpose, it is clear that mineral oil and gas are 'goods' and 'products' falling within GATT Article V.

Electricity might be the only product that is less obviously classifiable, because it is an intangible 'good' of a special nature – electrons moving in a circuit, for example, inside a conductor such as a metal wire. Electricity is special in a physical sense, because it cannot be stored (except to a limited extent in batteries) and traded as a chemical compound. Despite these physical characteristics, many national and other legal systems understand electric power as a 'good',[69] and WTO literature supports the thesis that GATT 1994 applies when electricity crosses borders.[70]

[68] See, for a detailed discussion of GATT and ECT transit regulation, Ehring and Selivanova, 'Energy transit'.

[69] The European case law goes back to Case 6/64, *Costa* v. *Enel*, [1964] ECR 585, where the European Court of Justice confirmed implicitly that electricity is a good.

[70] See P-X Pierros and S. Nüesch, 'Trade in electricity, spot on', *Journal of World Trade*, 34 (4) (2000), 95–124; M. M. Roggenkamp, 'Implications of GATT and EEC on network-bound energy trade in Europe', *Journal of Energy & Natural Resources Law*, 12(1) (1994), 59–82, at 64, 66; M. M. Roggenkamp, 'Transit of networkbound energy: a new phenomenon?', *World Competition* 19(2) (1995), 126; S. Charnovitz, 'Trade and climate: potential

GATS may apply simultaneously to the transmission and distribution of electrical power. However, as the WTO jurisprudence has made clear on several occasions,[71] GATT 1994 and GATS are not mutually exclusive.

In addition, in the *context* of Article V, one could refer to tariff commitments inscribed in WTO Members' schedules. A number of WTO Members have tariff commitments on energy products, including on 'tariff item number 2716.00.00, electrical energy' (e.g. the European Community and the United States).[72]

WTO Members' schedules of tariff commitments and the Harmonized System both confirm that the terms 'goods' and 'products' in the GATT 1994 cover energy products, including electrical power.[73] It is therefore difficult to substantiate an argument that the GATT cannot be applied to the transit of energy and energy products, given that neither may be able to be considered a 'good'.[74]

With respect to transportation modes, it has been argued that *pipelines* do not fall under GATT Article V, as they are not expressly mentioned in that article. GATT Article V:7 contains the only explicit exception, regarding aircraft in transit. Considering the broad definition of 'traffic in transit' in GATT Article V:1, an express reference to

conflicts and synergies' in J. E. Aldy et al., *Beyond Kyoto* (Arlington, VA: Pew Center on Global Climate Change, 2003), pp. 141, 147. In the case of *alternating* current (AC), which is the dominant form, the direction of the movement changes so frequently that the electrons do not travel very far before they switch direction. In other words, electrons move (back and forth) across the border only within a short piece of cable (meaning that the electrons may be imported or exported, but they usually do not travel far enough to transit through a whole country). However, this is different in the case of *direct* current (DC), which is sometimes used for long-distance transportation, and in general it would seem wrong to overemphasise the relevance of the difference between AC and DC for questions of legal classification under the GATT 1994. A consumer of electricity pays for the extraction of energy, as do the users of other forms of energy. This, and the fact that different forms of energy can be transformed into each other, also works in favour of treating all forms of energy in the same manner. See Ehring and Selivanova, 'Energy transit'.

71 Appellate Body Reports, *EC – Bananas III*, para. 221; *Canada – Autos*, para. 160; *Canada – Periodicals*, p. 19; Panel Report, *Canada – Periodicals*, paras. 5.14–5.19.

72 These tariff commitments gain legal significance through Article II of the GATT 1994, but it would be odd to apply Article II:1(a) of the GATT 1994 to trade ('commerce') in electrical energy but not those GATT provisions that refer to 'goods' or 'products', such as Articles I:1, II:1(b), III and V of the GATT 1994.

73 Chapter 27 of the Harmonized Commodity Description and Coding System lists headings for all these energy products, including coal, petroleum, gas *and* electrical power.

74 B. Clark, 'Transit and the Energy Charter Treaty: rhetoric and reality', *Web Journal of Current Legal Issues* (1998), http://webjcli.ncl.ac.uk/1998/issue5/clark5.html (accessed 15 March 2015).

cover particular transport modes such as pipelines is not necessary. Whether the GATT negotiators specifically thought of including one or the other specific modes of transportation should not lead to any doubt that pipelines are covered by the abstract terms of GATT Article V:1.[75] It should not be forgotten that oil pipelines already existed well before the conclusion of GATT 1947.[76]

II Freedom of Transit

According to GATT Article V:2, '[t]here shall be freedom of transit through the territory of each contracting party, via the routes most convenient for international transit'. This provision sets out a substantive right to freedom of transit through the territory of each Member for transport to or from the territory of other Members.[77] This means that a WTO Member must not prevent transit, that is, it must grant access to the existing transport infrastructure. The main controversies concern access to the fixed infrastructure when this is privately owned and when there are substantial capacity constraints.

1 Ownership of Infrastructure

The fixed infrastructure used for transportation of energy products is often privately owned. The GATT obligation of 'freedom of transit', however, is imposed on Members. One can of course raise the question whether measures taken by private companies could be attributed to the respective WTO Member according to principles of general international law. Sometimes, the government's involvement in these actions may also be sufficient to allow a finding of a direct breach. GATT Article XVII, which deals with the conduct of state enterprises and enterprises with special or exclusive rights, applies to purchases and sales involving imports or exports. It cannot therefore be invoked to enforce transit rights or rights of access to fixed infrastructure in

[75] This is also the position of the Energy Charter Secretariat in its publication *Trade in Energy: WTO Rules Applying under the Energy Charter Treaty*, para. 94, and of M. M. Roggenkamp, 'Transit of networkbound energy: the European experience' in T. W. Wälde (ed.), *The Energy Charter Treaty. An East–West Gateway for Investment and Trade* (Alphen aan den Rijn: Wolters Kluwer, 1996), pp. 499, 518.

[76] Ehring and Selivanova, 'Energy transit'.

[77] See the Note by the Secretariat, Article V of GATT 1994 – Scope and application, para. 11.

general. However, it has been argued that GATT Article V applies directly to companies that can be classified under Article XVII.[78]

On one hand, the WTO Agreement is an international treaty among subjects of public international law that binds WTO Members and not private parties. On the other, this fact does not rule out the possibility that a WTO obligation under GATT Article V is binding upon Members, even where a private company owns the transport infrastructure.

The obligation for a Member to guarantee freedom of transit applies in any case, regardless of who owns the transportation infrastructure. It may simply mean that the government must take suitable measures vis-à-vis the private owner of the infrastructure in order to meet its WTO obligations. Of course, the question of private ownership of the transport infrastructure was not so pertinent when the GATT was negotiated, since such networks were generally state-owned in the Contracting Parties of the GATT. Private ownership of transportation networks has become more common since the energy sector reforms carried out in most countries.

In conclusion, even when private owners control transit infrastructure, WTO Members are still under the obligation to guarantee the freedom of transit. In other words, GATT Article V contains an obligation of a *result*.[79] The opposite conclusion could lead to WTO obligations becoming ineffective where private companies own fixed infrastructure. Therefore, even though the WTO obligations ensure that freedom of transit applies only to WTO Members, the government can fail to meet its international obligations if private owners impede transit.

2 Capacity Constraints

As already mentioned, because of the large capital investments involved in fixed transport infrastructure projects, pipelines are not constructed with large extra capacities, but rather to serve the needs of specific projects. The existing transport infrastructure will therefore often turn out to be insufficient to meet all demands.

The decision on the *construction* of transport infrastructure on a country's territory is typically a matter of that country's sovereignty – which could, however, be restricted by international commitments. Questions have been raised as to whether GATT contains such an

[78] Roggenkamp, 'Implications of GATT and EEC on networkbound energy trade in Europe'.

[79] Ehring and Selivanova, 'Energy transit'.

obligation. There is nothing explicit in the transit provisions of GATT obliging WTO Members to allow infrastructure construction or to expand existing transit capacity.[80] The first sentence of GATT Article V:2 ('via the routes most convenient') seems to indicate that freedom of transit exists only within existing capacities. Indeed, only 'routes' that already exist can be 'convenient', and only one of several existing routes can be the 'most convenient'.[81]

What would happen if the lack of capacity is such that it effectively denies any 'freedom of transit'? In this case, one may argue that 'freedom of transit' should imply the possibility to create new infrastructure. In particular, it has been argued that if an investor offers to undertake the construction of the necessary additional infrastructure, the government's opposition to the project could be argued to constitute a denial of 'freedom of transit'.[82]

Another important question concerns the access to *existing* limited transport infrastructure where there is congestion. Arguably, a WTO Member cannot apply MFN treatment and refuse to grant access for all goods in transit while reserving the route in question for its domestic transportation or import/export traffic; the WTO Member must, arguably, allocate scarce transport capacities in such a way that transit is possible.

If capacity allocation is considered either a regulation or a formality in connection with transit, the obligation of MFN treatment in GATT Article V:5 applies. In this respect, the question is whether distinctions between transit and domestic (or import/export) transportation are permitted. This is uncertain, although it is clear that a distinction between transiting goods based on the foreign/domestic ownership of the goods or the means of transport would be in violation of GATT Article V.[83]

F Concluding Remarks

Because significant amounts of renewable energy have to be able to connect to networks over the next decades, long-term investments in energy infrastructure and interconnections are necessary. Regulation of

[80] This argument is made in Roggenkamp, 'Implications of GATT and EEC on network-bound energy trade in Europe', pp. 73 and 82, and 'Transit of networkbound energy', p. 510, noting that 'means of transport shall be deemed to be in transit'. Note, however, that means of transport are deemed in transit only when they pass across the territory, which is not the case with fixed infrastructure.

[81] See Ehring and Selivanova, 'Energy transit'. [82] Ibid. [83] Ibid.

access, on reasonable terms, to transport and distribution networks will be crucial for the development of cross-border energy trade.

Energy trade's dependence on fixed infrastructure implies that traditional trade disciplines in themselves might not be enough to ensure free trade in energy. While WTO rules have traditionally focused on market access commitments, these might not be sufficient to ensure liberalised energy trade. Indeed, trade in energy is often obstructed by problems in gaining access to transportation infrastructure. Typical barriers include excessive access fees and discriminatory allocation of network capacities by the transmission system operator. These problems are aggravated by non-transparent terms and conditions for calculating access fees and usage conditions.

Moreover, energy transportation networks are mostly controlled not by governments that undertake obligations under international or regional treaties, but by powerful private companies. These companies often claim lack of capacity and charge transportation fees exceeding the cost of services rendered.

WTO rules and commitments ensure market access and non-discrimination, and tackle quantitative measures. However, these rules do not currently address the anti-competitive practices that are common among incumbents controlling different segments upstream and downstream along the energy value chain. Additional measures have to be taken to guarantee the availability of fixed infrastructure and timely access to transportation pipelines/transportation networks, distribution systems and the like.

Issues linked to fixed infrastructure could possibly be addressed in the context of GATS negotiations. Moreover, in the context of the Trade in Services Agreement (TISA) negotiations, proposals have been made to address energy services specifically. Unfortunately, these proposals do not address the issue of energy transportation and access to fixed infrastructure.

In the GATS context, additional commitments in either an annex to GATS on Energy Services or a Reference Paper could be used to address competition issues and third-party access to fixed infrastructure, using an approach similar to the additional disciplines for the telecommunications sector contained in the Reference Paper. In this respect, the following issues need to be addressed to secure an effective pro-competitive framework for energy trade including TPA to fixed infrastructure:

- providing non-discriminatory TPA to and interconnection with energy networks and grids, and other essential infrastructure, whether dominated by government entities or privately owned companies;
- creating an independent regulator separate from and not accountable to any supplier of energy services;
- ensuring non-discriminatory, objective and timely procedures for the transportation and transmission of energy;
- maintaining appropriate measures for preventing certain anticompetitive practices in the sector;[84]
- ensuring transparency in the formulation and implementation of rules, regulations and technical standards;
- requiring the provision of non-discriminatory and timely information on data relevant to the transportation and transmission of energy, such as prices and transmission capacity;
- security of supply and non-interruption of energy transportation;
- expeditious dispute settlement (as the interruption of energy transportation services can have drastic consequences for consumers).

In addition, WTO transit provisions should be used more effectively to ensure that energy transit via fixed infrastructure is not obstructed. Although GATT Article V establishes a general rule of freedom of transit, some key issues pertinent to energy transit are not covered.

Freedom of energy transit is linked not just to the non-discriminatory use of existing infrastructure, but also to the possibility of creating additional transit capacity if needed. Considering the large investments needed for the construction of energy transport infrastructure, such networks rarely contain large spare capacities, but rather are tailor-made to serve specific projects with predefined capacity. GATT does not address the issue of new transit capacity construction. For this, an effective investment framework is needed to ensure that additional networks can be constructed, should existing ones fail to provide necessary capacity.

Moreover, there are no rules in GATT Article V for resolving congestion management problems. It is not clear how exactly freedom of transit is guaranteed when the capacity of existing infrastructure does not allow for transit of additional volumes of energy materials and products. Finally, the energy sector is dominated by monopoly enterprises that

[84] WTO, Council for Trade in Services, Committee on Specific Commitments, Communication from the United States, S/CSC/W/27.

can obstruct access to transit infrastructure, and their behaviour is not, in all cases, effectively tackled under the general WTO rules.

Since the general transit provisions of GATT Article V have not been designed specifically for dealing with energy transit (although they of course cover such transit, just like the transit of any other good), additional rules can help ensure energy transit on reasonable terms based on the balance between the sovereign interests of states and the need for security and stability of transit.

In the context of trade facilitation negotiations, a number of WTO Members have demonstrated an interest in devising some more specific rules related to energy transit. The inclusion of an explicit mention that the definition of traffic in transit covers goods moved via fixed infrastructure such as pipelines was proposed, as was an 'upgraded' non-discrimination requirement to require that the treatment accorded to transit must be no less favourable than that accorded to own export or import traffic. Finally, there was a proposal to discipline the behaviour of companies that operate transit infrastructure. Such companies would have to act consistently with the provisions on transit and solely in accordance with commercial considerations. However, agreement was not reached on these issues and the final text on trade facilitation does not address energy transit specifically.

Bibliography

Bhattacharyya, S. C. (2011). *Energy Economics: Concept, Issues, Markets and Governance*. London: Springer.

Cameron, P. D. (2007). *Competition in Energy Markets: Law and Regulation in the European Union*. Oxford: Oxford University Press.

Charnovitz, S. (2003). Trade and climate: potential conflicts and synergies. In *Beyond Kyoto*. Arlington, VA: Pew Center on Global Climate Change.

Chatzivasileiadis, S., Ernst, D. and Andersson, G. (2013). The Global Grid. *Renewable Energy*, **57**, 372–83.

Clark, B. (1998). Transit and the Energy Charter Treaty: rhetoric and reality. *Web Journal of Current Legal Issues*, http://webjcli.ncl.ac.uk/1998/issue5/clark5.html (accessed 15 March 2015).

Cossy, M. (2011). Energy services. In Yulia Selivanova (ed.), *Regulation of Energy in International Trade Law: WTO, NAFTA and Energy Charter*. Alphen aan den Rijn: Kluwer Law International, pp. 149–80.

Ehring, L. and Selivanova, Y. (2011). Energy transit. In Yulia Selivanova (ed.), *Regulation of Energy in International Trade Law: WTO, NAFTA and Energy Charter*. Alphen aan den Rijn: Kluwer Law International.

Energy Charter Secretariat. (2002). *Trade in Energy: WTO Rules Applying under the Energy Charter Treaty*, www.energycharter.org/what-we-do/trade-and-transit/trade-and-transit-thematic-reports/trade-in-energy-wto-rules-applying-under-the-energy-charter-treaty-2002/ (accessed 30 November 2016).

Evans, P. (2002). *Liberalizing Global Trade in Services*. Washington, DC: AEI Press.

(2003). Strengthening WTO Member commitments in energy services: problems and prospects. In Aaditiya Mattoo and Pierre Sauvé (eds.), *Domestic Regulation and Service Trade Liberalization*. Washington DC: World Bank and Oxford University Press, pp. 167–89.

Gamberale, C. (2011). Energy services. In WTO Secretariat, *Guide to the GATS: An overview of Issues for Further Liberalization of Trade in Services*. Alphen aan den Rijn: Kluwer Law International, pp. 259–94.

IEA staff. (2004). The role of regulatory authorities. *OGEL*, **2**(3), 3.

International Energy Agency. (2001). *Regulatory Institutions in Liberalized Electricity Markets*. Paris: IEA.

Melly, C. (2003). Electric power and gas market reform and international trade in services, note 2, in *Energy and Environmental Services: Negotiating Objectives and Development Priorities*. Geneva: United Nations Conference on Trade and Development (UNCTAD/DITC/TNCD/2003/3), pp. 164–77.

Mernier, A. (2008). Setting the rules of energy trade. In *Fundamentals of the Global Oil and Gas Industry*. London: Petroleum Economist.

Parker, D. and Van Siclen, S. (2003). OECD perspectives on competition policy and regulatory reform in the electricity sector. *OGEL*, **1**(3), 11.

Pierros, P-X and Nüesch, S. (2000). Trade in electricity, spot on. *Journal of World Trade*, **34**(4), 95–124.

Poretti, P. and Rios-Herran, R. (2006). A reference paper on energy services: the best way forward? *OGEL*, **4**(4), online.

Roggenkamp, M. M. (1994). Implications of GATT and EEC on networkbound energy trade in Europe. *Journal of Energy & Natural Resources Law*, **12**(1), 59–82.

(1995). Transit of networkbound energy: a new phenomenon? *World Competition*, **19**(2), 119–146.

(1996). Transit of networkbound energy: the European experience. In Thomas W. Wälde (ed.), *The Energy Charter Treaty: An East–West Gateway for Investment and Trade*. Alphen aan den Rijn: Wolters Kluwer, pp. 499–519.

Selivanova, Y. (2007). The WTO and energy: WTO Rules and Agreements of relevance to the energy sector. *International Centre for Trade and Sustainable Development, Issue Paper No. 1.*

(2010). Managing the patchwork: challenges for multilateral agreements in trade and investment. In Andreas Goldthau and Jan Martin Witte (eds.), *Global Energy Governance: The New Rules of the Game.* Washington, DC: Brookings Institution Press, pp. 49–72.

(2015). *Clean Energy and Access to Infrastructure: Implications for the Global Trade System*, E15Initiative (Geneva: International Centre for Trade and Sustainable Development (ICTSD) and World Economic Forum), http://e15initiative.org/wp-content/uploads/2015/01/E15_Clean-Energy_Selivanova_FINAL.pdf (accessed 29 November 2016).

Thomas, S. and Hall, D. (2006). *GATS and Electricity and Water Sector.* University of Greenwich: Public Services International Research Unit.

Waelde, T. (2000). *Access to Energy Networks: A Precondition for Cross-Border and Energy Services Trade.* Dundee: Centre for Energy, Petroleum and Mineral Law and Policy.

Waelde, T. W. and Gunst, A. J. (2003). International energy trade and access to energy networks. In *Energy and Environmental Services: Negotiating Objectives and Development Priorities.* Geneva: United Nations Conference on Trade and Development (UNCTAD/DITC/TNCD/2003/3), pp. 118–63.

World Bank. (2001). 'Power and gas regulation – issues and international experience', draft working paper. Washington, DC: World Bank.

WTO, Note by the Secretariat, Article V of GATT 1994 – Scope and application.

Council for Trade in Services, Committee on Specific Commitments, Communication from the United States, S/CSC/W/27.

Council for Trade in Services – Special Session – Energy Services – Information Note by the Secretariat, JOB(05)/204.

Report of the Appellate Body, *European Communities – Regime for the Importation, Sale and Distribution of Bananas*, WT/DS27/AB/R, 9 September 1997.

Report of the Appellate Body – Canada, WT/DS139/AB/R, WT/DS142/AB/R, 31 May 2000.

Council for Trade in Services, Committee on Specific Commitments, Communication from the United States, S/CSC/W/27, 18 May 2000.

12

Preferential Connection and Integration of Electricity Generators Using Renewable Energy Sources: Critical Assessment of the EU Rules

KAROLIS GUDAS

A Introduction

In the 1990s, the European Union (EU) decided to gradually liberalise electricity markets. Liberalisation was followed by the adoption of the climate and energy package.[1] The closely linked policy objectives of electricity liberalisation and climate change mitigation, however, can be conflicting, as commitments to a low-carbon economy are often implemented at the cost of market-based trade in both conventional and green electricity.

Besides promotion of renewable energy (RE) on the production side, the EU has introduced measures which allow EU Member States to provide preferential treatment to RE-sourced electricity over conventional power generators at the stages of connection, access to the grid and dispatch of electricity on the grid.[2] On one hand, grid-related incentives have the potential to facilitate the connection and integration of RE electricity generators into electricity systems. They may eliminate grid-related barriers, such as long waits to obtain authorisation for grid connection and discriminatory and non-transparent access to the grid procedures adopted to connect and integrate conventional power plants.[3]

[1] Under the climate change and energy package, the European Union undertook the commitment to reduce greenhouse gas emissions by 20 per cent from 1990 levels, to raise the share of energy consumption produced from renewable resources to 20 per cent and to improve energy efficiency by 20 per cent. The EU's present goals include increasing the share of renewable energy by 27 per cent by 2030, in order to contribute towards climate change mitigation. See European Commission, *Climate Change and Energy Policies*, http://ec.europa.eu/clima/policies/package/index_en.htm (accessed 10 October 2014).

[2] Article 16(2) and 16(3) of the RES Directive.

[3] See Á. R. Paraja, *Priority Access for Renewable Energy into the Grid: The Experience of a Regulator* (2010), www.energy-regulators.eu/portal/page/portal/FSR_

On the other hand, grid-related incentives may undermine the economics of conventional power technologies to the extent that they are no longer profitable.[4] Grid-related incentives may also hinder cross-border trade in conventional and RE electricity by creating loop flows that limit cross-border transmission capacity,[5] or potentially induce the shutdown of certain thermal power plants, the start-up costs of which can be carbon dioxide (CO_2)-intensive.[6]

This chapter addresses part of the technological bias problem that occurs in the technologically diverse electricity sector. It suggests that electricity-generating installations should be treated based on their ability to supply electricity. To do so, the chapter explains the different ways in which electricity-generating technologies are treated at the levels of production of electricity and grid operation. It further provides an overview of problems related to the implementation of rules giving priority to renewables, and an overview of the EU's rules tackling the issue of discrimination against electricity-generating installations at the stages of connection, access and dispatch of electricity on the grid.

B Overview of the Priority Rules under EU Law

Preferential treatment of RE-sourced electricity dates back to 1996, when the EU adopted the first legislative package of liberalisation of gas and electricity markets, which included Directive 96/92/EC concerning common rules for the internal market in electricity (1996 Electricity Directive). Articles 8(1) and 8(2) of the 1996 Electricity Directive placed the transmission system operators in charge of dispatching the generating installations based on objective criteria, to be published and applied

HOME/ENERGY/Policy_Events/Workshops/2010/EU_Energy_Law_Policy/MRP.pdf (accessed 10 November 2014).

4 G. Wynn, 'Europe's wind and solar grid challenge', *Reuters* (2013), www.businessspectator.com.au/article/2013/9/19/renewable-energy/europes-wind-and-solar-grid-challenge (accessed 10 October 2014).

5 European Wind Energy Association, *EWEA Position Paper on Priority Dispatch of Wind Power: A Paper from the EWEA Large-Scale Integration Working Group*, 2013, p. 13, www.ewea.org/fileadmin/files/library/publications/position-papers/EWEA_position_on_priority_dispatch.pdf (accessed 10 October 2015).

6 O. Rosnes, *Subsidies to Renewable Energy in Inflexible Power Markets*, Norwegian University of Life Sciences, www.ifn.se/BinaryLoader.axd?OwnerID=dfe1cad2-2ef5-4894-b64e-14648d3949c4&OwnerType=0&PropertyName=File1&FileName=Rosnes%20Subsidies%20to%20Renewables%20in%20Inflexible%20Power%20Markets.pdf (accessed 10 October 2014).

in a non-discriminatory manner. Article 8(3) of the 1996 Electricity Directive, in particular, allowed Member States to require the system operator to prioritise generating installations using renewable energy sources (RES) or waste or producing combined heat and power.

In 2001, priority rights to RE-sourced electricity-generating installations were extended with the introduction of Directive 2001/77/EC on the promotion of electricity produced from renewable energy sources in the internal electricity market (2001 RES Directive). Notably, Article 7(1) of the 2001 RES Directive introduced the requirement for Member States to take the necessary measures to ensure that transmission system operators (TSOs) and distribution system operators (DSOs) in their territory guarantee the transmission and distribution of electricity produced from RES. In addition, the 2001 RES Directive: (i) allowed Member States to give electricity produced from RES priority access to the grid system and (ii) required TSOs to give priority to generating installations using RES when dispatching generating installations, insofar as was permitted by the operation of the national electricity system.

Grid-related incentives were expected to help new technologies and players enter the electricity markets. They were also intended to address market failures, such as the dominance of vertically integrated utilities, inadequate infrastructure and the burdensome authorisation procedures for planning, building and operation of grids that existed at Member State level.[7] To further advance those objectives, in Directive 2009/28/EC on the promotion of the use of energy from renewable energy sources (RES Directive), which amended and subsequently repealed the 2001 RES Directive, the EU revised and amended its preferential treatment of RE-sourced electricity-generating installations.

I Priority Rules in the RES Directive and Their Implementation

The RES Directive maintains the 2001 RES Directive requirement for Member States to ensure that TSOs and DSOs in their territories guarantee

[7] See further European Commission (2007), 'Communication from the Commission to the Council and the European Parliament. Renewable energy road map. Renewable energies in the 21st century: building a more sustainable future' (COM/2006/0848 final); European Commission (2006), 'Communication from the Commission to the Council and the European Parliament Green Paper follow-up action report on progress in renewable electricity' (COM/2006/0849 final); European Commission (2013), 'Communication from the Commission. Delivering the internal electricity market and making the most of public intervention' (C(2013) 7243 final).

the transmission and distribution of electricity produced from RES.[8] It also leaves intact EU Member States' obligation to 'ensure that when dispatching electricity generating installations, [TSOs would] give priority to generating installations using renewable energy sources' insofar as is permitted by the electricity system.[9] Besides the above, the RES Directive of 2009 strengthens the priority regime for RE electricity generators in several ways when compared to the RES Directive of 2001.[10]

First, in light of the fact that EU electricity systems accommodate the further development of electricity production from RES, the RES Directive stipulates a requirement for Member States to 'take the appropriate steps to develop transmission and distribution grid infrastructure, intelligent networks, storage facilities and the electricity system, in order to allow the secure operation of [the system]'.[11] This general duty explicitly imposes a requirement to further develop the systems not only for conventional electricity-generation needs, but also for electricity generation from renewables, which features such challenges as forecasting and dependence on climatic conditions. However, the RES Directive neither provides any guidelines on the implementation of the rule nor explains the meaning of 'appropriate steps' to be taken to develop such infrastructure in order to accommodate a growing share of RES electricity production.

Second, the RES Directive imposes an obligation on EU Member States to 'provide for either priority or guaranteed access to the grid-system'. According to Recital 61 of the RES Directive, (i) *priority access* requires Member States to provide an assurance to connected generators of electricity from RES that they will be able to sell and transmit the electricity generated from RES in accordance with the connection procedures at all times, whenever the source becomes available and (ii) *guaranteed access* requires Member States to provide an assurance that electricity will obtain access to the grid, allowing the use of a maximum amount of electricity from RE electricity generators connected to the grid.[12] Similar to the 2001 RES Directive, it makes the implementation of priority or guaranteed access rules subject to requirements relating to

[8] Article 16.2(a) of the RES Directive. [9] Article 16.2(c) of the RES Directive.

[10] Article 16 of the RES Directive stipulates and clarifies, among other things, a number of rights related to connection and access to the electricity grids that are not covered by this chapter (as this chapter's intention is more to discuss the rationale of the regime than particular aspects of it).

[11] Article 16(1) RES Directive.

[12] Note that priority access and guaranteed access are similar measures: guaranteed access applies to RE electricity directly sold on the market, while priority access applies to

maintenance of the grid's reliability and safety, and therefore provides some flexibility to Member States regarding implementation of the rule.

Third, besides priority access and dispatch of electricity, the RES Directive also (i) introduces the option for Member States to provide for priority connection or reserved connection capacity for new electricity-generating installations using RES[13] and (ii) allows Member States to require TSOs and DSOs to bear, in full or in part, the costs related to RES-e generators' connection and integration, including costs of 'technical adaptations, such as grid connections and reinforcements, improved operation of the grid and rules on the non-discriminatory implementation of the grid codes'.[14] Notably, 'connection' and 'access' to the grid have different meanings in EU law. According to the Court of Justice of the European Union, 'connection' to the grid is related to the physical connection of the electricity generator to the grid, whereas 'access' to the grid is related to the terms and conditions of supply of electricity.[15]

The extent to which priority measures are implemented in the Member States of the EU varies widely, from many perspectives. Most of the Member States have different connection charge regimes as well as different distribution cost regimes.[16] Many EU Member States apply non-discriminatory treatment in terms of the connection of RE electricity generators, but give preferential access and dispatch terms to RE electricity generators. Some Member States give RE generators priority in terms of connection, access and dispatch of electricity on the grid. Others apply non-discriminatory treatment at the levels of connection, access and dispatch of electricity on the grid, irrespective of the technology used for the generation of electricity (except for feed-in tariffs or related schemes, which *de facto* ensure priority for renewables).[17]

For instance, France and Sweden do not give priority to RE-sourced electricity in terms of either connection or dispatch. Ireland reserves connection capacity for RES-e. Belgium and Spain give priority to RE electricity generators for connection, access and dispatch. The United Kingdom provides guaranteed access for electricity produced from all

purchase contracts with TSOs (European Wind Energy Association, *EWEA Position Paper on Priority Dispatch of Wind Power*, p. 6).

[13] Recital 61 of the RES Directive. [14] Article 16(3) and 16(4) of the RES Directive.

[15] Case C-239/07, *Sabatauskas and Others*, Court of Justice of the European Union, judgment of 9 October 2008, paras. 36–44.

[16] Ibid. [17] Ibid.

types of generators, whereas Germany requires the grid operator to connect RE electricity generators first and to ensure that they are given priority access.[18] In 2013, of sixteen states that took part in a survey by the Council of European Energy Regulators, (i) seven of them had different connection charge regimes for RES-e plants than for conventional plants; (ii) one had a different access charge regime for RES-e plants; (iii) five had imposed a 'deep' connection charges model and (iv) five had imposed a 'shallow' connection charges model.[19]

II Priority Rules in the EE Directive

Directive 2012/27/EU on energy efficiency (EE Directive) also allows the provision of preferential treatment to RE at the stages of connection, access and dispatch of electricity on the grid. The EE Directive establishes a framework for the promotion of energy efficiency within the EU at all stages of the energy chain and aims to remove barriers and overcome market failures by addressing the rules relating to transmission, distribution,

[18] RES LEGAL Europe Database, *Legal Sources on Renewable Energy* (2012), www.res-legal.eu/compare-grid-issues/ (accessed 10 October 2014). Also see Council of European Energy Regulators, *Status Review of Renewable and Energy Efficiency Support Schemes in Europe* (2013), www.ceer.eu/portal/page/portal/EER_HOME/EER_PUBLICATIONS/CEER_PAPERS/Electricity/Tab2/C12-SDE-33-03_RES%20SR_25%20June%202013%20revised%20publication_0.pdf (accessed 16 November 2015).

[19] Notably, depending on the connection to the grid cost components, different connection charging models are categorised into the 'deep connection cost charging model', 'shallow connection cost charging model' or 'deep', 'semi-deep', 'semi-shallow' and 'shallow' models. A 'deep' connection cost charging model imposes upon the power station's developer several grid infrastructure-related costs, such as grid connection, grid reinforcement and extension. A 'shallow' connection cost charging model imposes upon the power station's developer only the costs of the connection to the grid, not the costs of reinforcement and extension. Combinations of the 'deep' and 'shallow' connection models are categorised as 'semi-deep' or 'semi-shallow' (for further information see Council of European Energy Regulators, *Status Review of Renewable and Energy Efficiency Support Schemes in Europe in 2012 and 2013* (2015), pp. 32–6, www.energy-community.org/portal/page/portal/ENC_HOME/DOCS/4154396/33476CB022773277E053C92FA8C0B7A8.pdf (accessed 13 November 2015). For instance, the RES LEGAL Europe Database provides that in the Netherlands, the costs of connection to the grid are borne by the plant operator. In Poland, the fee for connection of RES of no more than 5 MW is half of the fee calculated on the basis of the actual costs. Slovenia exempts RES electricity generators from payment for necessary electricity grid reinforcements resulting from their connection (see RES LEGAL Europe Database, *Legal Sources on Renewable Energy*).

access to and dispatch of electricity on the grid.[20] Similarly to the RES Directive, in the EE Directive implementation of grid-related incentives is made subject to the reliability and safety of the grid – notably, the EE Directive requires consideration to be given to the rules on transmission, distribution, access and dispatch, as set out in Directive 2009/72/EC concerning common rules for the internal market in electricity (Electricity Directive),[21] but without prejudice to the rules on connection, access and dispatch contained in the RES Directive.[22]

In particular, the EE Directive requires Member States to: (i) guarantee the transmission and distribution of electricity from high-efficiency cogeneration; (ii) provide priority or guaranteed access to the grid for electricity from high-efficiency cogeneration and (iii) provide priority dispatch of electricity from high-efficiency cogeneration insofar as the secure operation of the national electricity system permits.[23] In the context of application of priority rules, Article 15(5) of the EE Directive enables Member States to set rankings 'as between, and within different types of, renewable energy and high-efficiency cogeneration' and requires EU Members to ensure 'in any case . . . that priority access or dispatch for energy from variable renewable energy sources is not hampered'.[24]

Therefore, the question of non-discrimination within the framework of the EE Directive is even more complex when compared to the RES Directive. First, unlike the RES Directive, the EE Directive aims to ensure grid access priority not only to RE electricity generators, but also to some conventional combined heat-and-power technologies. Thus, the EE Directive raises the question of discrimination between RE-sourced electricity and conventional electricity. Second, it introduces the term 'variable renewable energy sources' and raises the question of discrimination between 'variable' and 'non-variable' RES. However, there is no legal definition of 'variable renewable energy sources' or 'non-variable renewable energy sources' at the EU level.[25] Third, promotion of 'variable' over 'non-variable' renewable sources may hamper 'non-variable' renewable sources, irrespective of any negative environmental or market effects.

[20] Article 15(5) of the EE Directive.
[21] Article 15 of the Electricity Directive.
[22] Article 16(2) of the EE Directive.
[23] Article 15(5)(a–c) of the EE Directive.
[24] Article 15(5) of the EE Directive.
[25] S. Reith, T. Kolbel and P. Schlagermann, *Legal Conditions for Grid Access* (Brussels: Geolec, 2012), pp. 12–14.

C Grid Operation Neutrality

There are several reasons why RE-sourced electricity should not *per se* be given special treatment at the stages of connection, access and dispatch of electricity on the grid. First, the requirement to feed in electricity produced from a specific source is not associated with any special characteristics of the electricity, but rather with the outcomes at the production stage. From a system operation point of view, some of the RES-e generators are comparable to conventional electricity generators. Second, at the stage of connection to the grid, some conventional electricity generators face the same challenges as renewable electricity-generating installations. Finally, renewables are granted priority by Member States regardless of the nature of challenges (technical or market) related to renewables integration in the electricity systems. Priority rules are implemented regardless of alternative means to integrate renewables through certain market arrangements.

I System Operation Objectives and Preferential Treatment of RES-e Generators

Electricity demand and electricity systems reserves change continually. Therefore, the central task of the system operators is to match production with consumption. This task is performed through dispatching, which entails the central coordination of producers and large consumers. Coordination of electricity systems is typically carried out with due regard to the use of interconnectors.[26]

In light of the objective to match electricity supply with demand, the elimination of obstacles related to RE connection and integration through the introduction of priority rules is arguably excessive and unjustifiable from the standpoint of transmission and distribution systems operation. A number of reasons should be outlined here besides the fact that RE connection and integration can be facilitated with grid adaptation and market arrangements.[27] First, some renewable electricity-generating technologies are comparable with conventional electricity-

[26] B. Barton, L. K. Barrera-Hernandez, A. R. Lucas and A. Ronne, *Regulating Energy and Natural Resources* (Oxford: Oxford University Press, 2006), p. 174.

[27] See for instance European Network of Transmission System Operators for Electricity, *Developing Balancing Systems to Facilitate the Achievement of Renewable Energy Goals* (2011), www.entsoe.eu/fileadmin/user_upload/_library/position_papers/111104_RESBalancing_final.pdf (accessed 15 November 2015).

generating installations. RE comprises a heterogeneous class of technologies which, over differing time scales, can be variable and unpredictable, variable but predictable, constant or controllable.[28] Some of the RE electricity-generating technologies, therefore, have a similar ability to supply electricity as conventional electricity-generating technologies (i.e. gas turbines and solar; coal and biomass, solar thermal with storage).[29]

Second, implementation of priority rules is not directly related to market barriers specifically linked to RES connection and integration challenges. RE connection and integration barriers vary from one Member State to another and include issues related to connection costs, capacity limits, lack of transparency, public opposition, administrative inefficiency, capacity speculation and lack of financial incentives.[30] Therefore, in some EU states introduction of priority rules will not necessarily contribute to the objectives of RE connection and integration. For instance, in the Netherlands, the challenges related to renewables integration are related to a lack of sufficient grid capacity; in Slovakia, barriers are related to capacity speculation and delays during the connection to the grid process; and in Portugal, the main challenges identified are related to complicated licensing procedures and environmental impact assessment.[31]

Furthermore, these and similar challenges are also pertinent to conventional electricity-generating installations. However, the EU rules do not require the establishment of a causal link between priority rules and market failures. Nor does the European Commission suggest phasing out the priority rules, irrespective of the circumstance that priority rules will not tackle the specific challenges related to RE-electricity generators' connection and integration in the grid.[32]

[28] Intergovernmental Panel on Climate Change, *Renewable Energy Sources and Climate Change Mitigation* (Cambridge: Cambridge University Press, 2012), p. 8.

[29] They may provide base load power (e.g. nuclear, coal versus biomass, solar thermal with storage, hydropower), intermediate load power (e.g. coal versus geothermal) or peak load power (e.g. gas turbines versus solar). See B. Harack, *How Can Renewables Deliver Dispatchable Power on Demand?*, www.visionofearth.org/industry/renewable-energy/renewable-energy-review/how-can-renewables-deliver-dispatchable-power-on-demand (accessed 10 October 2014).

[30] Reith, Kolbel and Schlagermann, *Legal Conditions for Grid Access*, pp. 12–14. [31] Ibid.

[32] See European Commission, 'Communication from the Commission. Delivering the internal electricity market and making the most of public intervention', https://ec.europa.eu/energy/sites/ener/files/documents/com_2013_public_intervention_en_0.pdf

II Discrimination against Electricity-Generating Installations

The problem of discrimination against electricity-generating installations that was a paramount issue back in 2001 is no longer of such relevance since the 2009 adoption and implementation of the third package of legislative proposals for gas and electricity markets.

Neither in 1996 (when the first package for liberalisation of electricity and gas markets was introduced) nor in 2001 (when the first renewable energy directive was introduced) were there strong institutional mechanisms in place to coordinate and control electricity markets. Back then, electricity markets were dominated by vertically integrated undertakings. Rules for access to the grid at the Member State level were largely discriminatory from substantive and procedural points of view.[33] They provided much flexibility to system operators to decide which electricity-generating installations to connect first, and for which of them to ensure access and dispatch of electricity on the grid.[34]

The liberalisation of the electricity sector demonstrates that electricity generation and supply can be legally, conceptually and commercially separated from the operation of transmission and distribution systems, and that discrimination against different electricity-generating installations can be tackled through a number of regulatory means (notably, if renewables are provided non-discriminatory access to the grid, their low operating costs will place them before conventional power producers in the dispatch merit order).[35]

(accessed 10 November 2015). Notably, the European Commission states that 'rules for the responsibility for grid balancing, priority dispatching and financial responsibility for network development . . . with the development of the open and competitive electricity markets . . . may lose their justification once the internal electricity market is completed'. However, grid connection, access and priority dispatch regulatory tools are still widely used in EU states that promote RE electricity, irrespective of the fact that EU Members have fully implemented measures that effectively tackle with discrimination of electricity generators at the stages of connection, access and dispatch of electricity on the grid.

33 See K. Gudas, 'Intersection of WTO and EU law in international electricity trade: analysis of the regulatory environment for electricity systems development', PhD thesis (2017), University of Bern.

34 Ibid.

35 The European Commission itself notes that 'when renewables producers have equal access to the market, their low operating costs (particularly for wind and solar power production) place them before conventional power producers in the merit order. As such, systems with centralised dispatch fade and priority dispatch rules become less relevant for renewable energy technologies active in the market'. See European Commission, 'Communication from the Commission. Delivering the internal electricity market and making the most of public intervention'.

For a long time, the European Commission argued that new generators' connection and integration in the electricity markets was hampered mainly by vertically integrated companies. Unbundling was therefore designed to eliminate the dominance of the vertically integrated utilities and to create an open and transparent market with non-discriminatory access to electricity networks. Unbundling transmission and distribution from electricity generation and supply has essentially removed the conflict of interest between generators (suppliers) and the operators of transmission systems in most EU Member States.[36] As a result, the transmission system operators were separated from the vertically integrated companies acting in the areas of generation and supply and the operators of distribution systems were granted a high degree of independence, regardless of whether or not a particular distribution system operator has remained part of a vertically integrated company.[37]

Second, the regulatory third-party access regime introduced in 2009 with the Electricity Directive restricted the denial of access to the grid only to situations in which the respective system operator lacks the necessary capacity. If access is denied, Article 32(2) of the Electricity Directive requires the provision of reasons duly substantiated by the respective transmission system operator or distribution system operator and based on objective and technically and economically justified criteria.[38] Therefore, if an RE electricity generator is denied access on discriminatory grounds, it is provided regulatory means to defend its commercial interests.

Third, the Electricity Directive imposes on Member States a requirement to establish competent bodies with the function of regulatory authorities, independent from the interests of the electricity industry. The regulatory authorities are assigned the tasks of both market monitoring and market regulation. They also have the power, if necessary, to compel TSOs to modify their terms and conditions. Therefore, if TSOs create discriminatory obstacles that prevent connection, access and dispatch of electricity on the grid, the national independent regulatory authorities may require them to modify these rules and eliminate the barriers.[39]

[36] See European Commission (2011), 'The entry into force of the EU third energy package', www.learneurope.eu/files/7113/7483/5945/Third_energy_package_en.pdf (accessed 10 October 2014).

[37] Articles 24–26 of the Electricity Directive.

[38] Article 32(2) of the Electricity Directive.

[39] Article 36(e) of the Electricity Directive.

Fourth, the EU's general non-discrimination rules and the competition laws established under the Treaty on the Functioning of the European Union restrict the transmission system operators and distribution system operators from abusing their dominant position.[40] Introduction of the network codes in the EU will further implement the non-discrimination principle and mitigate any risks related to discrimination against electricity generators at the stages of connection, access to and use of electricity grids.[41] Network codes will, to some extent, harmonise connection to the grid and market rules.[42]

D Concluding Remarks

While grid-related incentives may indeed have the potential to increase the amount of RES in the electricity sector, it is evident that from an economic, technical and legal point of view, these measures may have negative side effects for trade and the operation of electricity systems.

The priority rules may be applied by Member States irrespective of the lack of a causal link between the objectives of the measures and their effects, or the availability of less trade-restrictive measures. The promotion of RE through grid-related incentives does not necessarily ensure a contribution to climate change mitigation. Grid-related incentives may create unplanned power flows that limit cross-border transmission capacity, undermining the economy of conventional power generators as well as leading to their unexpected shutdown.

Instead of priority rules, the regulation of connection to the grid and operation of the grid should be subject to rules aimed at ensuring neutrality across technologies through grid adaptation and market

[40] Violation of the EU's competition rules may result in a fine of up to 10 per cent of total annual income. See further on third-party access and competition rules, A. Kotlowski, 'Third-party access rights in the energy sector: a competition law perspective', *Utilities Law Review*, 16(3) (2007), 102–18.

[41] See Gudas, 'Intersection of WTO and EU law'.

[42] Also, the European Commission encourages Member States to ensure that national rules which are not harmonised are non-discriminatory across technologies and calls for further harmonisation of market rules with the idea to allow all producers to participate in the intraday, balancing and ancillary services markets. See European Commission, 'Communication from the Commission: Delivering the internal electricity market and making the most of public intervention'.

arrangements. Several reasons justify this approach. First, some conventional and renewable electricity-generating technologies are comparable from the system operation point of view. Second, the major obstacles to RE connection and integration can be addressed through institutional control, implementation of a regulated third-party access regime and other legal remedies available under EU law.

13

The Treatment of Restrictions and Financial Charges on Imports and Exports of Electricity under EU and International Law

ILARIA ESPA

A Introduction

Cross-border electricity trade is intensifying in all regions of the world as global demand for electricity grows and intermittent renewable energy sources account for a progressively higher share of countries' electricity mix. Yet, cross-border electricity flows remain exposed to a wide array of practices capable of limiting the importation and/or exportation of electricity. These measures may be necessary to preserve or restore the security of electricity supply in light of the physical peculiarities of trade in electricity and, in particular, renewably produced electricity, or may be dictated by national interests or reasons of political economy. Such measures are also used by European Union (EU) Member States, irrespective of the construction of the EU internal electricity market.

In as much as electricity trade gradually expands on a larger scale and over long distances, it is important that the use of practices amounting to import and export restrictions on electricity is adequately disciplined at the international level in order to avoid unnecessary disturbances to cross-border electricity trade. In the international legal framework, the World Trade Organization (WTO) and the Energy Charter Treaty (ECT) rulebooks do contain provisions on import and export restrictions that are applicable to international trade in electricity but, contrary to the case in the EU legal framework, none of them is electricity-specific.[1] The

[1] In both regimes, electricity is conventionally considered a good. WTO rules on import and export restrictions are therefore also applicable to trade in electricity to the extent that they are applicable to trade in goods. WTO Panel Report, *Canada – Measures Relating to the Feed-in Tariff Program (Canada-FIT Program)*, circulated on 19 December 2012, WT/DS426/R, para. 7.11, footnote 46. The same holds true for ECT rules on import and export

question is therefore whether such provisions can, even so, effectively foster long-distance electricity trade as part of a cross-border electricity-trading regime. This chapter aims to answer this question by adopting a comparative approach: first, it illustrates the scope and coverage of EU rules applicable to import and export restrictions on electricity; second, it assesses the adequacy of generally applicable WTO rules on import and export restrictions against the EU electricity-specific 'model'.[2]

The chapter will thus be organised as follows. Section B will give an overview of the types of practices affecting cross-border electricity flows that amount to restrictions on importation and/or exportation. The analysis will be limited to the types of measures implemented in the European Union.[3] Section C will examine the EU's primary and secondary law instruments for disciplining the use of such measures. Section D will take stock of the existing WTO and ECT rules on import and export restrictions relevant to electricity trade and assess whether they are adequate to enable and foster long-distance electricity trade. Finally, Section E will provide some concluding remarks on the main strengths and weaknesses of existing international law rules applicable to electricity trade in the perspective of the gradual integration of national, regional and, eventually, inter-regional grids.

B Overview of Import and Export Restrictions Affecting Cross-Border Electricity Trade

A wide range of measures may affect cross-border electricity trade either directly or indirectly. These measures may be imposed on the importation and/or exportation of electricity. They may be price measures such as import or export fees imposed on cross-border electricity flows, or quantitative restrictions affecting the volume of electricity being transmitted cross-border. In some cases, trade restrictions are imposed to ensure the security and quality of supply. In others, countries may apply

restrictions, as electrical energy is explicitly listed in the Annex EM to the Energy Charter Treaty among the 'Energy Products and Materials' covered by ECT rules.

2 Electricity is considered a good also in the EU legal framework. According to existing European Court of Justice (ECJ) case law, imports and exports of electricity fall within the scope of the EU Treaty rules on the free movement of goods. See Cases C-393/92, *Almelo* (1994) ECR I-1477 and C-158/94 *Commission* v. *Italy* (1997) ECR I-5789.

3 Albeit limited in scope, this selection is still representative considering that the EU internal electricity market constitutes what is so far the most advanced experience of regional integration in the electricity sector. See Chapter 1 and Chapter 6.

trade-restrictive measures for reasons of political economy or to pursue non-economic public policy goals. Finally, import or export restrictions may be imposed temporarily to address emergency situations.

I Import and Export Quantitative Restrictions on Cross-Border Electricity Trade

1 Import Prohibitions or Restrictions of Electricity Flows

The most immediate way to inhibit cross-border electricity trade is to physically impede the flow of electricity at the border. Although it seems counterintuitive for countries to actually ban the importation of electricity, especially if 'short' (i.e. countries with a domestic generating capacity inferior to domestic consumption), specific circumstances linked to the physical properties of electricity flows may induce countries to resort to import restrictions. This is the case of Poland, which has recently introduced power flow-controlling devices known as phase-shifting transformers aimed at regulating and, if necessary, interrupting the importation of green electricity coming from Germany, following the example of countries such as the Netherlands and Belgium.[4] This practice followed the rapid and extensive development of wind farms in the northern part of Germany and solar photovoltaic systems in the southern part of Germany in the past few years, often producing electricity from decentralised generators (i.e. generators connected directly to the distribution network so that the German transmission system operator (TSO) cannot control them in its balancing exercise).[5] Due to the inherent intermittency of wind and solar electricity, the use of phase-shifting transformers is meant to preserve the security of supply in the national network of neighbouring countries, which could otherwise be exposed to severe imbalances resulting from excess exports and/or unscheduled and uncontrollable loop flows at peak times,[6] or even drastic shortages of

[4] G. Parkinson, 'Poland builds electronic wall to keep out German renewables' (18 November 2013), http://reneweconomy.com.au/2013/poland-builds-electronic-wall-to-keep-out-german-renewables-72084 (accessed 15 March 2015).

[5] VDE, *The 50.2 Hz Problem* (2014), www.vde.com/en/fnn/Pages/50-2-hz.aspx (accessed 15 March 2015).

[6] For a description of the physics of loop flows, see O. Daxhelet and Y. Smeers, 'Inter-TSO Compensation Mechanism' (7 November 2005), www.hks.harvard.edu/hepg/Papers/Daxlet_Smeers_11.07.05.pdf (accessed 15 March 2015). In the case of Germany, in particular, the overload produced in the northern area was originally meant to be discharged to the southern part of the country, and vice versa. Yet, in real time, the national network is not

supply in extreme cases of over-frequency, causing decentralised generators to shut down.[7]

Although these devices *de facto* inhibit the physical flows of electricity from a country into an interconnected neighbouring country, two factors need to be considered. First, phase shifters aim at ensuring the security of supply vis-à-vis potential overload situations altering the production–consumption balance, and thus putting the stability of national electricity systems at risk. Second, by regulating electricity injections into the interconnectors and protecting power lines from overload, these devices actually do not hamper trade *ab absoluto*; rather, they create the conditions for increasing cross-border electricity transactions by making available and 'schedulable' additional cross-border transmission capacity.[8]

2 Export Prohibitions or Restrictions of Electricity Flows

Limiting or stopping the electricity supply available for exportation could also impede cross-border electricity flows. Electricity exports are usually banned in emergency situations, potentially creating electricity shortages. In such sudden crises, national TSOs may intervene to restore the production–consumption balance and therefore preserve the security and the stability of supply.

At the time of writing, the most recent case illustrating how countries can concretely limit the electricity supply available for exports is the 2012 winter crisis experienced by several Balkan States (Albania, Bulgaria, Greece, Romania, the Former Yugoslav Republic of Macedonia, and Serbia). These countries faced a particularly severe winter, leading to increased domestic demand for electrical power but also to reduced

able to retain the surplus of green electricity produced at peak times. Neighbouring countries 'flooded' with German electricity flows would thus need to shutdown active high-power stations in real time to compensate for unexpected surpluses of energy and avoid a blackout. See M. Duchamp, 'Poland plans to ban German solar and wind energy imports' (6 December 2011), http://motls.blogspot.ch/2011/12/poland-plans-to-ban-german-solar-and.html (accessed 15 March 2015).

[7] This is known as the '50.2 Hz problem'. It describes the risk that generators located in a same geographical area and connected to low-voltage distribution networks would all cease to operate at once in extreme weather circumstances, as they are programmed to shut down when the main frequency exceeds 50.2 Hz. In such a scenario, the TSO could not intervene, as decentralised electricity generators are not connected to the transmission system. VDE, *The 50.2 Hz Problem*.

[8] W. L. Kling, D. A. M. Klaar and G. H. Schuld, 'Phase shifting transformers installed in the Netherlands in order to increase available international transmission capacity', *Cigré C2-207 Working Paper* (2003), www.transform.ru/articles/pdf/sigre/c2-207.pdf (accessed 15 March 2015).

availability of thermal production plants and hydropower owing to the low temperatures. This situation generated a domino effect, as some of the countries experiencing a lack of sale offers started to be affected by the prohibition of electricity exports from neighbouring countries, and thus decided to also introduce similar measures in order to protect their own network.[9]

The measures used by the Balkan countries affected by the winter crisis to limit or totally ban the exportation of domestically generated electricity were implemented at various levels by intervening in and altering the process of allocation of available interconnector capacity.[10] In other words, countries restricted or prohibited export electricity supplies by limiting the availability of interconnection capacity until the conditions necessary to preserve the functionality and the safety of the national electricity system were restored.[11] This result could be achieved by simply suspending the market-based procedures for allocating cross-border capacity based on auctioning and reallocating it according to national priorities, by limiting electricity exports based on bilateral procedures or by instructing traders to set their export schedules to zero. In some cases, however, TSOs still continued to allocate cross-border market capacities through auctioning in order to avoid restricting electricity export for transit; in others, TSOs also forbade the practice of arbitrage.[12]

The limitation of electricity supply available for exportation via the curtailment of available cross-border transmission capacity directly affects the traders active in the affected countries, as well as actors such as public providers, distribution companies and end suppliers relying on undelivered supply. The traders have three choices:

[9] Energy Charter Treaty News, *Restrictions on Export of Electricity in the Balkan Region: Updates* (22 February 2012), www.encharter.org/index.php?id=21&id_article=267&L=0 (accessed 15 March 2015).

[10] Electricity can be traded cross-border (i.e. it can be bought from a generator located in country A and sold to a consumer situated in country B) insofar as it can be transported through an interconnector, that is, a cable connecting two separate markets. As with any transmission line, however, an interconnector only has a limited capacity. Such capacity has to be allocated by the TSOs in order to ensure the availability of the interconnection's maximum capacity, while still avoiding congestion problems in accordance with safety standards of secure network operation. IEA (2013), *Cross-Border Trade in Electricity and the Development of Renewable-Based Electric Power: Lessons from Europe*, www.oecd-ilibrary.org/trade/cross-border-trade-in-electricity-and-the-development-of-renewables-based-electric-power_5k4869cdwnzr-en (accessed 15 March 2015), p. 50.

[11] Discussions with representatives of various power trading companies.

[12] Ibid.

i. they could try to fulfil their supply obligations by buying the electricity from other generators situated in other neighbouring countries at very high prices;
ii. they could suffer the costs arising from having an unbalanced schedule;[13]
iii. they could claim *force majeure* suspension of deliveries to escape these costs.

In such a situation, the public providers, distribution companies and end suppliers are forced to announce new tenders to ensure security of supply at much higher prices.[14]

3 Import or Export Restrictions Arising out of the Process of Cross-Border Capacity Allocation

TSOs are responsible for managing the cross-border allocation of interconnector capacity. Import and export transactions, and thus international electricity trade, directly depend upon the availability of cross-border transmission capacity. To ensure maximum transfer capacity, TSOs have developed standardised definitions of allocable capacities based on the assessment of various factors, such as weather forecasts (i.e. temperature), congestion of national transmission lines, security margins and planned maintenance of interconnectors.[15] Accordingly, out of the total transfer capacity of an interconnection,[16] TSOs only allocate the so-called net

[13] TSOs are responsible for constantly ensuring the balance between supply and demand in the national network. They do so by managing the so-called 'balancing markets', either by running another competitive auction aimed at fixing the balancing price of electricity for under- and over-scheduling operators or by buying and selling electricity to balance the system and then charging an imbalance penalty fee to unbalanced traders. IEA, *Cross-Border Trade*, p. 43.

[14] As economic operators driven by profit, they would choose the least expensive option depending on the conditions under which they were unable to supply contracted electric power to their clients. However, under emergency situations placing high pressure on the network, the prices incurred by traders and suppliers to make up for lack of electricity supply are generally 'devastating'. See Y. Selivanova, 'Emergency situation in South Eastern Europe in winter 2012', presentation to the Energy Charter Secretariat, Brussels, 23 May 2012.

[15] European Transmission System Operators, *Definitions of Transfer Capacity in Liberalised Electricity Markets*, Final Report April 2001, www.entsoe.eu/fileadmin/user_upload/_library/ntc/entsoe_transferCapacityDefinitions.pdf (accessed 15 March 2015).

[16] Total Transfer Capacity (TTC) is 'the maximum exchange programme between two areas compatible with operational security standards applicable at each system if future network conditions, generation and load patterns were perfectly known in advance'. Ibid, p. 6.

transfer capacity (NTC), which is defined as 'the maximum exchange programme between two areas compatible with security standards applicable in both areas and taking into account the technical uncertainties on future network conditions'.[17] The calculation of the NTC made by national TSOs is critical for exploiting the full potential of the interconnections to safely develop cross-border electricity trade. It is thus important that TSOs assess the NTC in a transparent way and that the calculations are not manipulated in such a way as to *de facto* restrict cross-border electricity trade.

The NTC is allocated in different time frames. The part of NTC 'that remains available, after each phase of the allocation procedure for further commercial activity' is known as available transfer capacity (ATC).[18] The ATC announced by TSOs is then divided into different maturities and sold on a yearly, monthly, weekly and daily basis.[19] Market-based allocation methods such as implicit (for coupled electricity markets) and explicit (for uncoupled electricity markets) auctioning are now widely used and are generally more efficient than the non-market based system based on bilateral agreements such as the first-come-first-served or pro rata allocations, as they guarantee that cross-border transmission capacity is allocated to those users who are willing to pay the highest price.[20] Differences in cross-border allocation methods and in gate closure times in uncoupled electricity markets may indirectly affect international trade in electricity because the lack of coordination causes inefficient price signalling and potentially higher prices, discouraging cross-border electricity transactions.[21]

Finally, cross-border electricity trade is also affected by transaction curtailment procedures affecting the availability of already allocated cross-border transmission capacity. Adjustments to the use of allocated

[17] Ibid, p. 7. The NTC is the TTC minus the so-called Transmission Reliability Margin (TRM), that is, a security margin that copes with the uncertainties on the computed TTC values.

[18] Ibid., p. 7.

[19] Currently, only a few countries also allocate transfer capacities intraday. See IEA, *Cross-Border Trade*, p. 30 and p. 51.

[20] Ibid., pp. 52–3. While this is true in general, more integrated networks such as as the EU internal electricity market abide by more advanced rules, which prohibit non market-based methods. See Commission Regulation (EU) 2015/1222 of 24 July 2015 establishing a guideline on capacity allocation and congestion management: http://eur-lex.europa.eu/legal-content/EN/TXT/PDF/?uri=CELEX:32015R1222&from=EN (accessed 15 September 2015).

[21] Ibid., pp. 24–5 and 30–1.

capacity may be required to ensure congestion management in emergency and *force majeure* situations that would not allow TSOs to wait for re-dispatching or counter-trading.[22] They may also be needed to cope with cross-border balancing needs caused by unexpected surplus or shortages (e.g. caused by intermittent renewables) when all programmed transmission capacity is allocated at the day-ahead stage.[23]

II Import and Export Fees for Cross-Border Electricity Flows

Cross-border electricity flows may be affected by the imposition of fees on the importation or exportation of electricity. These measures are normally applied at the transmission level and *de facto* raise the costs incurred by domestic electricity consumers (in the case of an import fee) or generators (in the case of an export fee). They do so by exposing such consumers and generators to passed-on network charges in addition to the domestic transmission fees imposed by their TSO for accessing and using the national network.[24] This duplication of costs is generally known as the 'pancaking' effect and ultimately results in high trade costs for cross-border electricity transactions.[25]

Import and export fees were widely applied in the past, when both interconnected countries used to apply national transmission fees for cross-border trade, thereby charging on electricity traders two different

[22] European Transmission System Operators, *Definitions of Transfer Capacity in Liberalised Electricity Markets*, pp. 4–5. The economic consequences in terms of firmness and potential obligation of a TSO or auction office to compensate vary depending on the different circumstances. In conditions of normal operation, no curtailment of previously allocated transmission capacity is allowed, but buyback voluntarily is permissible. See, e.g., Coordinated Auction Office in South East Europe (SEECAO), 'Auction Rules for Capacity Allocation in SEE CAO' (July 2014), www.energy-community.org/portal/page/portal/ENC_HOME/DOCS/3350143/0633975ADBA37B9CE053C92FA8C06338.PDF (accessed 15 March 2015).

[23] IEA, *Cross-Border Trade*, p. 25. Indeed, various representatives of power trading companies have denounced a significant reduction in the NTC and ATC values made available on a yearly basis by national TSOs (as well as more frequent curtailments of allocated capacity on a weekly and even daily basis). They attribute this to the increasing share of intermittent renewables in the countries' electricity mix.

[24] ENTSO-E, *Overview of Transmission Tariffs in Europe: Synthesis 2013*, www.entsoe.eu/fileadmin/user_upload/_library/Market/Transmission_Tariffs/Synthesis_2013_FINAL_04072013.pdf (accessed 15 March 2015).

[25] M. Shahidehpour and M. Alomoush, *Restructured Electrical Power Systems: Operation: Trading, and Volatility* (Basel: CRC Press, 2001), p. 47.

transmission fees (i.e. one importing fee and one exporting fee) for each across-the-border transaction.[26] At the beginning of the 2000s, the EU introduced the 'inter-transmission system operator compensation mechanism' for the costs of hosting cross-border flows of electricity, including the cost of providing cross-border access to the interconnected network. Its aim was to create a common regulatory approach in order to eliminate trade distortions resulting from the co-existence of several different systems for transmission charging imposed at the national level.[27] Yet, some EU Member States have continued to charge additional fees on cross-border electricity flows (i.e. Romania, Bulgaria and Greece).[28] Technically, these countries do not announce the introduction and/or maintenance of 'import' or 'export' fees on cross-border electricity flows, but rather extensions of the tariff models applied at the national level by their TSOs (i.e. the feed-in fees and/or take-up fees applied to domestic users of the national network) to either importers or exporters.[29] However, feed-in fees and take-up fees are usually inconsequently applied, so that the final result is to treat national consumers of imported electricity and national generators delivering electricity abroad (namely, imports and exports of electricity) like national generators and consumers,

[26] A. Merlin, 'European power market design', keynote speech given at the San Antonio-CIGRE-IEEE Symposium, 5–7 October 2005.

[27] Regulation (EC) No 1228/2003 of the European Parliament and of the Council of 26 June 2003 on conditions for access to the network for cross-border exchanges in electricity, Articles 3 and 4, www.energy-community.org/pls/portal/docs/36276.PDF (accessed 15 March 2015); Regulation (EC) No 714/2009 of the European Parliament and of the Council of 13 July 2009 on conditions for access to the network for cross-border exchanges in electricity and repealing Regulation (EC) No 1228/2003, Articles 13 and 14, http://eur-lex.europa.eu/LexUriServ/LexUriServ.do?uri=OJ:L:2009:211:0015:0035:EN:PDF (accessed 15 March 2015); Commission Regulation (EU) No 838/2010 of 23 September 2010 on laying down guidelines relating to the inter-transmission system operator compensation mechanism and a common regulatory approach to transmission charging, http://eur-lex.europa.eu/LexUriServ/LexUriServ.do?uri=OJ:L:2010:250:0005:0011:EN:PDF (accessed 15 March 2015).

[28] C. Urbanke, 'Power market issues in Romania', presentation given at the EFET Workshop on Wholesale Power Trading in the CEE and SEE Regions, Brussels, 9 July 2013; P. Popov, 'Case study: Bulgaria', presentation given at the EFET Workshop on Wholesale Power Trading in the CEE and SEE Regions, Brussels, 9 July 2013; A. Siri, 'Adverse effects on cross border trading across Greek interconnections', presentation given at the EFET Workshop on Wholesale Power Trading in the CEE and SEE Regions, Brussels, 9 July 2013.

[29] M. Garbers, 'Why export and import fees are illegal', presentation given at the EFET Workshop on Wholesale Power Trading in the CEE and SEE Regions, Brussels, 9 July 2013.

respectively – thus discriminating against different groups of generators/consumers. In the former case, countries applying feed-in fees extend them to domestic consumers importing electricity, thereby privileging national consumers buying electricity from a national generator and/or supplier. In the latter case, countries charging domestic consumers at the national level apply such take-up fees to exporting generators as well, but not to domestic generators selling electricity for national consumption. In addition, such fees are also generally applied to traders seeking to import or export electricity into or out of the country concerned, although they clearly do not generate and/or consume power in the domestic network.[30]

Depending on the transmission tariff model adopted by each country, the actual fee may have different components. Frequently, such fees include, along with a network access and/or a transport component, 'Combined Heat and Power (CHP)' and/or 'green' components to promote CHP or renewable energy.[31] In other cases, the fees include components for unrecoverable balancing costs.[32] These different components are either variable in and of themselves or subject to considerable and frequent variations that are difficult to predict, thus creating uncertainties among traders and ultimately increasing trade frictions linked to cross-border electricity flows.[33]

Import and export fees charged at the transmission level have accordingly been under strong scrutiny for their discriminatory effects against different groups of generators and/or consumers and for discouraging electricity traders, which do not directly produce or consume electricity, from participating in the wholesale market. These measures make cross-border electricity flows more expensive, thus deterring electricity trade to the advantage of domestic generation or consumption. As price measures not aimed at addressing specific crises but

30 Urbanke, 'Power market issues in Romania'; Popov, 'Case study: Bulgaria', Siri, 'Adverse effects on cross border trading across Greek interconnections'.

31 Urbanke, 'Power market issues in Romania'; Popov, 'Case study: Bulgaria'.

32 Siri, 'Adverse effects on cross border trading across Greek interconnections'.

33 Urbanke, 'Power market issues in Romania'; Popov, 'Case study: Bulgaria', Siri, 'Adverse effects on cross border trading across Greek interconnections'. In particular, wholesale electricity traders decide how much yearly cross-border transmission capacity rights to buy based on the announced import/export fees in place in the countries in which they are operating. Variations in such fees, however, occur on a more frequent basis than this. In the case of Bulgaria, for instance, changes in export fees are frequently announced on a monthly basis (discussions with representatives of various power trading companies).

rather imposed on any cross-border electricity transactions, moreover, import and export fees are difficult to justify as instrumental to ensuring the security and stability of the supply. Rather, they reveal an underlying political economy rationale (e.g. protection of incumbent utilities' market positions and maintenance of price differentials serving the economic interests of national generators or consumers).[34] Accordingly, the first informal complaint regarding Bulgaria and Romania's use of import and export fees for cross-border electricity flows was brought before the Directorate-General for Energy of the EU Commission by the European Federation of Energy Traders (EFET) in 2013. The complaint has been kept confidential so far and, at the time of writing, it is still pending. Following submission of the complaint, these fees have been decreasing in the countries concerned.[35]

C EU Law Applicable to Import and Export Restrictions on Electricity Flows

The EU legal framework encompasses both primary and secondary law instruments relevant and applicable to disciplining cross-border electricity trade and, in particular, restrictions on electricity imports and exports. Pertinent primary law rules include aspects of both trade in goods and trade in services regulation, to the extent that the costs entailed by such restrictions are also borne by wholesale traders involved in international transactions. While electricity is considered a good in the EU legal framework, wholesale electricity trading in fact qualifies as a mixed activity including aspects of both goods and services.[36] Relevant EU secondary law instruments include electricity-specific regulations and directives adopted to realise the EU internal electricity market.

[34] IEA, *Cross-Border Trade*, p. 44 ff.

[35] I. Peltegova, 'Bulgaria reduces electricity export tariff from 1 August' (30 July 2013), www.icis.com/resources/news/2013/07/30/9692393/bulgaria-reduces-electricity-export-tariff-from-1-august/ (accessed 15 March 2015); S. Udubasceanu, 'Doubt surrounds scrapping of Romanian export fee' (30 June 2014), www.icis.com/heren/articles/2014/06/11/9790683/power/edem/romania-scraps-tariff-to-export-electricity.html (accessed 15 March 2015). Romania seems to have eliminated such fees very recently due to market coupling with Hungary.

[36] Existing ECJ jurisprudence made clear that a single measure may be covered simultaneously by the provisions on goods and the provisions on services. See Case C-390/99, *Canal Satélit Digital SL* (2002) ECR I-607.

I Relevant EU Primary Law

As it affects not only imports and exports of electricity but also wholesale trading, existing price and quantity measures, the restriction of cross-border electricity flows may run counter to the principle of free movement of goods under Articles 34 and 35 of the Treaty on the Functioning of the European Union (TFEU) and the principle of freedom to provide services under Articles 56 and 57 TFEU.

According to Articles 34 and 35, quantitative restrictions on imports and exports, as well as all measures having equivalent effect, shall be prohibited between Member States. A substantial body of ECJ jurisprudence has consistently interpreted these principles in a broad manner as to prohibit not only all discriminatory barriers to trade but also genuinely non-discriminatory measures that nevertheless have the ultimate effect of impairing trade.[37] The import and export restrictions discussed above are thus very likely to constitute an obstacle to the free movement of goods within the meaning of Articles 34 and 35 TFEU, either as amounting to quantitative restrictions per se (e.g. import prohibitions through phase shifters, emergency export bans or cross-border transmission capacity curtailments) or as measures having equivalent effect (e.g. import and export fees).

Depending on the rationale underlying the use of such restrictions, however, it may in principle be possible to seek justification under Article 36 TFEU. According to Article 36, 'the provisions of Articles 34 and 35 shall not preclude prohibitions or restrictions on imports [and] exports ... justified on grounds of ... public policy or public security; the protection of health and life of humans, animals or plants ... or the protection of industrial and commercial property'. Based on the question whether a causal link could be credibly established between the measure restricting cross-border electricity trade and the need to preserve or restore the security of supply and/or the safety of the national network, the argument could be made that such measures fall under the scope of the 'public security' exception recognised under Article 36 TFEU. Measures aimed at stimulating the share of green electricity fed into the grid and combined heat and power (CHP) generation could in principle be defended under the 'public health' exception recognised in Article 36 or, more generally, by

[37] Case 120/78, *Rewe* v. *Bundesmonopolverwaltung für Branntwein* (1979) ECR 649; Cases C-267-8/91, *Keck et Mithouard* (1993) ECR 6097.

invoking environmental protection grounds,[38] provided that they can be proved to be proportionate and well-founded.[39] Finally, according to the last sentence of Article 36, even if justified under one of the exceptions listed in Article 36, any prohibitions or restrictions limiting cross-border electricity flows 'shall not constitute a means of arbitrary discrimination or a disguised restriction on trade between Member States'. In other words, a Member State cannot, under Article 36 TFEU, justify import and export restrictions directly or indirectly aiming at discriminating in favour of domestically produced electricity.[40]

Based on the ECJ's interpretation of Article 36 TFEU, the import and export fees currently applied by some eastern EU Member States seem unlikely to be defendable under the second part of Article 36 TFEU, irrespective of any associated environmental rationale. On the other hand, transmission capacity curtailments implemented for congestion management purposes and temporary export restrictions addressing specific emergency situations, and not affecting electricity flows related to transit, are more likely to be justifiable under Article 36 TFEU, final sentence, than are 'general' restrictions applicable to any cross-border electricity transfer.

By affecting wholesale traders, import and export restrictions on electricity also impinge on the EU primary law principle of freedom to provide services recognised under Article 56 TFEU.[41] This principle has

[38] Although 'protection of the environment' is not explicitly mentioned in Article 36 TFEU, EU Member States have increasingly resorted to environmental justifications, and a consistent body of ECJ jurisprudence has recognised it to be an overriding mandatory requirement under which it is possible to defend limitations to the principle of free movements of goods. See Case C-2/90 *Commission* v. *Belgium* (1992) ECR I-4431; Case 302/86 *Commission* v. *Denmark* (1988) ECR 4607; Case C-473 *Toolex* (2000) ECR I-5681; Case C-379/98 *PreussenElekta* (2001) ECR I-2099; Case C-573/12 *Ålands Vindkraft AB* v. *Energimyndigheten* (2014) EU:C:2014:2037.

[39] A Member State is required to prove the appropriateness and the proportionality of a restriction under Article 36 by showing that it is necessary to achieve the declared goal and the least restrictive among all the alternatives ensuring the same level of protection. For an overview of the ECJ case law on Article 36 see European Commission, *Free Movement of Goods: Guide to the Application of Treaty Provision Governing the Free Movement of Goods* (2010), http://ec.europa.eu/enterprise/policies/single-market-goods/files/goods/docs/art34-36/new_guide_en.pdf (accessed 15 March 2015), p. 28.

[40] Case 34/79 *Heen and Darby* (1979) ECR 3795; Cases C-1/90 and C-176/90 *Aragonesa de Publicidad Exterior and Publivìa* (1991) ECR I-4151.

[41] Under Article 57 TFUE, 'Services shall be considered to be "services" within the meaning of the Treaties where they are normally provided for remuneration, in so far as they are

also been interpreted in an expansive way by the ECJ as prohibiting not only all discriminatory barriers to trade, but also genuinely non-discriminatory measures that ultimately inhibit trade.[42] Accordingly, the above-mentioned considerations apply, and the import and export restrictions on electricity discussed above are likely to constitute an obstacle to the freedom to provide services as recognised under Article 56 TFUE. On the basis of Article 62 TFEU, however, the exceptions based on grounds of public policy, public security or public health that are laid down in Article 52 TFEU also apply to the freedom to provide services. They may therefore prove relevant in defending at least those restrictions addressing specific and sudden crises, which put at risk the integrity of national electricity systems.

II Relevant EU Secondary Law

Various electricity-specific instruments of EU secondary law are relevant and applicable to import and export restrictions on cross-border electricity flows. An important source is Regulation 714/2009/EC on conditions for access to the network for cross-border exchanges in electricity.[43] Article 14, in particular, disciplines the use of charges for access to networks. It requires Member States to ensure that network operators charge generators and/or consumers for the national network access in a non-discriminatory manner (Article 14 (1)), and excludes cross-border traders that do not generate or consume power in the network from paying these charges on individual transactions (Article 14 (2)). Existing import and export fees imposed on cross-border electricity flows thus seem very likely to contravene Article 14 of Regulation 714/2009/EC. In fact, they comprise a 'network access' component which is applied inconsistently among different groups of generators and consumers.

not governed by the provisions relating to freedom of movement for goods, capital and persons. "Services" shall in particular include: . . . (b) activities of a commercial character'. As a commercial activity through which imports and exports of physical electricity occur in exchange for remuneration, wholesale trading of electricity falls also under the scope of the EU provisions regulating trade in services.

42 Case 275/92, *Schindler* (1994) ECR I-1039; Case C-384/93, *Alpine Investment* (1995) ECR I-1141; Case C-272/94, *Guiot* (1996) ECR I-1905; Case C-3/95, *Reisebüro Broede* (1996) ECR I-6257. Public security or public health exceptions are also applicable to the free movement of services by means of Article 52 (1) TFUE.

43 Regulation (EC) No 714/2009 of the European Parliament and of the Council of 13 July 2009 on conditions for access to the network for cross-border exchanges in electricity and repealing Regulation (EC) No 1228/2003, http://eur-lex.europa.eu/legal-content/EN/ALL/?uri=CELEX%3A32009R0714 (accessed 15 March 2015).

They are generally applied to wholesale traders, although the mere trading of electricity, without generating or consuming, cannot be subject to network fees.

While Member States are pre-empted from imposing additional network charges for the use of cross-border transmission capacity (Article 14 (4)), Article 16 of the Regulation 714/2009/EC allows them to apply additional cross-border charges to declared imports or exports 'with non-discriminatory market-based solutions' for congestion management purposes (Article 16 (1)). Member States may thus increase the price of the implicit or explicit auctions within the process of allocating interconnector capacity to address network congestion problems. Article 16 also regulates transaction curtailment procedures. The general rule is that the maximum available cross-border transmission capacity shall be made available to market participants (Article 16 (3)). Allocated capacity can thus be curtailed only 'in emergency situations where the transmission system operator must act in an expeditious manner and re-dispatching or countertrading is not possible', provided that market participants are compensated. The only exception is when allocated capacity is restricted under *force majeure* circumstances. Based on the language of Article 16, existing import and export fees still remain illegitimate as discriminatory additional network charges, but quantitative restrictions on the import or export of electricity applied for congestion management purposes, including in *force majeure* cases, may be consistent with Article 16 insofar as they address specific emergency situations.

Directive 2009/72/EC concerning common rules for the internal market in electricity may be relevant for justifying otherwise unlawful import and export restrictions on cross-border electricity flows.[44] Under Article 3 of Directive 2009/72/EC, in particular, Member States may seek justification for so-called 'public service obligations' imposed on undertakings operating in the electricity sector, which relate, inter alia, to 'environmental protection, including energy efficiency, energy from renewable sources and climate promotion', provided that such obligations 'clearly defined, transparent, and non-discriminatory' (Article 3 (2)). Among the import and export restrictions on electricity trade that have been discussed, import and export fees including a 'green' or a 'CHP'

[44] Directive 2009/72/EC of the European Parliament and the Council of 13 July 2009 concerning common rules for the internal market in electricity and repealing Directive 2003/53/EC, http://eur-lex.europa.eu/legal-content/EN/TXT/PDF/?uri=CELEX:32009L0072&from=EN (accessed 15 March 2015).

component may in principle be defended under Article 3 of Directive 2009/72/EC to the extent that they pursue non-economic public policy goals such as renewable energy or CHP development.[45] However, they remain difficult to justify under Article 3 of Directive 2009/72/EC to the extent that they treat different groups of generators and/or consumers inconsistently.

Finally, Directive 2009/72 is also relevant for justifying any type of restrictions on electricity trade in emergency situations. Article 42 (1) in fact allows Member States to take temporary safeguard measures 'in the event of a sudden crisis in the energy market and where the physical safety or security of persons, apparatus or installations or system integrity is threatened', provided that such measures 'cause the least possible disturbance in the functioning of the internal market' and are 'not wider in scope than is strictly necessary to remedy the sudden difficulties which have arisen' (Article 42 (2)). Ad hoc emergency measures such as those applied during the winter crisis in the Balkan region in 2012 may thus fall under the scope of Article 42 to the extent that they are discontinued immediately after a crisis ends. Member States have nevertheless to provide notification of the restrictions to the other Member States and to the Commission, so that the latter may require the Member State to amend or abolish them 'insofar as they distort competition and adversely affect trade in a manner which is at variance with the common interest' (Article 43 (3)). These procedures are evidently conceived to prevent abuses of state of emergency exceptions and limit the losses incurred by traders operating in the countries affected by the restrictions.

D Existing International Rules on Import and Export Restrictions Applicable to Cross-Border Electricity Trade

The analysis of the EU rules relevant and applicable to import and export restrictions on electricity reveals that any of the existing measures

[45] These measures generally fall within the remit of, respectively, Directive 2009/28/EC of the European Parliament and of the Council on the promotion of the use of energy from renewable sources and amending and subsequently repealing Directives 2001/77/EC and 2003/30/EC, http://eur-lex.europa.eu/legal-content/EN/TXT/PDF/?uri=CELEX:32009L0028&from=EN (accessed 15 March 2015), and Directive 2004/8/EC of the European Parliament and of the Council of 11 February 2004 on the promotion of cogeneration based on a useful heat demand in the internal energy market and amending Directive 92/42/EEC, http://eur-lex.europa.eu/legal-content/EN/TXT/PDF/?uri=CELEX:32004L0008&from=EN (accessed 15 March 2015).

affecting cross-border electricity flows are covered by EU primary law and secondary, electricity-specific legislation, so that unnecessary disturbances in the functioning of the internal electricity market should in principle be avoided. At the same time, the use of emergency measures limiting cross-border electricity trade is also regulated under specific exceptions, thus leaving Member States with sufficient policy space to ensure the security and stability of the supply.

Relevant international law rules applicable to trade in electricity are not targeted to such peculiar needs. WTO rules on trade in goods and trade in services are neither energy- nor electricity-specific. ECT rules regulate trade in energy materials and products but do not specifically address cross-border electricity flows. It is therefore important to assess, on the one hand, whether existing measures practically restricting cross-border flows of electricity may fall under the scope of the generally applicable WTO and ECT rules regulating the use of import and export restrictions and, on the other, whether existing WTO and ECT exceptions are available and applicable to justify the implementation of emergency measures restricting cross-border electricity trade.

I WTO Law

1 Relevant GATT Rules on Import and Export Quantitative Restrictions

In the WTO legal framework, electricity is considered a good, and is accordingly disciplined under the General Agreement on Tariffs and Trade (GATT). Article XI:1 GATT, in particular, outlaws any import and export 'prohibitions or restrictions other than duties, taxes or other charges, whether made effective through quotas, import or export licences or other measures'. WTO dispute settlement bodies have consistently interpreted the expression 'other measures' in connection to the term 'restriction' as a 'broad residual category',[46] including not solely a category of measures that may be considered formal quantitative restrictions – such as quotas – but also a whole variety of means, irrespective of their *de iure* or *de facto* nature,[47] insofar as they impose 'a limiting

[46] WTO Panel Report, *India – Measures Affecting the Automotive Sector*, adopted on 5 April 2002, WT/DS/146/R and WT/DS/175/R, para. 1.14.

[47] GATT Dispute Settlement Report, *Japan – Trade in Semi-Conductors*, L/6309 – 35S/116, adopted on 4 May 1988; WTO Panel Report, *Argentina – Measures Affecting the Export of*

condition or regulation' capable of inhibiting imports or exports.[48] In *Colombia – Ports of Entry*, the Panel opined that 'measures which create uncertainties and affect investment plans, restrict market access for imports or make importation prohibitively costly' also fall under the scope of Article XI:1 GATT.[49] In *China – Raw Materials*, it further added that any measures having 'the very potential to limit trade . . . constitute[s] a "restriction" within the meaning of Article XI:1 of the GATT 1994'.[50]

In light of the existing WTO jurisprudence interpreting Article XI:1 GATT, any of the above-discussed 'quantitative' measures affecting cross-border electricity flows seem likely to qualify as a 'measure' amounting to a 'restriction' on the importation or exportation of electricity within the meaning of Article XI:1 GATT. It is clear that such measures practically reduce the volume of electricity that can be 'transported' cross-border, either by controlling power flows (i.e. through phase shifters) or by interfering in the process of allocation of cross-border transmission capacity. They should thus be removed, as per Article XI GATT, unless they fall under the scope of any of the specific exceptions admitted under Article XI:2 GATT.

Based on the recurring rationales associated with quantitative restrictions on imports and exports of electricity, Article XI:2 (a) is particularly relevant, as it admits '*export* prohibitions or restrictions temporarily applied to prevent or relieve critical shortages of foodstuffs or other products essential to the exporting contracting party'. The language of Article XI:2 (a) allows ad hoc export bans and restrictions which address state-of-emergency situations putting at risk the security of supply, such as in the case of the winter crisis affecting the Balkan region in 2012. However, import bans and/or restrictions do not fall under its scope. This situation reflects the 'general' nature of GATT provisions, and fails

Bovine Hides and the Import of Finished Leather, WT/DS155/R, adopted on 16 February 2001.

48 WTO Panel Report, *India – Quantitative Restrictions on Imports of Agricultural, Textile and Industrial Products*, WT/DS90/R, adopted on 22 September 1999, para. 5.128.

49 WTO Panel Report, *Colombia – Indicative Prices and Restrictions on Ports of Entry*, adopted on 20 May 2009, WT/DS366/R, para. 7.240.

50 WTO Panel Report, *China – Measures Related to the Exportation of Various Raw Materials*, adopted on 22 February 2012, WT/DS394/R, WT/DS395/R and WT/DS398/R, para. 7.1081. For a thorough analysis of the WTO jurisprudence on Article XI:1 GATT, see I. Espa, *Export Restrictions on Critical Minerals and Metals: Testing the Adequacy of WTO Disciplines* (Cambridge: Cambridge University Press, 2015), pp. 169–75.

to take into account that, in the electricity sector, a crisis may arise from whatever imbalance there is between supply and demand (i.e. any variation occurring on the demand or supply side which affects scheduled flows). This is one of the main differences from the relevant EU electricity-specific secondary legislation (i.e. Article 42 of Directive 2009/72/EC or Article 16 of Regulation 714/2009/EC), which is drafted in such a way as apply to imports and exports of electricity equally.

2 Relevant GATT Rules on Import and Export Price Measures

Article XI:1 GATT expressly excludes 'duties, taxes or other charges ... on the importation [or] on the exportation' from its scope of application. A specific legal regime applies to import duties under Article II GATT. WTO Members are required to bind progressively lower levels of import tariffs for each product into their Schedules of Concessions through subsequent rounds of negotiations.[51] Under Article II:1 (b) GATT, in particular, WTO Members shall refrain from imposing import tariffs in excess of those set forth in the Schedules, as well as '*all other duties or charges of any kind* imposed on or in connection with the importation' (emphasis added). Article II:2 GATT, however, allows WTO Members to impose 'charges equivalent to an internal tax' insofar as the legal treatment imposed on imports is consistent with the national treatment rule under Article III:2 GATT.[52] In addition, WTO Members are also required to respect the most favoured nation (MFN) clause.[53]

No parallel legal framework is provided under the GATT for binding export tariffs. Article II:1 (a) leaves open the possibility for Members to negotiate commitments other than import tariffs 'in the appropriate Part of the appropriate Schedule annexed to this Agreement'. No WTO Member, however, has included export duty

[51] See Article XXVIII *bis* GATT.

[52] The national treatment rule provides that a Member cannot discriminate between imported goods and domestically produced like products by imposing on the former less favourable legal treatment than that borne by the latter.

[53] According to most favoured nation treatment, 'any advantage, favour, privilege or immunity granted by any contracting party to any product originating in or destined for any other country shall be accorded immediately and unconditionally to the like product originating in or destined for the territories of all other contracting parties'.

concessions on electricity in its GATT Schedule in a way similar to import tariffs. A selected number of newly acceded WTO Members have assumed country-specific obligations on the use of export duties under individual provisions of their WTO accession protocol. These export duty commitments are quite uneven in scope and coverage, but only in a handful of cases (e.g. China, Montenegro, Tajikistan) is electricity among the products for which export taxes are regulated.[54] All other WTO Members, by contrast, remain free to introduce and/or maintain export taxes on electricity based on standard GATT obligations.

On the import side, not every WTO Member has included electricity among the products subject to import duty concessions;[55] currently, however, only thirty-one countries apply tariffs on the importation of electricity, and such tariffs are generally quite low.[56]

Based on the language of Article II GATT, it is uncertain whether the import and export fees applied on cross-border electricity flows could be considered import and export 'tariffs' for the purpose of GATT regulation. On the import side, the argument could in principle be made that the fees on electricity flows may qualify as 'other charges of any kind' imposed in excess of scheduled commitments for electricity, and thus be condemned under Article II:1 (b) GATT inasmuch as feed-in fees be extended to electricity imports only. This avenue would however be available to challenge exclusively those WTO Members whose import fees on electricity result in the imposition of a charge that is in excess of what would be allowed according to Article II:1 (b) GATT. On the export side, no WTO Member would in any case breach the Article II GATT obligation.

To the extent that existing import and export fees for cross-border electricity flows are applied at the transmission level and include a network access and/or transport component, which is charged to compensate TSOs for hosting cross-border flows of electricity in their

[54] Espa, *Export Restrictions on Critical Minerals and Metals*, pp. 145–63. For the systemic implications entailed by the fragmentation of WTO disciplines on export duties, see ibid., pp. 194–208.

[55] In the Harmonised System, electricity is attributed the optional heading 2716.00 within subsection 27 ('Mineral fuels, mineral oils and products of their distillation; bituminous substances; mineral waxes') of section v ('Mineral products').

[56] IEA, *Cross-Border Trade*, p. 27.

networks, the argument could be made that these measures may fall under the scope of Article VIII GATT. According to Article VIII:1 (a), in particular,

> All fees and charges of whatever character (other than import and export duties and other than taxes within the purview of Article III) imposed by contracting parties on or in connection with importation or exportation shall be limited in amount to the approximate cost of services rendered and shall not represent an indirect protection to domestic products or a taxation of imports or exports for fiscal purposes.

The availability of Article VIII GATT to import and export fees applied to electricity flows would be important for two reasons. First, Article VIII:1 (a) prohibits fees and charges applied both to importation and to exportation. Second, Article VIII applies to all fees and charges (other than internal taxes) imposed at the border, even if the product concerned is not subject to tariff binding. Article VIII, however, refers to fees and charges levied in correspondence with a rendered 'service', which was defined by the panel as any '*government* activities closely enough connected to the processes of customs entry',[57] such as the 'consular transactions', 'statistical services' and 'analysis and inspection' indicated in the illustrative list in paragraph 4 of Article VIII.[58] It seems unlikely, therefore, that import and export network fees imposed on cross-border electricity flows at the transmission level by national TSOs could be covered by Article VIII GATT. Even if this were so, they could hardly be considered to correspond to a 'service' rendered by TSOs. Such charges are in fact also imposed on wholesale traders who are evidently neither producing nor consuming electricity in the network, and are borne by domestic *and foreign* generators and/or consumers, causing pancaking effects.

Finally, to the extent that the various components of the transmission fees imposed on cross-border electricity flows are also imposed at the domestic level on electricity consumption and/or production, it could be claimed that these charges are inconsistent with the national treatment rule under Article III GATT because of their inherently discriminatory nature. In fact, the types of fees currently applied to cross-border electricity flows treat different groups of generators and/or consumers inconsistently, ultimately making cross-border electricity flows more

[57] WTO Panel Report, *US – Customs Users Fee*, BISD 35S/245, para. 77 (emphasis added).
[58] Ibid., para. 76.

expensive – to the advantage of domestic generation or consumption – and discouraging electricity traders that do not produce or consume electricity from participating in cross-border transactions. A national treatment-based claim could however not serve to condemn export fees applied to cross-border electricity flows inasmuch as the national treatment rule applies uniquely to imports.[59]

3 Relevant GATT Exceptions

GATT-inconsistent import and export restrictions on electricity could be justified under the general exceptions recognised under Article XX GATT. Among these, the exceptions that may be relevant for electricity trade are Article XX (b), Article XX (g) and Article XX (j). The first two are the so-called environmental exceptions, and justify, respectively, 'measures necessary to protect human, animal or plant life or health' and 'measures relating to the conservation of exhaustible natural resources if such measures are made effective in conjunction with restrictions on domestic production or consumption'. The third exception allows WTO Members to derogate from GATT obligations when applying measures

> essential to the acquisition or distribution of products in general or local short supply; *Provided* that any such measures shall be consistent with the principle that all contracting parties are entitled to an equitable share of the international supply of such products, and that any such measures, which are inconsistent with the other provisions of the Agreement shall be discontinued as soon as the conditions giving rise to them have ceased to exist.

The environmental exceptions could in principle be invoked to justify the 'green' or 'CHP' components of import fees should the latter be found in breach of either Article II GATT or Article III GATT, and, in principle, those import or export restrictions implemented through the curtailment of cross-border transmission capacity, which are meant to cope with unexpected surplus or shortages caused by intermittent renewables. Article XX (j) proposes a shortage-of-supply exception available for violations of any GATT provisions and not just for Article XI:1-inconsistent measures, as in the case of Article XI:2 (a). Yet, it remains inapplicable to import restrictions on

[59] S. Defilla, 'Energy trade under the ECT and accession to the WTO', *Journal of Energy and Natural Resources Law*, 21 (2003), p. 431.

electricity owing to its express reference to situations of *short* supply. Article XX (j) therefore does not offer much help in filling the gap left by the absence of a carefully balanced exception which would adequately reflect the fact that the security and stability of the electricity system (and, thus, of the electricity supply) could equally be threatened by shortages of supply *and* congestion problems (i.e. oversupply) altering the balance between production and consumption.

Finally, similarly to the exceptions provided for in the EU context, Article XX requires that measures provisionally falling under any of the exceptions be non-discriminatory. The introductory paragraph of Article XX GATT states in fact that only measures which are 'not applied in a manner which would constitute a means of arbitrary or unjustifiable discrimination between countries where the same conditions prevail, or a disguised restriction on international trade' shall be justified despite being inconsistent with GATT obligations.

II ECT Law

The ECT contains specific trade provisions applicable to trade in energy materials and products that basically refer to GATT. According to Article 4 ECT, in particular, 'nothing in this Treaty shall derogate, as between particular Contracting Parties which are parties to the GATT, from the provisions of the GATT and Related Instruments as they are applied between those Contracting Parties'. Article 29 contains provisions applicable to trade between ECT Contracting Parties at least one of which is still a WTO Member. The ECT integrates by reference most WTO law, including the GATT provisions relevant for regulating import and export restrictions on cross-border electricity flows: Article I, Article III, Article XI and Article XX GATT.[60] It does not cover, however, trade in energy services. Accordingly, the ECT rules neither add to nor resolve any of the gaps and/or inconsistencies arising from the WTO regulation of import and export restrictions when applied to electricity.

[60] Energy Charter Secretariat, *Trade in Energy: WTO Rules Applying under the Energy Charter Treaty* (2001), www.encharter.org/fileadmin/user_upload/document/WTO_Rules_applying_to_the_ECT_-_2002_-_ENG.pdf (accessed 15 March 2015).

E Conclusions

Cross-border electricity flows are still subject to a wide range of restrictions on the importation and/or exportation. While some of them are discriminatory in nature and respond to political economic rationales running counter to the spirit of increased integration between electricity markets (e.g. import or export fees), other types of restrictions may be necessary to preserve the security and stability of supply across the (interconnected) national networks when faced with congestion problems (e.g. import restrictions implemented through phase shifters, cross-border transmission capacity curtailments or additional charges imposed on auctions) or to restore the production–consumption balance during emergency and *force majeure* situations threatening the integrity of the system (e.g. export bans or restrictions, limiting either physical flows or interconnection capabilities).

This chapter has shown that the relevant disciplines applicable to import and export restrictions on electricity are significantly more developed in the EU context than in the WTO and ECT legal frameworks. This situation is a direct consequence of the advanced stage of implementation of the EU internal electricity market, which was made possible through the entry into force of electricity-specific EU secondary legislation targeting potential disturbances in cross-border electricity trade, on both the physical and the commercial sides.

The WTO legal framework and the Energy Charter Treaty, by contrast, do not specifically address trade in electricity, but generally regulate, respectively, trade in energy goods and services and trade in energy materials and products. This situation has direct implications for the relevance and adequacy of existing international rules applicable to import and export restrictions on electricity. On the one hand, all existing quantitative restrictions applied on the importation or the exportation of electricity are outlawed by GATT, but only those export restrictions applied to prevent or relieve critical electricity shortages could seek justification under GATT-relevant exceptions. On the other, the disciplines applicable to price measures such as the import and export fees imposed on cross-border electricity flows are overall deficient and, in any case, uniquely regulate those measures applied on the import side. In the electricity sector, however, quantitative forms of restrictions may be needed to manage the system in accordance with safety standards for secure network operation and under emergency situations, while discriminatory transmission fees applied to cross-border transactions serve no purpose for the electricity system and ultimately decrease overall welfare, to the advantage of selected groups of domestic generators or consumers.

The analysis of the electricity-specific mechanisms, procedures and rationales subsumed under the use of import and export restrictions suggests that relevant but generally applicable international rules such as those available in the WTO and ECT legal frameworks may not be well equipped to enable, facilitate and/or encourage long-distance electricity trade. With a view to promoting inter-regional trade in electricity, international disciplines on import and export restrictions should incorporate provisions that take into account the peculiarities of cross-border electricity trade and transmission, similarly to the EU example. From an electricity-specific perspective, aspects requiring further attention include the deficiency of GATT disciplines on export duties and other charges, the need for specific flexibilities allowing for congestion management and state-of-emergency responses, and the limited reach of GATS disciplines for regulating the effects of import and export restrictions on wholesale electricity traders. While existing provisions under either the WTO or the ECT legal frameworks could in principle be expanded and/or integrated to this end, a first-best solution would consist of including electricity-specific provisions regulating the use of import and export restrictions as one of the building blocks of a comprehensive sectoral Agreement on Electricity.

Bibliography

Daxhelet, O. and Smeers, Y. (2005). *Inter-TSO Compensation Mechanism*, 7 November, www.hks.harvard.edu/hepg/Papers/Daxlet_Smeers_11.07.05.pdf (accessed 15 March 2015).

Defilla, S. (2003). Energy trade under the ECT and accession to the WTO, *Journal of Energy and Natural Resources Law*, **21**, 428.

Duchamp, M. (2011). Poland plans to ban German solar and wind energy imports, 6 December, http://motls.blogspot.ch/2011/12/poland-plans-to-ban-german-solar-and.html (accessed 15 March 2015).

Energy Charter Secretariat. (2001). *Trade in Energy: WTO Rules Applying under the Energy Charter Treaty*, www.encharter.org/fileadmin/user_upload/document/WTO_Rules_applying_to_the_ECT_-_2002_-_ENG.pdf (accessed 15 March 2015).

ENTSO-E. (2013). *Overview of Transmission Tariffs in Europe: Synthesis 2013*, www.entsoe.eu/fileadmin/user_upload/_library/Market/Transmission_Tariffs/Synthesis_2013_FINAL_04072013.pdf (accessed 15 March 2015).

European Commission. (2010). *Free Movement of Goods: Guide to the Application of Treaty Provision Governing the Free Movement of Goods*, http://ec.europa.eu/enterprise/policies/single-market-goods/files/goods/docs/art34-36/new_guide_en.pdf (accessed 15 March 2015).

Garbers, M. (2013). *Why Export and Import Fees Are Illegal*. Presentation given at the EFET Workshop on Wholesale Power Trading in the CEE and SEE regions, 9 July 2013.

IEA. (2013). *Cross-border Trade in Electricity and the Development of Renewable-Based Electric Power: Lessons from Europe*, www.oecd-ilibrary.org/trade/cross-border-trade-in-electricity-and-the-development-of-renewables-based-electric-power_5k4869cdwnzr-en (accessed 15 March 2015).

Kling, W. L., Klaar, D. A. M. and Schuld, G. H. (2003). Phase shifting transformers installed in the Netherlands in order to increase available international transmission capacity, *Cigré C2–207 Working Paper*, www.transform.ru/articles/pdf/sigre/c2-207.pdf (accessed 15 March 2015).

Merlin, A. (2005). *European Power Market Design*, keynote speech given at the San Antonio-CIGRE-IEEE Symposium, 5–7 October 2005.

Parkinson, G. (2013). Poland builds electronic wall to keep out German renewables, 18 November 2013, http://reneweconomy.com.au/2013/poland-builds-electronic-wall-to-keep-out-german-renewables-72084 (accessed 15 March 2015).

Peltegova, I. (2013). Bulgaria reduces electricity export tariff from 1 August, 30 July, www.icis.com/resources/news/2013/07/30/9692393/bulgaria-reduces-electricity-export-tariff-from-1-august/ (accessed 15 March 2015).

Popov, P. (2013). *Case Study: Bulgaria*. Presentation given at the EFET Workshop on Wholesale Power Trading in the CEE and SEE Regions, 9 July.

Selivanova, Y. (2012). *Emergency Situation in South Eastern Europe in Winter 2012*. Presentation given at the Energy Charter Secretariat, Brussels, 23 May.

Shahidehpour, M. and Alomoush, M. (2001). *Restructured Electrical Power Systems: Operation: Trading, and Volatility*. Basel: CRC Press.

Siri, A. (2013). *Adverse Effects on Cross Border Trading Across Greek Interconnections*. Presentation given at the EFET Workshop on Wholesale Power Trading in the CEE and SEE Regions, Brussels, 9 July.

Udubasceanu, S. (2014). Doubt surrounds scrapping of Romanian export fee, 30 June, www.icis.com/heren/articles/2014/06/11/9790683/power/edem/romania-scraps-tariff-to-export-electricity.html (accessed 15 March 2015).

Urbanke, C. (2013). *Power Market Issues in Romania*. Presentation given at the EFET Workshop on Wholesale Power Trading in the CEE and SEE Regions, Brussels, 9 July.

VDE. (2014). *The 50.2 Hz Problem*, www.vde.com/en/fnn/Pages/50-2-hz.aspx (accessed 15 March 2015).

14

The Integration of Electricity from Renewable Energy Sources in the European Union Electricity Market: The Case for 'Smart Grids'

JOËLLE DE SÉPIBUS

A Introduction

The power sector is set to play a central role in the low-carbon economy. Analysis shows that electricity could be almost carbon dioxide (CO_2)-free by 2050, and such low-carbon electricity promises to partially replace the fossil fuels used in transport and heating.[1] In all the decarbonisation scenarios of the European Union (EU), renewable energy sources (RES) feature as a crucial part of the solution.[2]

Renewable energy resources are plentiful all over Europe and their share in the European energy mix of electricity production is constantly increasing. The current growth path, however, is increasingly challenged by insufficient grid capacity. Grids constitute major bottlenecks for the expansion of RES, with many transmission lines, especially cross-border interconnections, suffering from congestion and overload. Moreover, the intermittent availability of many renewables threatens the stability of the grids.

Despite the policies set in place by the EU under its 'energy and climate package', the annual growth of RES hence may not be sustained in the mid- to long term.[3] Recognising the need to upgrade and modernise its grids, the EU has called for the creation of a 'Smart Supergrid' that interconnects European grids at the continental level and makes them 'intelligent' through the addition of information and communication technology (ICT).[4]

1 European Commission, COM (2011) 112, final version.
2 European Commission, COM (2011b) 885, final version.
3 European Commission, COM (2012) 271, final version.
4 European Commission, COM (2011) 658, final version.

To implement this agenda, the EU has taken a leading role in coordinating research efforts and creating a common legislative framework for the necessary modernisation of Europe's grids within the context of a liberalised electricity market. This chapter is intended to provide both an overview and a critical appraisal of the measures taken so far by the EU to 'transform' the grids into the backbone of a decarbonised electricity system. It suggests that if competition is to play a significant role in the deployment of Smart Grids, the current regulatory paradigm will have to be reassessed.

The chapter is organised as follows. Section B sets out the case for the creation of a Smart Supergrid. In Section C, both the research agenda and the legislative framework on Smart Grids are critically assessed. Section D examines two of the most salient issues: the creation of common standards and regulatory incentives for the implementation of Smart Grids. The chapter ends with the outlook for a possible paradigm shift in electricity regulation.

B Electricity from Renewable Energy Sources in the European Union

I The Case for Renewable Energies in a Liberalised European Electricity Market

The integration of electricity from RES in the EU is taking place in a market that has been progressively liberalised since 1996.[5] In the fledgling stages of the development of its liberalisation agenda, the European legislator did not pay much attention to the deployment of RES. Clearly, concerns regarding economic efficiency and security of supply were the main drivers of the process.

With the adoption of the first European Climate Change Programme (ECCP) in 2000, a paradigm change was introduced. The latter was reinforced in 2007 with the adoption of the '20/20/20' target and the implementation of the so-called climate and energy package,[6] which

5 A. Johnston and G. Block, *EU Energy Law* (Oxford: Oxford University Press, 2012).

6 The climate and energy package is a set of legislative acts which aims to ensure that the EU meets its climate and energy targets for 2020. These targets, known as the '20-20-20' targets, set three key objectives: a 20 per cent reduction in EU greenhouse gas emissions from 1990 levels; raising the share of the EU's energy consumption that is produced from renewable resources to 20 per cent; and a 20 per cent improvement in the EU's energy efficiency. See http://ec.europa.eu/clima/policies/strategies/2020/ (accessed 25 November 2016).

led in particular to the revision of the internal market Directive (hereafter 'E-Directive').[7]

Since then, the promotion of RES has represented a major building block of the 2020 strategy for smart, sustainable and inclusive growth,[8] and it is one of the principal pillars of the Roadmap towards achieving a decarbonised economy by 2050.[9] In 2014 a new target was set for renewable energies: namely, that their share in overall energy consumption should reach at least 27 per cent by 2030.[10]

Despite the positive trends over recent years, it is unlikely that the planned increase in the share of RES will be achieved.[11] In particular, grids will have to be modernised and reinforced. Anchoring its approach in the European 2020 strategy for smart, sustainable and inclusive growth, the EU has thus recognised the need for a step change in the way grids are planned, constructed and operated[12] and has called for an upgrade of the networks with the aim of setting up a European Smart Supergrid.

This concept encompasses two different approaches.[13] The first aims at reinforcing cross-border transmission lines for the transport of electricity over long distances and is commonly known as the Supergrid. This is particularly important for the integration of large offshore power stations, which are generally located far away from areas of high demand. The second responds to the necessity to equip distribution grids with smart information technologies so that volatile supplies of intermittent energy sources, such as wind and photovoltaic systems, can be matched with demand. This approach is commonly referred to as the Smart Grid.

[7] Directive 2009/72/EC, *Official Journal* of 14 August 2009, L 211, 55–93.

[8] European Commission, COM (2010) 2020, final version.

[9] On 4 February 2011, the European Council stressed its commitment to achieving the long-term target of an 80–95 per cent reduction in domestic production of greenhouse gases and recognised that this requires nothing less than 'a revolution in our energy systems'. One month later, on 8 March 2011, the European Commission reiterated this communication in its *Roadmap for moving to a competitive low carbon economy in 2050*. See European Commission, COM (2011) 112, final version.

[10] See http://ec.europa.eu/clima/policies/2030/index_en.htm (accessed 25 November 2015).

[11] European Commission, COM (2012) 163, final version.

[12] See for instance the communication on Energy 2020 and its blueprint for an integrated European energy network. See also European Commission, COM (2010) 639, final version; COM (2010) 677, final version; European Commission, COM (2011) 658, final version.

[13] A. Battaglini, J. Lilliestam and G. Knies, 'The SuperSmart Grid – paving the way for a completely renewable power system' in H. J. Schellnhuber, M. Molina, N. Stern, V. Huber and S. Kadner (eds.), *Global Sustainability: A Nobel Cause* (Cambridge, MA: Cambridge University Press, 2010), p. 291.

Although realisation of the Smart Grid could to a certain extent reduce the necessity to build the Supergrid, the two approaches are by no means mutually exclusive. Combined intelligently, a Smart Supergrid could contribute to speeding up the decarbonisation of the European electricity system.[14] This is supported by economic arguments as well as the need to ensure security of supply.[15]

1 The Supergrid

With liberalisation, the development of the grids underwent a profound change. Decisions regarding the modernisation, reinforcement and expansion of grids were no longer controlled by generators, but became the responsibility of grid operators.[16] Accordingly, optimal coordination between generation and development became more difficult and new forms of cooperation had to be found to guarantee that future investments in generation were matched by investments in transmission, and vice versa.

When the process of liberalisation first began, the European legislator paid little attention to the effects the process had on grids. Over time, it became clear that to guarantee their optimal development, grid operators would have to be controlled by independent regulators that have at their disposal a minimal set of competences – in particular, the right to fix grid tariffs and to monitor interconnection capacity.[17]

To ensure a sufficient European focus, regulators were further required to cooperate with each other in the framework of the European Regulators' Group for Electricity and Gas (ERGEG), the energy regulators' advisory group to the European Commission. Moreover, the first Cross-Border Regulation was adopted to ensure the construction of new interconnection capacity and the development of fair criteria for allocating existing cross-border capacity.[18] A further step in the completion of this legislative framework was the Security of Supply Directive,[19] which invites Member States to define security standards for grids and seeks to increase interconnections between countries to enable effective competition.[20] Finally, guidelines for trans-European

[14] Ibid. [15] Ibid. [16] Directive 2009/72/EC, Article 10 (2) c.
[17] Directive 2009/72/EC, Article 23.
[18] Regulation (EC) No 1228/2003, *Official Journal* of 15 July 2003, L 176, 1.
[19] Directive 2005/89/EC, *Official Journal* of 4 February 2006, L 33, 22.
[20] European Commission, COM (2003) 740, final.

energy networks (TEN-E) were adopted to define projects which are eligible for financial assistance from the EU.[21]

Despite these improvements in coordination and support, investments in interconnection capacity have remained slow. Long and uncertain authorisation procedures delay the implementation of cross-border infrastructure projects and vertically integrated utilities continue to show little interest in developing new interconnection capacity,[22] as increased cross-border flows lead to more competitive pressure and may reduce the profitability of electricity generation in their 'home' markets.

To address these shortcomings, the European legislator has decided to step up European cooperation. Accordingly, the third energy package lays down the basic elements for a more European focus in grid planning. The revised Cross-Border Regulation provides new rules on cross-border compensation and investments under the guidance of the newly created Agency for the Cooperation of Energy Regulators (ACER).[23] Furthermore, transmission system operators (TSOs) have been required to elaborate regional and European ten-year grid development plans within the framework of the European Network of Transmission System Operators for Electricity (ENTSO).

This legal framework has been further completed by a thorough overhaul of the guidelines for trans-European Energy Networks (TEN-E).[24] In the new regulation on guidelines for trans-European energy infrastructure, the European Commission sets out how interconnections can be improved and the grids adapted to meet the new needs of a progressively decarbonised electricity system, addressing in particular authorisation and financing issues. Stressing the need to modernise as well as upgrade the grids to meet increasing demand, foster market integration and maintain existing levels of system security, it proposes the creation of so-called electricity highways that can accommodate increasing renewables-based electricity generation, connect these 'new generation hubs' with major storage capacities and consumption centres and cope with increasingly flexible and decentralised electricity demand and supply.[25]

[21] The first guidelines were set out in 1996. They were revised several times, in particular in 2003, 2006 and 2013. See Decision 1364/2006/EC, *Official Journal* of 22 September 2006, L 262, 1–23 and Regulation (EC) No 347/2013, *Official Journal* of 25 of April 2013, 39–75.

[22] European Commission, COM (2010) 677, final version.

[23] Ibid.

[24] Regulation (EC) No 347/2013, *Official Journal* of 25 April 2013, L 115, 39–75.

[25] Ibid.

Building on the first ten-year network development plan, strategic trans-European energy infrastructure corridors and areas, as well as new rules for the identification of projects of common interest (PCIs), are to be set up.[26] Financial aid to carry out the selected projects will come from a new institution, the Connecting Europe Facility. The picture is completed by a methodology and process for the elaboration of a harmonised cost–benefit analysis for PCIs.

2 The Smart Grid

Until recently, grids predominantly transported power in one direction: from large power stations, via the transmission and distribution systems, to the final consumers. In this situation, distribution grid operators assumed an essentially passive role, ensuring the delivery of unidirectional power flows to end-users. With the emergence of new technologies such as Flexible Alternating Current Transmission Systems (FACTS), the situation has changed: grid operators are now able to accommodate bi-directional power flows and can accept injections of power by distributed generators,[27] as well as taking into account demand response from consumers. This situation, however, entails a fundamental modification of their behaviour: grid operators will have to become far more active than they were in the past and will have to make distribution grids 'smart'.

What exactly is meant by 'smartening' grids defies a clear understanding. While transmission grids, controlled on the basis of reliable data, are already relatively 'intelligent', distribution grids are mostly 'as dumb as dumb can be'.[28] An element common to most definitions is that electricity grids become 'intelligent' through the addition of two-way digital communication technology to devices associated with the grid. Each device on the grid can be given sensors, managed by automation technology, to gather data, thus allowing the management of data flows from a central location. As a result, the adjustment and control of millions of devices becomes possible and the actions of all users connected to the grid – generators as well as consumers – can be integrated intelligently.[29]

[26] European Commission, COM (2013) 711, final version.

[27] Distributed generation stands for electricity that is fed in directly to the distribution grid.

[28] Distribution grids are medium and low-voltage grids. The quotation is from M. Kurth, 'Smart metering, smart grids, smart market design' in E. M. Noam, L. M. Pupillo and J. J. Kranz (eds.), *Broadband Networks, Smart Grids and Climate Change* (New York, NY: Springer, 2013), p. 11.

[29] See US Department of Energy, http://energy.gov/oe/technology-development/smart-grid (accessed 25 November 2015).

Several types of technology are generally distinguished. In upstream (generator) or downstream (consumer) markets, smart meters, measuring output or consumption in real time, play a crucial role.[30] For grid operators, communications instruments transmitting data on the status of the grid in real time constitute the principal assets operators can use to optimise the management of demand, decentralised production resources, storage and possibly also fleets of plug-in electric vehicles.[31]

Smart grid technologies offer a wide range of possibilities. Each country has its own view as to which market segment would gain most from Smart Grids.[32] Among the benefits most often stressed are the transformation of consumers into active players, the reduction of peak loads to avoid the necessity for grid reinforcements, the integration of renewables and energy storage, the promotion of innovation, the reduction and anticipation of outages and the development and storage of data flows.[33]

While the potential to improve the whole system is clear, its benefits will not accrue to all players to the same extent, and some may lose out. While Smart Grids may help consumers reduce their electricity bills, and generators may be able to adjust production to meet demand, grid operators can optimise traffic in a more economical manner and would have at their disposal more effective instruments to balance supply and demand, in particular thanks to selective load-shedding at peak hours.[34] The deployment of Smart Grid communications resources will also assist new players (aggregators or managers of intermittent generation resources) in devising new strategies in a reconfigured market.[35]

The number of applications that can be used in connection with the Smart Grid once the communications technology is deployed is growing fast. Also, whereas many technologies will remain invisible to the consumer, smart meters, which are devices that record the consumption of electricity, will have to be installed on the consumer's premises. They represent only the first step, however.[36] Eventually they will communicate with smart thermostats and other appliances, giving people a clearer view of their patterns of consumption of electricity.

Smart grid technologies are crucial for the deployment of large amounts of RES. As increasing quantities of intermittent RES imply a

[30] C. Clastres, 'Smart grids: another step towards competition, energy security and climate change objectives', *Energy Policy*, 39(9) (2011), 4.

[31] Ibid., 4. [32] Ibid., 6. [33] Ibid., 6. [34] Ibid., 8. [35] Ibid., 32.

[36] *The Economist*, 'Building the smart grid', www.economist.com/node/13725843 (accessed 25 November 2015).

greater need to manage variability and uncertainty, more flexibility is required both from the generation mix and from demand. Moreover, storage can help regulate the balance between supply and demand.[37] All this implies that advanced communications technologies, combined with smart electricity meters linked to control centres, should be installed and that generators and consumers should be informed in real time and given the appropriate incentives to adapt their behaviour. Furthermore, improved operational methods are necessary to determine the reserve required both to maintain the demand–generation balance and to ensure optimal generation scheduling.[38]

C Smart Grids in the European Union: From Innovation to Deployment

While the benefits of Smart Grids are widely recognised, their implementation prompts many questions. Unlike the United States, which engaged in legislative development efforts at an early stage, the EU initially mostly supported research and fostered government-backed deployment of Smart Grids.[39] Since 2009, however, it has paid increased attention to standardisation, privacy and market design issues. The following sections provide a brief overview of the EU policy framework and discuss some of the most salient issues.

I The European Research Agenda

EU interest in Smart Grids dates back to 2004, when the first international conference on the integration of renewable energy sources and distributed energy resources was held in Brussels. In 2005, spurred on by industrial stakeholders and the research community, the European Commission set up the European Technology Platform for the Electricity Networks of the Future (hereafter Smartgrids Platform), aiming at the creation of a joint vision and a research agenda for the deployment of

[37] IPCC, *Renewable Energy Sources and Climate Change Mitigation: Special Report of the Intergovernmental Panel on Climate Change* (Cambridge, MA: Cambridge University Press, 2012), 109.

[38] Ibid.

[39] Z. Zhang, 'Smart Grid in America and Europe: similar desires, different approaches (Part 1)', *Public Utilities Fortnightly*, 149(1) (2011), 47.

Smart Grids.[40] Representing a wide variety of stakeholders, this platform has produced a series of reports on how grids can become smart, in particular by defining a vision and setting out a strategic agenda for research up to 2035.[41]

With the adoption of the SET Plan in 2009, which was a response to declining R&D budgets for energy and the EU's desire to position itself at the forefront of research and innovation in low-carbon technologies, the EU has moreover created a valuable tool to foster research into low-carbon energy.[42] Based on the premise that the climate challenge may only be addressed if the entire energy system is transformed, with far-reaching implications for how energy is produced, transported, traded and used, the EU has decided to establish so-called 'European Industrial Initiatives' (EIIs).[43] The aims of the EIIs are to strengthen industrial participation in energy research and demonstration projects, boost innovation and accelerate the deployment of low-carbon energy technologies. One of the EIIs that has been created is the European Electricity Grid Initiative (EEGI), the scope of which includes promoting the establishment of electricity systems that will be able to integrate up to 35 per cent of electricity from distributed RES by 2020 and work towards completely decarbonised electricity production by 2050.

With the SET Plan, a novel kind of political process has been created, which combines top-down and bottom-up approaches. On the one hand, this encourages a broad variety of stakeholders to form 'platforms' to develop responses to new challenges and to act both as 'incubators' and lobbyists; on the other, it ensures an appropriate policy response. While participation in the platforms is usually open to all interested stakeholders, the political process is guided by a Steering Group, composed of a group of representatives from EU Member States, chaired by the European Commission. This framework is further complemented by the European Energy Research Alliance (EERA), which fosters the

[40] See European Distribution System Operators' Association for Smart Grids, www.edsoforsmartgrids.eu/index.php?page=smart-grids-european-technology-platform (accessed 25 November 2015).

[41] See European Technology Platform for the Electricity Networks of the Future, www.smartgrids.eu/ETP%20Documents (accessed 25 November 2015).

[42] The EEGI has adopted an implementation plan for the upgrade of networks, particularly at distribution level, and stressed the need for close collaboration between distribution and transmission operators. See European Commission, COM (2011) 202, final version.

[43] See SET-Plan, http://setis.ec.europa.eu/about-setis/set-plan-governance (accessed 25 November 2015).

cooperation of leading European research institutes, and SETIS, the Strategic Energy Technologies Information System, which monitors the progress of the SET Plan's implementation.

II The European Legislative Agenda

Despite strong support for research activities, the implementation of the Smart Grid did not accelerate as expected. Among the reasons identified for this delay were the prevailing consumer uncertainties regarding acceptance of the new technologies, the lack of clarity regarding future business models, the scale of the necessary investments and, to some extent, the technologies that would be needed for its deployment.[44]

While the EU initially adopted a relatively hands-off approach to legislative issues, this changed over time. To improve coordination and to speed up the legislative process at the European level, the European Commission set up a Smart Grid Task Force (SGTF), which is composed of a Steering Committee and four expert groups. Its role is to advise the Commission on the implementation of the Smart Grid. More specifically, it was tasked to 'produce a set of regulatory recommendations' to ensure the 'consistent, cost-effective, efficient and fair implementation of Smart Grids' throughout the EU.

The work of the SGTF has focused on several cross-cutting issues concerning, inter alia, the roles of the various stakeholders, standards, privacy and security challenges and infrastructure deployment.[45] Based on the outcomes of the SGTF's work, the European Commission issued a Communication in 2011 entitled 'Smartgrids: from innovation to deployment'. In this document, the European Commission identified four policy areas which require European cooperation: first, developing common European standards for Smart Grids; second, addressing data privacy and security issues; third, regulatory incentives for deployment of Smart Grids; and fourth, continuous support for innovation and its rapid application.

The following sections discuss two issues in more detail: the development of common standards, and regulatory incentives for deployment of Smart Grids.

[44] See Smart Grid Task Force, https://ec.europa.eu/energy/en/topics/markets-and-consumers/smart grids and meters/smart grids task force (accessed 25 November 2015).

[45] B. Covassi da Encarnaçao, 'A smart future? The EU digital agenda between broadband, the grid and energy efficiency' in E. M. Noam, L. M. Pupillo and J. J. Kranz (eds.), *Broadband Networks, Smart Grids and Climate Change* (New York, NY: Springer 2013), p. 96.

1 Developing European Smart Grid Standards

The deployment of Smart Grids requires the definition of hundreds of new standards describing the technical requirements of the devices associated with the grids and their interactions. The need to define so-called interoperability standards, specifying how actors and components throughout the Smart Grid interact with each other, was recognised at an early stage in the United States.[46] Although the country has a strong tradition of 'bottom-up' private sector standard-setting processes,[47] it acknowledged that the enormous scope and complexity of the Smart Grid would make federal involvement indispensable.[48]

A federally coordinated process thus started as early as 2007 under the leadership of the National Institute of Standards and Technology (NIST), which was given the 'primary responsibility to coordinate the development of a framework . . . to achieve interoperability of Smart Grid devices and systems'.[49] It works under the guidance of the Smart Grid Interoperability Panel (SGIP)[50] and is assisted by the Federal Energy Regulatory Commission (FERC), whose role is to adopt the standards established by the NIST if this is required to ensure the interoperability of the federal electricity system.[51] The scope of the standard-setting process is wide and encompasses, in particular, the creation of a common information framework.[52] So far, the FERC has refused to adopt mandatory standards at the federal level, essentially leaving the regulatory role to the states.

The EU's involvement in the standard-setting effort started much later than was the case in the United States, with the first mandate being given to the European standardisation organisations (ESOs) in 2009, for the interoperability of smart utility meters, and another in 2010, for chargers for electric vehicles.[53] Acknowledging the necessity for a broader range of standards, the European Commission eventually issued a mandate in 2011 to develop standards 'facilitating the implementation of high-level

[46] Zhang, 'Smart Grid in America and Europe', 47.
[47] J. B. Eisen, 'Smart regulation and federalism for the Smart Grid', *Harvard Environmental Law Review*, (2013) 129.
[48] Ibid., 132 ff.
[49] See Office of the National Coordinator for Smart Grid Interoperability, www.nist.gov/public_affairs/releases/upload/smartgrid_interoperability_final.pdf (accessed 25 November 2015).
[50] Eisen, 'Smart regulation and federalism for the Smart Grid', 138.
[51] Ibid., 137.
[52] Zhang, 'Smart Grid in America and Europe', 49.
[53] European Commission, COM (2012) 202, final version.

Smart Grid services and functionalities'.[54] It announced at the same time that if progress was insufficient, it would consider defining a 'network code'.[55]

Building on the work of NIST, the European standardisation process has since gained considerable speed.[56] In 2012, the Smart Grid Coordination Group (SG-CG) set up by the ESOS delivered reports on the main elements of the mandate: a technical reference architecture, the first set of standards supporting the exchange of information and the integration of all operators, sustainable standardisation processes and standards on information security and data privacy. It also produced a Framework Document, which provides an overview of all the activities related to the deployment of Smart Grids and describes how the different elements fit together.[57] The first stock-taking conference took place in January 2013.[58]

The methodology applied by the SG-CG for the standard-setting process is based on so-called high-level 'use cases', which represent narrative descriptions of how actors will interact.[59] Applying the principles of system engineering to standardisation, this methodology departs considerably from the way in which standardisation processes have so far been organised and will constitute the backbone of all the future standardisation work within the ambit of Smart Grids.

With this standardisation effort, the EU has set an important milestone for the future deployment of Smart Grids. Interaction between the thousands of devices and actors will, however, not be possible without the creation of common interoperability standards. Standards do more than make interaction possible, though: they promote investments in

[54] For a definition of these functionalities see European Commission, SEC (2011) 463, final version.

[55] European Commission, COM (2012) 202, final version. European network codes are sets of rules whose aim is to provide effective access to the transmission grids across borders and to ensure the coordinated evolution of the transmission system. The European network codes are not intended to replace the necessary national network codes for non-cross-border issues.

[56] See the summary made by Commissioner Oettinger at the Conference in January 2013, http://ec.europa.eu/energy/gas_electricity/events/20130128_smartgrids_standardization_en.htm (accessed 25 November 2015).

[57] See European Telecommunications Standards Institute, www.etsi.org/news-events/news/646-2013-01-smart-grid-results (accessed 25 November 2015).

[58] See, for a full account of the Conference, http://ec.europa.eu/energy/gas_electricity/events/20130128_smartgrids_standardization_en.htm (accessed 25 November 2015).

[59] See Eisen, 'Smart regulation and federalism for the Smart Grid', 127.

technology, eliminate market obstacles and foster competition.[60] If it is true that a standardisation process also entails some risk of suppressing innovation, this risk is probably minor compared to the consequences of building Smart Grids without foundational standards.[61] Also, if designed properly, standards may well allow for future upgrades, ensuring their fluid evolution.

There is not yet enough experience to determine whether an expanded role for the EU in the development of standards is advisable. Following the example of the Internet, a flexible approach would probably yield better results than dictating mandatory standards at an early stage.[62] The risk of balkanisation of standards will however have to be closely monitored to allow the establishment of network codes in the field of Smart Grids, if need be.

2 Regulatory Incentives for Smart Grid Deployment

Prior to 2009, the EU had not given much guidance to Member States regarding the implementation of Smart Grids. This changed with the adoption of the revised RES Directive[63] (RES Directive) and a new Energy Efficiency Directive (EE Directive),[64] which set framework conditions for structural changes to the grids to ensure a larger uptake of renewables and to provide incentives for energy efficiency.

The Uptake of Large Quantities of Renewable Energy Sources by the Smart Grid According to the RES Directive, Member States are required to take all appropriate steps to develop transmission and distribution grid infrastructure, intelligent networks and storage facilities to absorb the expected growth in RES.[65] Particular emphasis is placed on the acceleration of authorisation procedures and better coordination between grid infrastructure approval, planning and administrative procedures.[66] These issues are crucial, as delays resulting from inefficient procedures hamper investments in RES.

Member States must further ensure that priority is given to RES when dispatching electricity 'in so far as the secure operation of the national electricity system permits' it.[67] To minimise possible curtailments, they

[60] Ibid., 124. [61] Ibid. [62] Ibid., 155.
[63] Directive 2009/23/EC, *Official Journal* of 16 May 2009, L 122, 6.
[64] Directive 2012/27/EU, *Official Journal* of 14 November 2012, L 315, 1.
[65] Directive 2009/23/EC, Recital 61, Article 16. [66] Ibid., Article 16 para. 1.
[67] Directive 2009/23/EC, Article 16 para. 2 (c).

have to ensure that 'appropriate grid and market-related operational measures are taken' and should require system operators to take corrective measures 'to prevent inappropriate curtailments'.

When it comes to connecting RES to the grid, the allocation of related costs is of the utmost importance.[68] So-called connection charges may either cover 'shallow costs', that is, the costs related to the single connection, or 'deep costs', which, in addition to connection costs, include costs related to the reinforcements of the core grid.[69] Whereas connection charges are generally a minor cost factor for large utilities, as reinforcement costs related to the transmission network are generally shared by all users of the network, this is not the case for distributed generation, for which connection charges may represent a significant cost factor.

While the RES Directive provides considerable leeway to Member States to choose how they want to allocate the connection costs for RES producers,[70] they must ensure that rules 'relating to the bearing and sharing of costs of technical adaptations, such as grid connections and grid reinforcements' are published and are based on objective, transparent and non-discriminatory criteria 'taking particular account of the costs and benefits associated with the connection of those producers to the grid'.[71] RES producers must also be appropriately informed and may, if Member States allow it, organise a call for tender for the connection work.[72]

In addition to connection charges, other fees, in particular those covering costs of transport and ancillary services, influence the uptake of RES. Although in most countries most of these costs are borne by consumers, a few countries allocate a non-negligible component of such tariffs to generators.[73] To avoid these rules placing a disproportionate burden on RES, the Directive posits that the overall charging of transmission and distribution tariffs may not discriminate against RES,[74] and

[68] See, for a thorough discussion of transmission costs and tariffs, THINK, *EU Involvement in Electricity and Natural Gas Transmission Grid Tarification*, Final Report (2012), www.eui.eu/Projects/THINK/Documents/Thinktopic/ThinkTopic6.pdf (accessed 25 November 2015).
[69] THINK, 'EU involvement', 22.
[70] According to the Directive, Member States may, but are not obliged to, pass connection costs on to the system operators. See Directive 2009/23/EC, Article 16 para. 4.
[71] Directive 2009/23/EC, Article 16 paras. 3 and 6.
[72] Directive 2009/23/EC, Article 16 para. 5.
[73] See, for a thorough discussion of these issues, THINK, *EU Involvement*, 21.
[74] Directive 2009/23/EC, Article 16 para. 7.

that tariffs have to 'reflect realisable cost benefits resulting from the plant's connection to the network, in particular benefits from the direct use of the low-voltage grid'.[75] Accordingly, Member States must make sure that grid tariffs reflect the benefits, such as a reduced need for grid reinforcements, which distributed RES provide to the grid.

The Roll-out of 'Smart Meters' 'Smart meters' are electronic devices that can measure the consumption of energy and provide bi-directional communication between the consumer and supplier or operator. Unlike traditional meters, which have to be read physically at their point of installation, smart meters make it possible to read and record energy consumption frequently. Among other advantages, they enable 'demand response' and 'time-of-use' pricing for electricity and the introduction of 'smart' household devices.

Smart meters are considered a prerequisite for the deployment of Smart Grids. The E-Directive has set out new requirements for the roll-out of smart metering systems.[76] Member States are required to install them by 2020 for at least 80 per cent of their customers where such a roll-out is assessed positively regarding its long-term costs and benefits. To assist Member States in carrying out these assessments, the European Commission has issued a recommendation in which it provides advice with respect to three key elements: data protection and security considerations, a methodology for the economic assessment of the long-term costs and benefits and minimum functional requirements.[77]

Despite the legislative framework set in place by the EU, numerous customers have not yet received appropriate information on how to regulate their energy consumption.[78] While not requiring a mandatory roll-out of smart meters, the Directive lays down new requirements aimed at improving smart metering systems' contribution to energy efficiency.[79] More specifically, the EE Directive requires that metering systems provide customers with information on actual time of use and measure the electricity exported to the grid from a consumer's premises. Furthermore, if no meters are installed, Member States must ensure that

[75] Directive 2009/23/EC, Article 16 para. 8.

[76] Directive 2009/72/EC, Article 3 para. 11; European Commission, COM(2014) 356, final version.

[77] European Commission, Recommendation 2012/148/EU, *Official Journal* of 13 March 2012, L 73, 9–22.

[78] Directive 2012/27/EU, Recital 32.

[79] Directive 2012/27/EU, Articles 9–11.

billing information is appropriate and based on actual use.[80] Finally, the Directive mandates that both billing information and access to consumption data for final customers are supplied free of charge.[81]

The Promotion of Energy Efficiency Another focus of the EE Directive is the reduction of so-called grid losses[82] related to the design and/or operation of the grids.[83] This objective may be achieved both through the replacement of equipment and through modification of the grid architecture and its management.[84] For instance, grid losses can be reduced if electricity is consumed locally. Accordingly, well-located RES can help increase the energy efficiency of the grids. This is true, however, only if the production can be adjusted to meet the demand, that is, if the production is flexible and/or backed up by storage.[85]

Particular emphasis is placed on the development of grid tariffs and regulations that incentivise grid users to implement energy efficiency measures.[86] Two main elements are relevant in that respect: the proportion of the grid tariffs that is charged to the grid user and whether there is a regulation in place that allows the system operator to recover costs related to investments for energy efficiency purposes.[87]

To ensure that Member States comply with their obligations, the EE Directive mandates that by 30 June 2015 they must have undertaken an assessment of the energy efficiency potentials of their grids and have identified the measures and investments necessary for cost-effective energy efficiency improvements, with a timetable for their introduction.[88] Member States are further required to remove those incentives in grid tariffs that are detrimental to the overall efficiency of the system and those that 'hamper participation of demand response, in balancing markets and ancillary services procurement'. Finally, Member States have to make sure that grid operators improve the design and operation of the

[80] Directive 2012/27/EU, Article 10. [81] Directive 2012/27/EU, Article 11.

[82] Grid losses are defined as the amount of electricity that is consumed by a power system for the transmission of electricity from generation to consumption. In the EU, the average losses are 1–2.6 per cent in transmission grids and 2.3–11.8 per cent in distribution grids. See Papaefthymiou et al., 'Incentives to improve energy efficiency in EU Grids', 2.

[83] Directive 2012/27/EU, Article 15 para. 1.

[84] Papaefthymiou et al., 'Incentives to improve energy efficiency in EU Grids', 4.

[85] Ibid., 5. [86] Directive 2012/27/EU, Article 15 para. 1.

[87] Directive 2012/27/EU, Annex XI; Papaefthymiou et al., 'Incentives to improve energy efficiency in EU Grids', 7.

[88] Directive 2012/27/EU, Article 15 para. 2.

infrastructure to increase its efficiency and allow suppliers to promote consumer participation in system efficiency.[89]

Mirroring the provisions on RES, the EE Directive requires that Member States guarantee the transmission and distribution of electricity from high-efficiency co-generation by providing it priority or guaranteed access and by determining special conditions for its connection to the grid.[90] Also, Member States have to ensure that demand-side resources, such as demand response, are encouraged 'alongside supply in wholesale and retail markets'. Accordingly, demand-response providers, including aggregators, have to be treated in a non-discriminatory manner with respect to balancing and ancillary services.[91] The picture is rounded off by similar provisions on the design of National Energy Efficiency Action Plans, reporting and monitoring to those laid down by the RES Directive.[92]

III Evaluation of the European Legislative Framework

The European regulatory framework sets out basic conditions for the deployment of Smart Grids according to the priorities set out by the 20/20/20 target aiming at a reduction of greenhouse gas emissions, the deployment of RES and an increase in energy efficiency. Overall broad discretion is left to Member States as to how to implement these goals. This takes into account the diversity of the power-generation mix, the market and the existing structure of Member States' electricity grids, and reflects the different levels of 'smartness' that are needed.

However, the rather vague character of many provisions also carries the risk that Member States will drag their feet in implementing appropriate changes. Also, the current economic downturn may tempt many Member States to reduce their support. Given the uncertainty regarding carbon prices after 2020, private parties may be reluctant to invest in infrastructure with benefits that are uncertain and that, even if they do invest, will materialise only in the mid- and long-term. Finally, it is important to keep in mind that some of the goals to be achieved by Smart Grids threaten the business models of some of the main players, who may well attempt to block or at least slow down the changes.

[89] Directive 2012/27/EU, Article 15 para. 4.
[90] Directive 2012/27/EU, Article 15 para. 7.
[91] Directive 2012/27/EU, Article 15 para. 8.
[92] Directive 2012/27/EU, Article 24.

While the setting of sometimes rather vague framework conditions reflects the unavoidable compromises entailed in the European decision-making process, it is also an expression of the regulatory puzzle posed by the deployment of Smart Grids. At present there is no consensus as to what type of regulatory framework will prompt a quick and efficient deployment of Smart Grids. While some experts argue that, given the characteristics of the industry, only its complete re-regulation could bring about the desired results,[93] others stress that a market-driven deployment of Smart Grids will contribute both to more innovation and to more cost-efficiency, and will eventually also address some of the problems of the liberalisation process.[94]

The more general question is therefore how policies should be framed to ensure that grids incorporate increased intelligent functionalities and make them accessible to the largest majority of users, while keeping costs as low as possible. This in turn raises questions regarding the form of regulations and who will be responsible for delivering the required functionalities.[95] While we do not attempt to provide a final answer to these questions, we will sketch out some of the decisions that Member States will have to make within the boundaries set by the European framework.

1 The Creation of New Incentives

One of the crucial questions that Member States will have to address is what needs to be done to incentivise grid operators to encourage both the integration of large quantities of distributed RES and greater energy efficiency. More specifically, how can electricity grid regulation move from a grid-cost-incentive based model towards a combined grid-cost-incentive, operational quality-incentive and sustainability-incentive-based one?

When it comes to defining regulatory incentives for Smart Grids' outputs, regulators face an unprecedented challenge. Grid operators' current revenue models are mostly volume-based and include various forms of price-control mechanisms, such as price and revenue caps, which incentivise short-term cost reductions, but not innovation and modernisation.[96] While volume-based remuneration discourages energy efficiency measures, cost-reduction incentives may prevent the development and deployment of distributed generation.

[93] Smart Grid Technology, http://smartgridtech.wordpress.com/the-case-for-regulated-monopoly-in-the-electric-grid/ (accessed 25 November 2015).
[94] See European Commission, COM (2011) 202, final version.
[95] Clasters, *Smart Grids*, 31. [96] THINK, *EU Involvement*, 11.

Under the new regulatory framework, three main goals have to be pursued: first, the integration of massive amounts of distributed generation, in particular RES; second, the integration of demand-response measures and storage; and third, the upgrade of transmission lines to absorb energy from large-scale RES.[97] All three targets require the development of new grid services, new technologies and the creation of incentives for grid users to participate actively in the management and development of the grid.

For instance, to encourage distributed generation, tailor-made services need to be developed and technologies installed that ensure the monitoring, communication and control capabilities of the distribution grid.[98] Incentivising demand response and installing storage requires active management of the grid: information has to be dispatched to all actors and their participation has to be encouraged with a view to optimising the functioning of the grid.[99] Finally, the integration of more significant quantities of large-scale RES requires general reinforcement of the transmission grids; the deployment of innovative technologies, such as high-voltage direct current for offshore grids; and producers' participation in the planning of new grids.[100]

Accordingly, to accelerate the deployment of Smart Grids, a fundamental reassessment of grid operators' revenue models will have to be undertaken to adjust the operators' incentives. An interesting approach in this respect is represented by the new RIIO model, which rewards the achievement of outputs that aim at the improvement of services for grid users, long-term cost reductions and incentives for energy efficiency.[101] However, many questions remain as to how it will accelerate the deployment of Smart Grids.

Although output regulation is theoretically superior to input regulation, as it leaves more flexibility to companies to choose how to achieve the goal, it also has its limitations. In practice, it can be difficult to define and measure appropriate performance targets and output parameters, in particular with respect to the numerous and sometimes conflicting requirements of Smart Grids.[102] Also, a reform of grid tariffs is unlikely

[97] L. Meeus et al., 'Smart regulation for Smart Grids', *EUI Working Papers* RSCAS 45 (2010) 2.

[98] Ibid., 3. [99] Ibid., 3. [100] Ibid., 4. [101] THINK, 'EU involvement', 11.

[102] European Union Smart Grid Task Force, *Expert Group 3: Roles and Responsibilities of Actors involved in the Smart Grids Deployment*, http://ec.europa.eu/energy/gas_electricity/smartgrids/doc/expert_group3.pdf (accessed 25 November 2015).

to be able to address all challenges. Grid tariffs are not, for instance, an appropriate instrument to incentivise infant technologies that are still at the R&D stage.[103] Hence, other funding mechanisms should be envisaged, such as public–private partnerships and the imposition of public service obligations, particularly if goals are pursued that benefit not grid users but rather society at large. Finally, it is important to create an overall environment that is conducive to innovation processes.[104]

2 The Setup of the Information Infrastructure

Until recently, the distribution grid was considered a classical natural monopoly, which distributed electricity from point A to point B. With the inclusion of new services and the active management of bi-directional flows of both information and electricity, the nature of the grid business changes radically. Among the questions raised by this new situation are: which tasks should be performed by the system operators, and which can or should be 'outsourced' to third parties? Also, to the extent that third parties are entrusted with some of these tasks, the question arises as to whether this will affect the overall regulatory design.

We shall illustrate this question through the example of the information layer. One of the crucial tasks of a Smart Grid will be to provide an information infrastructure which both collects and dispatches information on the actual system conditions from and to the various grid users. Given this setup, the information infrastructure has been described as a multi-sided platform that is characterised by cross-network effects, which means that benefits to be realised by one group of customers crucially depend on the participation of others.[105]

The creation and operation of an information platform requires investment in the physical infrastructure (i.e. communication technologies and smart meters) and the mechanisms that allow the dispatch of the information.[106] Theoretically, these two elements can be carried out by two separate entities – which will, however, have to collaborate closely to ensure the effective functioning of the information platform. Given its

[103] Meeus et al., 'Smart regulation for Smart Grids', 8.

[104] D. Bauknecht, 'Incentive regulation and network innovations', *EUI Working Papers*, RSCAS 2 (2011).

[105] C. Brandstätt et al., 'Roles and responsibilities in Smart Grids: a country comparison', *IEEE Conference Publications: 9th International Conference on the European Energy Market (EEM)* (2012) 2, http://ieeexplore.ieee.org/stamp/stamp.jsp?arnumber=06254698&tag=1 (accessed 25 November 2015).

[106] Ibid., 2.

network effects, it is difficult to see how a business case can be built without some form of state intervention.

When it comes to allocating responsibility for the operation of the information platform, the national regulatory framework and culture plays an important role.[107] The United Kingdom, for instance, plans to entrust handling of the data to a data and communications company that will be selected through a competitive tendering process. The chosen company will be subject to the regulator's control and will recover its costs through regulated service charges. France, in contrast, has decided to charge the distribution system operator (DSO) with setting up the physical infrastructure and handling the information flows. Costs of the roll-out of smart meters are to be borne by society at large.[108]

As these two examples suggest, the information platform can be delivered by different actors, and this features both advantages and trade-offs.[109] The allocation of responsibility to the DSO will probably enhance the overall coordination of the system. It will, however, be more difficult to ensure the neutrality and non-discriminatory handling of the information flows. The independent company model provides better guarantees as to neutrality, but this might come at the expense of other grid services related to the enhancement of the system's stability.[110]

D Outlook: A Paradigm Shift for Electricity Regulation?

At present, electricity regulations focus on grid operators. To ensure that they do not abuse their monopoly position, the terms and conditions for connection and access to the grids are controlled by neutral regulators, whose core duty constitutes the fixing or approving of transmission and distribution tariffs or their methodologies.[111] While the emphasis placed on the regulation of grid tariffs may have been justified in the past, this approach becomes too narrow in the light of the requirements entailed by the deployment of Smart Grids. One of the main regulatory challenges will be to create a neutral platform that ensures the smooth coordination of an increasing set of heterogeneous grid users and markets.

[107] Ibid., 2. [108] Ibid., 4.

[109] N. Friedrichsen, C. Brandstätt and G. Brunekreeft, 'The need for more flexibility in the regulation of Smart Grids – stakeholder involvement', *Bremer Energie Institut Paper-series* No. 13 (2013), 10.

[110] Ibid., 10.

[111] Directive 2009/72/EC, Article 15 para. 1; T. Maes, 'The energy regulators' core competence: network tariff setting' in B. Delvaux, M. Hung and K. Talus (eds.), *EU Energy Law and Policy Issues Vol. 3*, ELRF Collection (Portland, OR: Intersentia, 2012), p. 123.

We argue that these new challenges justify a more fundamental reassessment of the current regulatory system than has been carried out to date. First, the current focus on grid tariffs has to give way to a broader view which encompasses all terms and conditions related to access to the facilities that are essential for the Smart Grid activities. Second, the regulator's role in the tariff-setting procedure should be challenged. The current system, which requires *ex ante* decisions on either the methodologies or the tariffs by the regulator, may indeed become increasingly inappropriate in a changing environment that values innovation, flexibility and speed. Third, the regulatory framework should take into account the increasing involvement of third parties in the grid business.

Pollitt suggests that a stronger regulatory focus should be placed on processes, rather than outcomes.[112] Along the same lines, some authors argue that more space should be given to negotiated settlements, which would guarantee more targeted outcomes, reduce regulatory intervention and provide protection against a 'regulator knows best' philosophy.[113] Such an approach does not make the regulator redundant, but would change its current role. Instead of fixing or approving access conditions, it would become an enabler of negotiations and 'an instance to fall back upon and provide arbitration when negotiations fail'.[114] While such a change of paradigm does not imply that grid-access regulation would become obsolete, it does imply that outcomes are shaped principally by stakeholders and that state intervention will be confined to tackling market failures and ensuring the pursuit of goals in the public interest.[115]

Some lessons in this respect may possibly be drawn from telecom regulations.[116] The premises of the Smart Grid resemble those of the telecommunication sector, where the regulator focuses on promoting competition for a large set of activities while securing non-discriminatory access to those facilities that are owned by the network operators and that cannot be duplicated.[117]

[112] M. G. Pollitt, 'The future of electricity (and gas) regulation in a low-carbon policy world', *The Energy Journal, Special Issue: The Future of Electricity: Papers in Honor of David Newbery* (2008), 76.

[113] N. Friedrichsen, C. Brandstätt and G. Brunekreeft, 'The need for more flexibility', 6.

[114] Ibid., 8. [115] M. G. Pollitt, 'The future of electricity (and gas) regulation', 76.

[116] M. G. Pollitt, 'Does electricity (and heat) network regulation have anything to learn from fixed line telecoms regulation?', *Energy Policy*, 38 (2010), 1360–71.

[117] See, for a comparison of the two sectors, M. G. Pollitt, 'Does electricity (and heat) network regulation have anything to learn?'

To guarantee these conditions, the telecom regulator is entitled to impose a range of obligations on network operators with 'significant market power'.[118,119] It may, for instance, impose obligations of transparency, of non-discrimination or of accounting separation. An even stronger remedy represents the right to impose an obligation to provide access to, and use of, specific network facilities. Furthermore, they must not withdraw access to facilities that has already been granted and must grant open access to technical interfaces that are indispensable for the interoperability of services. The most 'invasive' remedy, finally, is the right granted to the regulator to impose price controls that prevent excessive and predatory pricing.

When imposing this set of obligations, regulators must be guided by the objectives of the telecom framework, in particular the promotion of competition to the benefit of the customers.[120] The intervention has to be proportionate, that is, it must not exceed what is necessary to achieve the goal pursued.

Although the electricity industry clearly differs in many respects from the telecom sector,[121] the underlying philosophy of its regulatory framework may well guide future regulation of Smart Grids. The following ideas may be worth considering and exploring further. First, actors should in principle be able, within the boundaries of competition law, to freely agree on the terms of their interaction. This is important in a system where the boundaries between monopoly and commercial businesses are increasingly blurred. Second, while state intervention is justified, namely to rectify market failures and to pursue goals in the public interest, it should be done in a proportionate manner and with a view to encouraging the provision of Smart Grid services on a competitive basis.[122] Last but not least, the regulatory framework should be sufficiently flexible to adapt to the changing needs and requirements of a rapidly evolving industry.

118 See, for a discussion of this term, M. B. Nenova, *EC Electronic Communications and Competition Law* (London: Cameron May, 2007), 209.

119 Directive 2002/19/EC of the European Parliament and of the Council of 7 March 2002 on access to, and interconnection of, electronic communications networks and associated facilities (Access Directive). See Articles 9–15.

120 See M. B. Nenova, *EC Electronic Communications and Competition Law*, p. 209.

121 See, for a comparison of the sectors, Pollitt, 'Does electricity (and heat) network regulation have anything to learn?'

122 See M. B. Nenova, *EC Electronic Communications and Competition Law*, p. 326.

Bibliography

Battaglini A., Lilliestam J. and Knies G. (2010). The SuperSmart Grid – paving the way for a completely renewable power system. In H. J. Schellnhuber, M. Molina, N. Stern, V. Huber and S. Kadner (eds.), *Global Sustainability: A Nobel Cause*. Cambridge, MA: Cambridge University Press, pp. 289–306.

Bauknecht, D. (2011). Incentive regulation and network innovations. *EUI Working Paper*, RSCAS 2.

Brandstätt, C., Friedrichsen, N., Meyer, R. and Palovic, M. (2012). Roles and responsibilities in smart grids: a country comparison. *IEEE Conference Publications: 9th International Conference on the European Energy Market (EEM)*, pp. 1–8, http://ieeexplore.ieee.org/document/6254698/ (accessed 25 November 2015).

Burri Nenova, M. (2007). *EC Electronic Communications and Competition Law*. London: Cameron May.

Clastres, C. (2011). Smart grids: another step towards competition, energy security and climate change objectives. *Energy Policy*, 39(9), 5399–408.

Covassi da Encarnaçao, B. (2013). A smart future? The EU digital agenda between broadband, the grid and energy efficiency', in E. M. Noam, L. M. Pupillo and J. J. Kranz (eds.), *Broadband Networks, Smart Grids and Climate Change*. New York, NY: Springer, pp. 91–102.

Eisen, J. B. (2013). Smart regulation and federalism for the smart grid. *Harvard Environmental Law Review*, 37, 101–56.

European Commission. (2003). Proposal for a Directive of the European Parliament and of the Council regarding measures to safeguard security of electricity supply and infrastructure investment, COM (2003) 301, final version.

(2010a). A strategy for competitive, sustainable and secure energy, COM (2010) 639, final version.

(2010b). A strategy for smart and sustainable growth, COM (2010) 2020, final version.

(2010c). Energy infrastructure and beyond – a blueprint for an integrated European energy network, COM (2010) 677, final version.

(2011a). A roadmap for moving to a competitive low carbon economy in 2050, COM (2011) 112, final version.

(2011b). Energy roadmap 2015, COM (2011) 885, final version.

(2012a). 'Revised 1958 Agreement', COM (2012) 202, final version.

(2012b). Renewable Energy: a major player in the European energy market, COM (2012) 163, final version.

(2012c). Renewable energy: a major player in the European energy market, COM (2012) 271, final version.

(2011). Smart Grids: from innovation to deployment, COM (2011) 202, final version.

(2014). The final implementation report of the EU Internatl Security Strategy, COM(2014) 356, final version.
European Union Smart Grid Task Force. (2012). *European Task Force for the Implementation of Smart Grids into the European Internal Market Mission and Work Programme*, http://ec.europa.eu/energy/gas_electricity/smartgrids/doc/mission_and_workprogramme.pdf (accessed 25 November 2015).
(2011). *Expert Group 3: Roles and Responsibilities of Actors Involved in the Smart Grids Deployment*, http://ec.europa.eu/energy/gas_electricity/smartgrids/doc/expert_group3.pdf (accessed 25 November 2015).
Friedrichsen, N., Brandstätt, C. and Brunekreeft, G. (2013). The need for more flexibility in the regulation of smart grids – stakeholder involvement. *Bremer Energie Institut Paperseries* No. 13.
IPCC. (2011). *Renewable Energy Sources and Climate Change Mitigation: Special Report of the Intergovernmental Panel on Climate Change*. Cambridge, MA: Cambridge University Press.
Johnston, A. and Block, G. (2012). *EU Energy Law*. Oxford: Oxford University Press.
Kurth, M. (2013). Smart metering, smart grids, smart market design. In E. M. Noam, L. M. Pupillo and J. J. Kranz (eds.), *Broadband Networks, Smart Grids and Climate Change*. New York, NY: Springer, pp. 11–16.
Maes, T. (2012). The energy regulators' core competence: network tariff setting. In B. Delvaux, M. Hung and K. Talus (eds.), *EU Energy Law and Policy Issues*, vol. 3, ELRF Collection. Portland, OR: Intersentia, pp. 123–44.
Meeus, L., Saguan, M., Glachant, J.-M. and Belmans, R. (2010). Smart regulation for smart grids. *EUI Working Papers*, RSCAS 45.
Papaefthymiou, G., Beestermöller, C. and Gardiner, A. (2013). *Incentives to Improve Energy Efficiency in EU Grids*. Berlin: ECOFYS.
Pollitt, M. G. (2008). The future of electricity (and gas) regulation in a low-carbon policy world. *The Energy Journal, Special Issue: The Future of Electricity: Papers in Honor of David Newbery*, pp 63–94.
(2010). Does electricity (and heat) network regulation have anything to learn from fixed line telecoms regulation? *Energy Policy*, 38, 1360–71.
THINK. (2012). *EU Involvement in Electricity and Natural Gas Transmission Grid Tarification*. Final report, www.eui.eu/Projects/THINK/Documents/Thinktopic/ThinkTopic6.pdf (accessed 25 November 2015).
Zhang, Z. (2011). Smart grid in America and Europe: similar desires, different approaches (Part 1). *Public Utilities Fortnightly*, 149, 1, 46–50.

15

Markets for Adequacy and Flexibility: An Interoperability Perspective

CHRISTIAN KUNZE*

A Objective

The implementation of 'disruptive technologies' is transforming the structure of the electric utility industry.[1] The structural change results from a convergence of factors that include falling costs of distributed generation and other distributed energy resources (DER); an enhanced focus on development of new DER technologies; increasing customer, regulatory and political interest in demand-side management technologies (DSM); and government programmes to incentivise selected technologies, for example, with guaranteed feed-in tariffs. The combination of the aforementioned factors and their secondary effects are already having a dramatic impact on customers, employees and investors, as well as on the availability of capital to fund future investment.[2]

This chapter analyses the impact for a transmission system operator (TSO) of challenges that arise from the aforementioned changes in terms of market design, using the example of the Swiss electricity

* This chapter takes into account developments up to the end of 2015 in a fast-moving environment. Subsequent evolutions and policy changes by Swissgrid are not herein reflected. The chapter is published in the author's personal capacity and does not commit SwissGrid in any way.

1 Christensen coined the term 'disruptive technologies' when referring to technological changes that damage established companies. He outlines two important characteristics: 'First, they typically present a different package of performance attributes – ones that, at least at the outset, are not valued by existing customers. Second, the performance attributes that existing customers do value improve at such a rapid rate that the new technology can later invade those established markets'. J. L. Bower and C. M. Christensen, 'Disruptive technologies catching the wave', *Harvard Business Review*, 73(1) (1995), p. 45. See also C. M. Christensen and M. Overdorf, 'Meeting the challenge of disruptive change', *Harvard Business Review*, 78(2) (2000), 66–76.

2 P. Kind, *Disruptive Challenges – Financial Implications and Strategic Responses to a Changing Retail Electric Business* (Washington, DC: Edison Electric Institute, 2013).

system.[3] The objective is to develop and present an integrated, consistent concept that guarantees interoperability of all its components, rather than simply taking an information and communications technology (ICT) perspective. This would support the implementation of European governments' strategies to facilitate the transition towards low-carbon economic growth.

B Background

Electricity market design is defined as the process and the result of creating an interaction of laws and rules under which all electricity market participants conduct their business.[4] It creates performance incentives for individual market participants, which should foster behaviour that is in line with overall political targets in relation to energy.[5]

As already mentioned, all aspects of the electricity system are being impacted by fundamental changes in the energy mix. While efforts to establish and implement network codes and guidelines are progressing, with the aim to deliver the European Target Model for electricity markets, discussions on further enhancement of the market design have gained momentum throughout Europe. In many European countries, national discussions are ongoing and the first measures have already been implemented. For example, Great Britain has held its first capacity auction for delivery starting 2018/19, France is close to starting its own capacity market and German stakeholders are extensively discussing alternative designs, with Germany having published a Green Paper describing the main ideas.[6]

From the perspective of a TSO, further development of the internal energy market would not be mainly an issue of the frequently discussed

[3] Swissgrid, *Flexibility Markets - Discussion Paper for a Future Energy Market Design* (Laufenburg: Swissgrid AG, 2015). The later publication, *Contributions to the Market Design for the Swiss Energy Strategy 2050 - Consultation Paper* (Laufenburg: Swissgrid AG, 2015), is not discussed in this paper.

[4] A. Ockenfels, 'Article Marktdesign' in *Gablers Wirtschaftslexikon*, 12th edn, wirtschaftslexikon.gabler.de/Archiv/17927/marktdesign-v7.html (accessed 6 January 2017).

[5] A. Kühnke, *Kernthema Strommarktdesign, Schriftenreihe 'Neues Systemverständnis & Erneuerbare Energien'* (Frankfurt: ZVEI - Zentralverband Elektrotechnik- und Elektronikindustrie, 2013), p. 1.

[6] Federal Ministry for Economic Affairs and Energy, *An Electricity Market for Germany's Energy Transition*.

information technology system interoperability. The ability of systems, units or market forces to provide services to and accept services from other systems, units or forces, and to use the services so exchanged to enable them to operate effectively together, has to be taken into consideration but, furthermore, the social, political and organisational factors that impact system-to-system performance also have to be taken into account.[7]

As market design proposals need to take specific regional and/or national industry characteristics into consideration, the following section uses the example of the Swiss electricity market and outlines the core elements of the structure of the Swiss electricity industry.

C The Swiss Electricity Market

I Generation and Demand Characteristics

Switzerland's electricity production is dominated by hydropower and nuclear energy. In total, 56 per cent of the electricity consumed in Switzerland is generated in hydropower plants, and half of this electricity is produced in storage power plants. Nuclear power accounts for 39 per cent, new renewable energies contribute 2 per cent and other forms (e.g. fossil fuel power plants) make up 3 per cent of the electricity produced.[8]

In 2009, Switzerland consumed more than 61,800 GWh of electricity and produced a net total of 63,900 GWh. This surplus, however, is not distributed evenly over the course of the year. In winter, electricity production may not meet demand. Imports from abroad cover this supply shortfall.

Owing to its geographical location and the electricity supply and demand patterns of neighbouring countries, Switzerland is a transit country for electricity: 54,159 GWh was exported and 52,002 GWh imported during 2013.[9]

[7] M. Hura, *Interoperability – A Continuing Challenge in Coalition Air Operations* (Santa Monica, CA: Rand, 2000), p. 7; T. Slater, *What Is Interoperability?* (New York, NY: Network Centric Operations Industry Consortium, 2012).

[8] As hydropower installations have been an important renewable generation asset in Switzerland since the beginning of industrialisation, photovoltaic systems and wind power generation are referred to as 'new renewable energies'.

[9] Swissgrid AG (undated), *Energy Sources*, www.swissgrid.ch/swissgrid/en/home/reliability/energy_sources.html (accessed 16 September 2014).

II Renewable Energy Sources (RES): Challenges for the Swiss Electricity Market

Although the share of new renewable energy in overall electricity generation in Switzerland is relatively limited, the massive increase in electricity generation from RES in Europe is having an impact on the Swiss market.

From a physical perspective, the integration of European RES generation into short-term wholesale electricity trading creates a widening gap between commercial flows and physical electricity flows. This development is illustrated in Figure 15.1.

Owing to the physical flow characteristics of electricity within the European network, it is obvious, for example, that increasing the volume of electricity generated from renewables in Northern Germany has a direct impact on other European regions. Taking Poland as an example, it might therefore be necessary to take expensive re-dispatch measures, coordinated between the Polish TSO and a German TSO, to maintain system stability in the Polish transmission grid.[10] Furthermore, the loop flow characteristics, for example, of wind energy

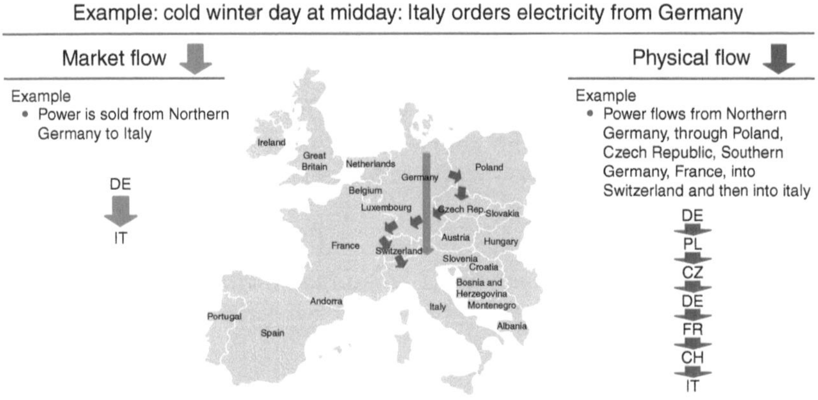

Figure 15.1 The increasing gap between market flow and physical flow
Source: Swissgrid

[10] R. Sverisson, 'TSO data confirms redispatch hiked Polish prices' (2014), www.montel.no/News/Story.aspx?id=539918 (accessed 16 September 2014).

produced at the Baltic Coast, also affect the availability of transmission capacity into and through Switzerland.

Despite the absolute growth of RES generation and existing prognosis errors for RES generation, there is an increasing need for sufficient flexibility to ramp up or down the volumes of electricity generated. However, generally decreasing wholesale market prices for electricity have implications for all non-subsidised generation technologies.[11] The flexibility of generation or demand does not have a specific 'value' in today's and tomorrow's electricity markets. This observation is in line with Woodhouse and Keane's opinion that there is an increasing need for flexibility in the electricity markets.[12] However, current market design does not value flexibility, and new market designs might continue to undervalue this feature.

III Market Design Challenges

In order to benefit from flexibility in the electricity system in terms of market prices, it is necessary to improve on the current design of the electricity market. Three main shortcomings of the current design are identified and summarised below.

1 Adapting the Market Design to Allow Integration of Renewables

Non-market-based incentives for the affordable deployment of new RES are in place in many European countries and electricity systems have become increasingly interdependent. Owing to the non-market-based incentives for renewable energy installations, many European electricity markets are currently in transition towards a system in which thermal plants must operate less, but with greater flexibility to respond when needed. This in turn requires improvements in terms of balancing incentives to ensure that flexible capacity (and demand-side) is appropriately

[11] J. Mayer, *Electricity Spot-Prices and Production Data in Germany 2013* (Munich: Fraunhofer Institute for Solar Energy Systems ISE, 2014), p. 3.

[12] S. Woodhouse and G. Keane, *Revealing the Value of Flexibility – How Can Flexible Capability Be Rewarded in the Electricity Markets of the Future?* (Oxford: Pöyry Management Consulting, 2014), www.poyry.com/sites/default/files/imce/files/revealing_the_value_of_flexibility_public_report_v1_0.pdf (accessed 17 September 2014).

rewarded for its contribution to the balancing of the electricity system. This trend is expected to continue in the years ahead.[13]

2 Regional Cooperation and Coordination of Energy Policies

The divergences between national energy policies limit the extent to which the full economic effect of an interlinked European electricity system can be realised. Currently, market designs follow different national approaches towards ensuring adequacy of generation. For example, Britain, France and Italy are implementing capacity markets, whereas Germany is discussing whether an extensive backup capacity mechanism is needed.[14] This raises the question of whether energy-only markets can provide sufficient long-term investment incentives.[15]

Various studies consider that capacity markets present considerable risks in terms of organisation and structure.[16] Furthermore, regulatory compliance with EU law prohibits taking measures that may negatively affect the internal market.[17] Ultimately, most capacity markets act to supplement the energy price, reducing the volatility of the residual energy price and legitimising the existence of 'missing money'. In an isolated market, effective incentives could still exist with separate capacity and energy markets, but the success of the European market-coupling arrangements requires that the spot price reflects the full value of electricity, not just part of it.

[13] S. Bössner, *Strengthening the European Electricity Market through Improved Franco-German Cooperation* (Paris: Jacques Delors Institute, 2015), p. 1.

[14] Negative evaluations of the potential implementation of capacity markets can be found in, for example, European Engine Power Plants Association (EUGINE), *Position Paper: Recommendations to Meet the 'Flexibility Challenge'* (Brussels: EUGINE, 2014), p. 1 and The Regulatory Assistance Project, *Power Market Operations and System Reliability: A Contribution to the Market Design Debate in the Pentalateral Energy Forum* (Berlin: Agora Energiewende, 2014), p. 9.

[15] P. Joskow, 'Capacity payments in imperfect electricity markets: need and design', *Utilities Policy*, 16(3) (2008), 159–70.

[16] Practical experience, e.g. in the United States, demonstrates that arriving at the correct capacity market design is a difficult undertaking that takes many years and may require many adjustments to rectify regulation errors. Furthermore, capacity markets result in higher system costs and also pose considerable risks in terms of over-complexity, mismanagement potential, inefficiency, irreversibility and path dependency. See Federal Ministry for Economic Affairs and Energy, *An Electricity Market for Germany's Energy Transition*, p. 46.

[17] DNV GL, *Potential Interactions between Capacity Mechanisms in France and Germany: Descriptive Overview, Cross-Border Impacts and Challenges* (Berlin: Agora Energiewende, 2015), p. 9.

3 European Perspective on Security of Supply

The European perspective on security of supply requires avoiding the suggestion of national solutions to transnational problems. Therefore, it is necessary to focus on the development of at least regional measures that guarantee security of supply in Europe, despite increasing intermittent generation and closures of conventional power plants. Potential solutions should be implemented in an 'energy-only' market design framework, which ensures that the market-coupling arrangements work effectively on spot prices that reflect the full value of electricity. This avoids increasingly prescriptive regulations targeted at ensuring resource adequacy. It would also allow for market-based approaches that would not overturn the market.[18]

D Markets for Flexibility and Adequacy: An Integrated Concept

As already noted, the true value of flexibility is not currently reflected in market prices for electricity. Therefore, the author proposes to strengthen the price signal for electricity in order to attribute a fair value to flexible generation or demand sources.

Furthermore, the following sections will demonstrate that the strengthened price signal and related scarcity prices will lead to the development of hedging products. Certain derivative products can then also be used to reform mechanisms that support adequacy in the European electricity system.

I Improving the Price Signal Incentivises Hedging and Demand-Side Integration

Figure 15.2 illustrates specific measures that the Swiss TSO proposed in order to improve the market price signal. The horizontal axis represents the chronology of market price quotations. It ranges from forward prices that result from long-term trading activities to real-time prices close to the physical electricity delivery. The grey-shaded areas represent already existing market segments. The numbered elements 1–5 represent the outcome of measures in order to develop sharp price signals and foster the market integration of renewable energies. They are:

[18] W. Hogan, *On an Energy Only Electricity Market Design for Resource Adequacy* (Cambridge, MA: John F. Kennedy School of Government, Center for Business and Government, 2005).

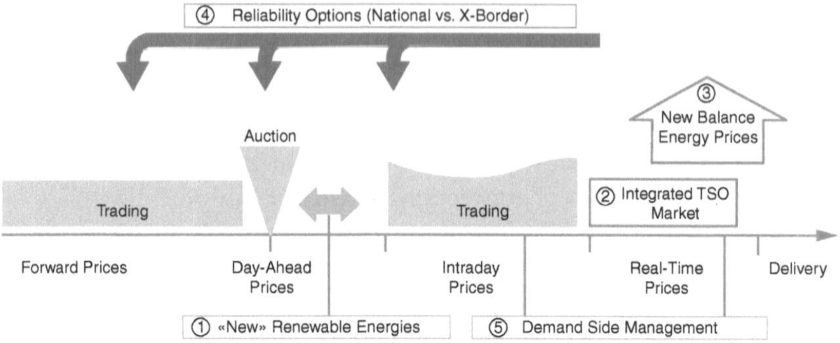

Figure 15.2 Proposed measures to strengthen the price signal
Source: Swissgrid

1. integration of 'new' renewables into the electricity market;
2. implementation of an integrated TSO market with real-time balancing prices;
3. application of scarcity pricing methods based on full marginal cost;
4. implementation of decentralised reliability options;
5. full integration of demand-side management.

1 Integration of 'New' Renewables into the Electricity Market

To ensure system security in view of the increasing levels of electricity-generation penetration by renewables in Europe, system operators need innovative approaches to deal with uncertainties. For instance, they may need to procure additional reserves or ancillary services, improve forecasting tools or invest in new cross-border transmission capacity.[19]

In addition, it is essential that the design of support mechanisms for renewable energies should minimise market inefficiencies and operational constraints. As a general principle, the legislative framework and market rules should stipulate that RES producers be bound by the same duties and responsibilities as all other electricity generators.[20]

[19] B. Johansson, 'Security aspects of future renewable energy system – a short overview', *Energy*, 61 (2013), 598–605.

[20] ECOFYS, *Design Features of Support Schemes for Renewable Electricity* (Utrecht, 2014), EU Project No. DESNL13116, p. 4; ENTSO-E, *Market Design Policy Paper* (Brussels: ENTSO-E, 2014), p. 3.

All mature technologies should be operationally exposed to wholesale market price signals, including balance responsibility. These general principles should be accompanied by additional measures that increase market efficiency. Although RES subsidies may be needed, they should be limited to periods in which the wholesale prices are negative, and producers of renewables should be expected to correctly forecast their feed-in and hedge their volatility in order to improve system security and economic efficiency.[21]

2 Creating Real-Time Balancing Prices by Implementation of an Integrated TSO Market

Many European markets are characterised by a separated, disaggregated market for tertiary balancing power, re-dispatch electricity and other comparable products. This leads to complicated markets with suboptimal price structures.

With the notion of the 'integrated TSO market' came the proposal to concentrate the procurement of tertiary, re-dispatch energy as well as further internationally exchanged TSO products into one single marketplace in a real-time timeframe. Concentration of all flexibility providers' offers in the same real-time marketplace could further increase liquidity.[22] This in turn could pave the way to introducing pricing on a 'pay as cleared' basis,[23] congruent with today's spot auctions in the wholesale market. The transparent procurement cost could provide market-based real-time prices for flexibility, on which imbalance prices can also be based.

The concept of integrated real-time markets, including a precise overview of real-time production, remotely compares to the possibilities of independent system operators in a nodal pricing system. Whereas appreciating the present zonal and balancing group-based system in the wholesale market for Europe, optimisation in the real-time timeframe can build on such concepts. In particular, system security can be better guaranteed by applying market processes. The transmission grid can be used more efficiently, ultimately leading also to increased cross-border capacities for TSO markets.

[21] ENTSO-E, *Market Design Policy Paper.*

[22] Some countries decided to further increase the liquidity of balancing markets by including all production capacities. Examples include France, Spain, Portugal, Belgium, the Netherlands and Italy.

[23] OFGEM, *Pay-as-Bid or Pay-as-Clear Pricing for Energy Balancing Services in the Balancing Mechanism* (London: OFGEM, 2014).

Cooperation between TSOs will be essential to reach this goal and the application of this concept can be fostered by regional implementation projects. The concept is in line with the Network Code's provisions on electricity balancing.[24]

3 Application of Scarcity Pricing Methods Based on Full Marginal Cost

Current regimes for balancing energy pricing in many European states reflect an average price, which is further diminished by the existence of separate reserve payments for capacity. On the one hand, this mechanism guarantees that there is sufficient capacity provision at every point in time to handle system imbalances. On the other, it leads to inefficient cost allocation. For example, in the case of negative imbalance of a balance group, the responsible party financially compensates the TSO exclusively for the electricity that is needed to balance its position. The cost of generators' capacity provision, that is, the expense of the TSO's option to request electricity at short notice in order to balance the system, is not directly allocated to the imbalanced market party. It becomes socialised via tariffs to all parties connected to the grid. Therefore, an efficient system for direct cost allocation is not currently in operation.

The implementation of a transparent real-time, full marginal cost mechanism can create an efficient scarcity price for balancing groups.[25] This price will provide an incentive for existing balancing groups to provide flexibility to the electricity system while also motivating investors to set up new, flexible or back-up generation (cf. adequacy), as this solves

[24] 'No later than one year after the entry into force of this Network Code, all TSOs shall develop a proposal for the pricing methods of each Standard Product for Balancing Energy. The pricing methods shall be based on marginal pricing (pay-as-cleared), unless TSOs complement the proposal with a detailed analysis demonstrating that a different pricing method is more efficient for European-wide implementation'. ENTSO-E, *Network Code Draft on Electricity Balancing, Version 3.0* (Brussels: ENTSO-E, 2014), p. 39.

[25] An example of the implementation of scarcity prices is the Texas market in the United States. Based on a study of the value of lost load, it allows scarcity prices of up to US$ 9.000/MWh. In this framework, an operating reserve demand curve explicitly adds a scarcity element to the real-time prices as the reserve margins become tight. With such mechanisms, scarcity pricing is ensured even in circumstances where regulatory constraints (perhaps to mitigate market power) limit market participants' ability to bid freely in the market. London Economics, *Value of Lost Load – Literature Review and Macroeconomic Analysis* (Boston, MA: ERCOT, 2013) (Public Utility Commission of Texas Project Number 40000).

the 'missing money' problem.[26] In all, this set-up underlines a consequent responsibility for balancing groups. Therefore, scarcity pricing is a promising measure to support the creation of system flexibility and to solve the adequacy challenge.

Great Britain expects to have implemented a full marginal cost scarcity price system in three phases by 2018.[27] Austria and Germany are discussing equivalent measures, whereas the Netherlands has implemented a system featuring real-time publication of balance regulation. [28] This enables internal balancing by balancing responsible parties (BRPs), which may reduce the size of system imbalances. The proposal to implement a full marginal cost mechanism is also supported by the draft of the Network Code on electricity balancing.[29]

4 Implementation of Decentralised Reliability Options

In a perfect market, the aforementioned scarcity pricing concept's implementation into the European electricity market would trigger market participants' interest in voluntarily hedging themselves against periodically very high prices. In this voluntary behaviour, or in the case of the introduction of a hedging obligation, decentralised reliability options or similar financial hedging products could be useful instruments.

5 Full Integration of Demand-Side Management

Today there is an obvious shift in the European electricity mix, from thermal, centralised generation towards renewable, distributed generation. In combination with the development of low-cost steering technology for industrial and household appliances, this offers the potential for demand-side response (DSR). DSR is load demand that can be actively changed by a

[26] A formal description of the missing money problem is given in C. Elberg, 'Cross-border effects of capacity mechanisms in electricity markets', *EWI Working Paper 14/11* (Cologne: EWI, 2014).

[27] OFGEM, *Balancing and Settlement Code (BSC) P305: Electricity Balancing Significant Code Review Developments* (London: OFGEM, 2015).

[28] The Austrian Federal Ministry of Agriculture, Forestry, Environment and Water Management recommends that market supervision should no longer consider temporary scarcity prices as the result of a misuse of market power. R. Haas, *Strommarktdesign für die Integration weiter steigender Anteile erneuerbarer Energie in Österreich und Europa, Endbericht* (Vienna: Bundesministerium für Land- und Forstwirtschaft, Umwelt und Wasserwirtschaft, 2015); Federal Ministry for Economic Affairs and Energy. *An Electricity Market for Germany's Energy Transition* (Berlin: Federal Ministry for Economic Affairs and Energy, 2014), p. 39.

[29] ENTSO-E, *Network Code Draft on Electricity Balancing, Version 3.0*, p. 61.

trigger. Demand-side management (DSM), meanwhile, is the utilisation of DSR for a purpose such as system security (i.e. balancing and congestion) or system adequacy.[30]

The value of potential investment in demand reduction or response will depend on the avoided cost of future energy purchases and on the short-run value of balancing services. Therefore, the introduction of full marginal cost real-time prices – complemented by the trading of hedging products such as options – will create an incentive for the broad implementation of DSR in Europe.[31]

6 Core Interoperability Challenges

Improvement of the price signal requires various activities to develop interoperability of the market design and deal with new challenges.

The integration of 'new' renewables into the electricity market is embedded in the reformulation of a general strategy on renewable energy. In Switzerland this is currently dealt with by the Federal Energy Office, which is mandated to answer questions on the most effective ways to integrate renewables into the market and to consider the social aspects of implementation of a strategy on renewables.

The implementation of an integrated TSO market requires interoperability of the IT systems used by national and foreign TSOs. It also has political implications, as it increases cooperation between neighbouring countries.

The application of scarcity pricing methods based on full marginal cost does not initially raise external interoperability issues. It requires, first and foremost, the interoperability of new system components (to calculate scarcity prices) with existing system components.

Most EU countries do not intend to implement decentralised reliability options in the market. The evolution of these products is considered to be

[30] ENTSO-E, *Demand Side Response Policy Paper* (Brussels: ENTSO-E, 2014). A similar definition is provided by ACER, where demand side is defined as 'Changes in electric usage by end-use consumers from their normal load patterns in response to changes in electricity prices and/or incentive payments designed to adjust electricity usage, or in response to the acceptance of the consumer's bid, including through aggregation'. See ACER, *ACER Framework Guidelines on Electricity Balancing* (Brussels: ACER, 2012). Certain publications focus on the utilisation of DSR for short-term system optimisation. However, its future role in integration into futures and forward markets should also be discussed more thoroughly.

[31] An overview of the implementation of DSM in Europe is published by Cambridge Economic Policy Associates, *Demand Side Flexibility – The Potential Benefits and State of Play in the European Union* (Cambridge: Cambridge Economic Policy Associates, 2014).

the natural outcome of market interactions between generators, consumers, traders, suppliers and exchanges. However, once the products have reached the development stage, it will be necessary to ensure that they are aligned with the physical and financial characteristics of TSO requirements. This needs to be guaranteed in order to ensure that market participants will be in the position to completely hedge their TSO exposure.

Finally, the full integration of DSM into the electricity market is likely to create the biggest interoperability challenges. First, social factors need to be taken into consideration to motivate future DSM providers to exchange their currently unused service potential on electricity markets. Second, interoperability decisions on a political level need to create compliant rules and guidelines for the integration of DSM into existing structures. Furthermore, it is obvious that the TSO–DSO interface needs to be defined more precisely in the future to enable the full potential of DSM to be utilised. In particular, this will require interoperability to be set up in a way that allows for the provision of flexibility on the DSO level, and the acceptance of these services on the TSO level.

E Connecting National Approaches to Security of Supply

The discussions on adequacy recently gave rise to national adequacy monitoring and the introduction of national capacity mechanisms. However, the author is of the opinion that adequacy should be handled by adopting a regional and market-oriented approach. This means that adequacy monitoring and the resolution of critical adequacy situations should be handled on a regional level. The benefits of integrated networks include the sharing of back-up capacity as well as efficient short-term dispatch. Furthermore, market participants have heavy responsibilities for guaranteeing adequacy. However, cross-border adequacy solutions, such as the cross-border participation of foreign generators in capacity mechanisms that might be implemented in some European countries, should be enhanced.

I Regional Adequacy Approach with Adequacy Responsibility for the Market

The adequacy situation should always be determined from a regional perspective.[32] This requires the exchange of necessary data and cooperation

[32] In this context, the term 'adequacy' refers to an analysis that identifies and assesses the level of risk faced by the power system during critical periods. These critical periods can

between authorities and TSOs, especially in terms of calculations. Joint approaches, such as those seen in the Pentalateral Energy Forum (PLEF) adequacy study, should therefore be enhanced.[33] TSOs could also provide adequacy reports which focus at a more national level and make sure that they are aligned with the reports of neighbouring countries, using common assumptions on weather patterns and system stress conditions, in particular.

The author supports the approach of regional cooperation in order to help ease national adequacy problems by sharing available national capacity. Also, bilateral or regional TSO contracts should be coordinated, or at least be applicable to solve critical regional adequacy situations. An important step towards this objective was the joint work on and publication of the first regional generation adequacy report.[34]

In addition, a market-oriented approach which gives clear adequacy responsibilities to the market participants is recommended, especially concerning the hedging of end-customer demand with sufficient supply. 'Hammock' effects, which allow market participants to pass on the responsibility for emergency cases to TSOs or authorities, should therefore be avoided.

A suitable measure here could be the previously outlined increased responsibility placed on balancing responsible parties, with full marginal cost pricing for imbalance. This would oblige them to provide adequacy to their balance perimeter. In this way, they are incentivised to mitigate volatility of generation and load level with both a short- and a long-term focus. In return, balance groups keep the freedom to optimise their balancing portfolio themselves and to identify the most efficient solutions. This might involve investing in relevant technologies, mixing different volatilities (diversification) and hedging with suitable options. Alternatively, it might involve purchasing energy on the short-term markets, such as the intraday market, and the implementation of a full marginal cost mechanism. Furthermore, the creation of new sources of flexibility, for example, through increased DSR activities, plays a crucial role.[35]

be a result of several individual or combined factors, such as weather conditions, lack of renewable generation or forced or planned outages of thermal units. The probabilistic assessment of the PLEF study calculates three typical criteria: loss of load expectation (LOLE), expected energy not served (EENS) and loss of load probability (LOLP). PLEF, *Pentalateral Energy Forum Support Group 2 Generation Adequacy Assessment* (Brussels: PLEF, 2015) p. 18.

[33] Ibid. [34] Ibid. [35] Ibid., p. 6.

Recent announcements made by market parties indicate that, in the short term, new products will be available on the wholesale electricity market to assist the balancing responsible parties in taking on increased balancing requirements and hedging for the related risks.[36]

II Enabling Cross-Border Trade in Adequacy: Efficient Investment to Be Guaranteed

The implementation of national capacity mechanisms is being discussed in many European countries and there is growing support for cross-border participation in these mechanisms. However, so far only one country, the United Kingdom, has devised a concrete solution which will allow interconnectors from other countries to participate in the national capacity mechanism. This includes, for example, the interconnector capacity from Norway, which will participate in the second capacity mechanism auction.[37] The other countries in which capacity mechanisms have been installed or are being discussed have so far not proposed concrete recommendations for measures that allow the participation of cross-border capacity in national mechanisms.

The discussions mainly focus on the following three models:

1. *Implicit participation.* The volume which needs to be purchased through the national capacity mechanism is reduced by the imports expected from foreign capacities. Foreign capacities receive payments in the 'energy-only market' but do not receive payments for their participation in the capacity mechanism.
2. *Interconnector participation.* The owner of the interconnector participates directly in the foreign capacity market with the capacity of the interconnector and receives the payments from the capacity market.[38]
3. *Direct generator participation.* Subject to the available interconnection capacity, foreign generators participate directly in the capacity market with their available capacity and receive payment for this.

From the author's point of view, all these solutions lack appropriate investment signals. The implicit participation model sends no investment

[36] EEX, *Working Paper Energy Turnaround Products* (Leipzig: EEX, 2015).
[37] Department of Energy and Climate Change, *Interconnectors to Participate in the Capacity Market from 2015* (London: Department of Energy and Climate Change, 2014).
[38] This solution will be applied for the border between the UK and Norway.

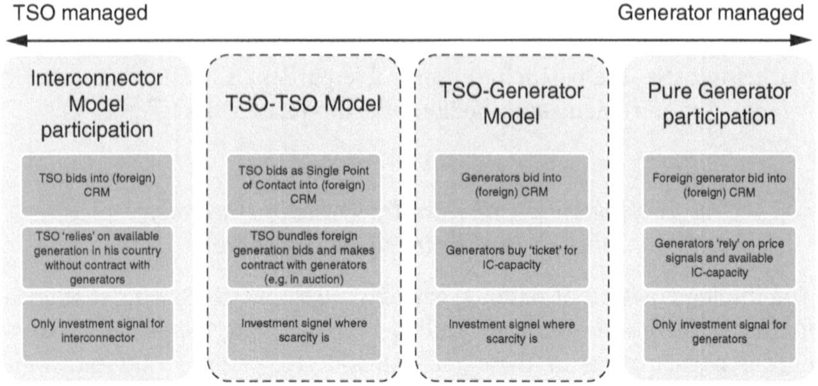

Figure 15.3 General models to enable cross-border adequacy
Source: Swissgrid

signals at all from the capacity market, but only from the 'energy-only market' (this is also the case for a market without cross-border participation). The other two options cannot react flexibly to scarcity signals from interconnection capacity or generation capacity, but focus only on one of the two (e.g. between Britain and Norway). They can therefore be applied to situations where only one scarcity signal is required. For highly meshed grids within Europe, however, it is difficult to determine where the scarcity is.

Therefore, the author recommends a model which reacts flexibly to scarcity signals from generators and interconnection capacities. Basically, two options are possible:

1. *TSO-generator model.* Generators participate in the foreign capacity market but pay the interconnection owner for the guarantee that interconnection capacity will be available (e.g. in a 'ticket' auction).
2. *TSO–TSO model.* Interconnection owners participate in the foreign capacity market but pay the generators to provide the necessary generation capacity (e.g. in an internal capacity auction).

The two models can be described as hybrids of (pure) interconnector participation and (pure) participation of generators, as Figure 15.3 shows.

Both models react flexibly to different scarcity situations and deliver appropriate investment signals. However, they also require suitable contracts between the interconnection provider and the participating generators. Furthermore, it is essential to ensure that TSOs remain responsible for their control area. The certification and monitoring of foreign generators should always be carried out by the domestic TSO in the home country of the generator.

III Core Interoperability Challenges

The connection of national approaches to security of supply is a highly sensitive political topic. Market frameworks and requirements, together with political decisions, lead to particular design alternatives being chosen in different countries. Therefore, the connection of national approaches to security of supply needs to take into consideration the interoperability of political philosophies that are reflected by the capacity scheme models.

At a later stage, system interoperability needs to allow the provision of services to and the acceptance of services from other national technical capacity management systems. In this case, one of the key questions will be: who should operate a national system that allows for participation in foreign capacity markets?

F Optimisation of Cross-Border Trading

The present efforts towards coupling of the pan-European day-ahead and intraday markets are important steps towards efficient cross-border trading. However, particularly in the intraday time frame, the present projects only demarcate the starting point for further developments. The author sees the need for further development, particularly in the following three areas.

I Harmonisation of National and Cross-Border Gate Closure Times

On intraday markets, price and volume volatility usually increases significantly as gate closure approaches. This increase in prices and volatility is an indication of system balancing challenges in the market and is most noticeable in the last hour before delivery time. However, in many cases it is currently not possible for flexible generators, or the demand side, to support system balancing on the intraday market in neighbouring countries.

This may be because nomination lead times for cross-border transactions are long in comparison to nomination lead times inside a country.[39] A lack of harmonisation of the intraday products (e.g. granularity) between neighbouring countries can also be observed. These restrictions limit the opportunity to optimise the social welfare of intraday cross-border transactions for the countries involved. The Agency for the Cooperation of Energy Regulators (ACER) aims to develop a framework to run intraday trading in Europe until the beginning of the balancing energy gate closure time.[40]

II Evaluation of Intraday Auctions

At present, the dominant trading mode in the intraday timeframe is 'continuous trading', in which trades are continuously closed. For cross-border trades this principle is combined with the 'first-come-first-served' principle, in which cross-border capacity (along with the traded energy) is explicitly or implicitly awarded to the first requester.

With increasing liquidity and trading frequency in the intraday market, the efficiency of the 'first-come-first-served' principle is being called into question. With an increasing demand for capacity on the intraday capacity platforms, a decreasingly satisfactory capacity allocation is noted: capacity is allocated not to the party with the highest economic benefit, but to the party with the fastest bidding infrastructure.

The introduction of auctions seems a particularly promising concept for reacting to the intraday market's increasing significance for system optimisation. The proposed market design measure is in line with the proposal in the draft network code on capacity allocation and capacity management.[41] It explicitly mentions the option to implement 'complementary regional intraday auctions ... within or between bidding zones'.[42]

[39] For example, the nomination lead times within Germany and Switzerland are approximately fifteen minutes, whereas the cross-border nomination lead time between these countries is forty-five minutes 'over the counter' (OTC) and sixty minutes via the coupled intraday platforms of EPEX SPOT.

[40] Related requirements are also covered in the CACM Network Guideline, but considerable effort will still be need to implement them.

[41] ENTSO-E, *Draft Document: Establishing a Guideline on Capacity Allocation and Congestion Management (CACM)* (Brussels: ENTSO-E, 2014).

[42] Ibid., p. 60.

III Optimisation of Cross-Border Forward Markets

The implementation of the Forward Capacity Allocation Guideline has proved successful. For example, the switch from physical to financial transmission rights (FTRs) is an important step towards efficient capacity allocation, leaving the maximum capacity to the (ideally cross-border coupled) spot auctions. Also, the increased harmonisation of auction rules is expected to increase overall efficiency.

Developments in national forward energy and derivative markets show that the present focus on deterministic baseload products may be shifting. In an environment of volatile consumption as well as volatile (renewable) production patterns, the value of such products decreases and new products emerge.[43] As products on national forward markets start evolving, the products sold across the border will also need a basic re-evaluation. Cross-border forward products may need to be developed in three areas:

1. *Volatility products.* These are products to transfer the volatility in production or consumption from one country to the next, so allowing for a volatility–flexibility hedge cross-border in the forward market.
2. *Reliability options.* These are products to transfer the risk of high balancing market prices across the border.
3. *Adequacy products.* These are products to transfer the security of supply for end-consumers across the border.

IV Core Interoperability Challenges

In terms of the harmonisation of national and cross-border gate closure times, organisational factors have an impact on the interoperability of national and neighbouring market system-to-system performance. Current cross-border processes have developed over time; the results of cross-border allocation and nomination procedures are the starting point for many internal TSO processes. Therefore, close cooperation between all the TSOs involved is initially required in order to evaluate a newly designed system-to-system interaction's impact on existing organisational processes. Once this evaluation is complete, the operational interoperability would have to be implemented on an ICT level in order to enable systems to operate together across national borders.

[43] EEX, 'Energy turnaround products', *Working Paper* (Leipzig: EEX, 2015); SQ Consult, *Capacity Remuneration Markets in Europe* (Utrecht: SQ Consult, 2013).

The proposed evaluation of intraday auctions implies that the design of interoperability procedures needs to start at the political level. Key decision-makers in the countries involved must agree on the redesign of existing processes. This is also likely to require an analysis of the social effects of interoperability, that is, it will be necessary to conduct social welfare analyses of the auctions' effects in the countries involved. Once again, positive results of this evaluation will be the starting point for the system-to-system interoperability of the TSOs that are involved in the new auction processes.

Furthermore, the optimisation of cross-border forward markets will initially require regulatory and political decisions to be made. The challenges relating to interoperability reach from the structuring of regulation for participation in capacity mechanisms in neighbouring countries to the implementation of local hedging products. Following a general decision on the acceptance of neighbouring electricity systems' capacity services in national capacity schemes, specific product design needs to take place. In this context, it is expected that market players, such as trading entities and exchanges, will agree on harmonised cross-border products and implement them. This will pose challenges in relation to ICT and organisational interoperability that will need to be solved in order to create efficient market structures.

G Conclusion

This chapter outlines how the Swiss electricity market is affected by increasing DER generation in Europe even though Switzerland itself generates only a limited amount of 'new' green energy. The Swiss energy mix would allow a contribution featuring flexible, renewable generation to balance supply and demand in Europe.

In order to fully utilise existing potential and incentivise the development of additional flexibility potential, a number of interconnected measures are required. These include the market integration of renewable energies, clarification of the role of the parties responsible for balancing and measures to increase the price signal in the short-term timeframe. Topics currently under discussion include the development of decentralised capacity mechanisms, the notion of full cost of balancing, the development of flexibility market products based on options and the role of flexibility itself in the European internal market.

Short-term measures to optimise the utilisation of Swiss flexibility potentials in Europe would require intraday markets' gate closure times to be set closer to the delivery time for the electricity. Also, international power flow nomination lead times should be harmonised.

The introduction of market design structures for flexibility and adequacy should foster the fair valuation of flexibility of generation assets in a European context. The related market design activities are complex, as interoperability challenges are not linked in a straightforward way to ICT questions. An interoperability-oriented market design approach also needs to take political and organisational factors into consideration at a very early stage of the planning. In TSOs' market design environment, interoperability has much wider implications than is the case in other industries. Therefore, for future TSO projects, a promising approach might be to subdivide them into work packages that directly address particular aspects of interoperability.

PART IV

Levelling the Playing Field: Regulatory Challenges

16

ASCM Disciplines and Recent WTO Case Law Developments: What Space for 'Green' Subsidies?

LUCA RUBINI*

A Introduction

Much has been written on the topic of 'green' subsidies, especially lately. Many of the points I make in this chapter are not new. Numerous scholars, including myself, have made them before.[1] It is worthwhile, however, to consolidate a few ideas and build on them. Undoubtedly the area is complex and is evolving fast – many issues, including legal ones, are still wide open, and thus need continuous analysis in order for an answer to be found.

This chapter attempts, in particular, to highlight that the key question of policy space with respect to public support for electricity generated from renewable energy sources cannot be answered by clinically isolating legal rules from their factual background – constituted by technological innovations, market developments and the broader political economy. Viewed from this perspective, the legal analysis must start when the technological, commercial and political considerations expounded in the other contributions to this volume are fully internalised.

The second contribution made by this chapter is more conceptual, and is founded on the analysis of two ideas that are considered very useful in visualising the question of (green) policy space and subsidy laws: the notions of point of balance and trade-off.

* I would like to thank the participants in the World Trade Forum 2014 (27 September 2014, Berne), and in particular J. Kreier, R. Howse and T. Cottier, for useful comments on a preliminary version of this chapter.

[1] See, e.g., L. Rubini, 'Ain't wastin' time no more: subsidies for renewable energy, the SCM Agreement, policy space and law reform', *Journal of International Economic Law*, 15(2) (2012), 525–79; L. Rubini, 'The good, the bad, and the ugly: lessons on methodology in legal analysis from the recent WTO Litigation on renewable energy subsidies', *Journal of World Trade*, 48(5) (2014), 895–938.

The facts and background of the issue of green policy autonomy are analysed in Section B. Section C then examines what the law says about government policies in support of green energy and in particular electricity. Section D is the conceptual core of the chapter, revolving around the above-mentioned notions of point of balance and trade-off. Section E shifts the focus from the laws to the courts, as the prominent actors called to interpret and apply them. Particular reference is made to the WTO Panels and the Appellate Body, and to the most recent jurisprudence with an impact on the issue of green policy space, that is, the *Canada – Renewable Energy/FIT* dispute. Section F concludes, offering a few remarks on law reform.

B Facts and Law

This section provides an easy heuristic to understand the key determinants of the issue of green policy space.

The question of 'policy space' or autonomy for green subsidies is as much factual as it is legal. Good lawyers and judges are those who first of all master the facts. Taking a step further, and quoting the Chicago sociologist Charles Henderson, Louis Brandeis famously held that a lawyer who has not studied sociology or economics is very apt to become a public enemy.[2] This chapter does not go as far as requiring radical changes in law schools' curricula or in legal training. But I want to emphasise that the question of policy autonomy heavily depends on the factual – that is, the *technology, market, political economy* conditions – that prevail in the green electricity sector or will do so in the future.

From the legal perspective, the main contribution made by the World Trade Forum from which this book derives is to explain to us lawyers what is and what will soon be possible in terms of trade in electricity. This is the precondition for scenarios in which support for the generation of electricity, which is then traded, can possibly give rise to causes for taking legal action. In what is certainly a very simplified chain, the technology comes first. When the technology becomes a possibility, the focus then markedly shifts to the market and the government as key actors, often working in partnership to make the technology as widely available as possible. If the public hand can sustain market opportunities, it can also cause distortions. In an internationally competitive market, these distortions can give rise to diplomatic rows. Finally, given the right

[2] L. Brandeis, 'The living law', *Illinois Law Review*, 10(7) (1916), 470.

political conditions, these claims can trigger legal remedies and complaints. The arsenal used to fight these disputes is the prevailing regulatory framework. It is against this simplified, but nonetheless complex, background that one can ask and answer questions about the policy autonomy available to governments wishing to support green electricity.

Hence, the key questions lawyers need to have answered before making any final assessment on the degree of policy autonomy are the following:

- What technology is used for the generation, storage, transmission and distribution of electricity? What does, or will, it permit?
- What types of cross-border trade can we think of?
- How fast will the markets respond to these technological possibilities?
- What role will public support play?
- Are the conditions in place for harm to be done?
- Who will complain about subsidies granted to electricity (that is, for its generation, storage, transmission and distribution)?
- What are the conditions under which disputes arise? Do they exist?

We can use a few stylised scenarios to further elucidate the framework of this inquiry. In the following section, we analyse the regulatory framework in more detail and hint at the technical issues that may affect the viability or otherwise of a subsidy law claim.

I Green Technology Subsidies

From the trade law perspective, this is so far the most common scenario. Public support for green technology (e.g. solar panels and wind turbines) has been significant. Since this technology, in the shape of final products or inputs, is extensively traded, public support can clearly have an impact on international trade. This may lead to subsidy claims being filed with the WTO dispute settlement system or to unilateral remedies (i.e. countervailing duties) being imposed in the affected jurisdictions. Most of the disputes that triggered the recent 'trade war' relate to this category of support.[3]

[3] For an overview of these cases, see, e.g., M. Wu and J. Salzman, 'The next generation of trade and environment conflicts: the rise of green industrial policy', *Northwestern University Law Review* (2014), 108(2), 401–74. With specific respect to trade remedies (including countervailing duties), see the chapter by E. Vermulst and M. Meng in this volume.

II Green Energy Subsidies

Public support for green energy (e.g. in the form of support to R&D or production) raises two questions, again from a trade law perspective.

First, we need to know whether there is any international trade in the commodity. There is certainly trade in gas and fuels (both biofuels and fossil fuels), but the issue is still more open with respect to electricity. On the one hand, we certainly have cross-border trade at the regional level but, quite often, the commonality of the participants (think for example of intra-EU electricity trade) makes challenges difficult. On the other hand, the idea of a 'global grid' is still speculative, with different arguments (mainly technological viability versus investment magnitude) being made. Owing to the specificity of the electricity sector, we will deal with this as a separate scenario below.

Second, one can have a trade law claim that does not necessarily target public support for the green generation of energy, but rather focuses on other aspects or conditions of that policy of support which have an impact on other (although not necessarily) green markets. The main economic and political mischief of the Ontario Feed-in Tariff (FIT) Programme that was subject to review in the *Canada – Renewable Energy/FIT* case was not the FIT in itself but rather the various types of local content requirements attached to it. This leads us back to the first scenario – that is, subsidies for green *technology.*

This second consideration leads to an important point. At least in principle, we can have a trade law, and particularly a subsidy law, issue, *even if* there is no cross-border trade with respect to a significant aspect of the policy. In the Ontario case, there was no trade in electricity but there was certainly trade in the inputs subject to the localisation requirement.

Furthermore, since support can also be granted indirectly, one might also wonder whether, through the direct support given to energy generation, one government is effectively (and indirectly) also supporting the producers of the equipment and inputs used to generate green energy. (The same consideration obviously applies to green *electricity.*) To put it another way, various markets (and obviously countries) can be affected by a subsidy.[4]

[4] One key fact which masterfully shows the indirect and transnational impact of subsidisation in the green sector was often repeated during the World Trade Forum: the Chinese green technology (e.g. solar panel) industry was created by German subsidies aimed to promote green electricity generation.

The law does reflect this economic reality through the doctrine of 'pass-through' of the benefit of the subsidy.

III Green Electricity Subsidies

This scenario can encompass support to the generation of green electricity but also to its transmission and distribution, or storage. While, as we are about to see below, claims with respect to transmission, distribution or storage are more difficult to visualise, one could certainly have claims focusing on a government's support for production. The key questions are then:

- Is there any cross-border trade in the generated electricity?
- Who will complain about the trade-distorting support measure?
- Are there competing conventional energy producers?
- Are there other (less subsidised) competing green energy producers?[5]

Clearly, the answers to these questions are, in the main, not legal. We must interrogate the technological and market possibilities, as well as the political economy.

IV Subsidy to Support Transmission, Distribution (i.e. the Grid) or Storage

Governments may also invest in infrastructure, be it for the storage of electricity or for its long- or short-distance transportation. It is clear that in some cases, such as cross-border transmission, we have by definition a cross-border element. This should not therefore be an obstacle to the application of trade laws.

[5] Given the widespread public support for the green energy sector, and the co-existence of various national support programmes, it is difficult to find one national industry that is completely 'clean' in this respect ('I have not received any support, but my foreign competitor has!'). The economic and competition issue is often one of 'net effects' ('My competitor gets more aid than me!'). Interestingly, this is exactly what the United Steelworkers Union (USW), which filed a section 301 petition with the US government in September 2010, claimed. The American union complained that the Chinese producers had received 'more than twice as much as the US spent in the sector'. See K. Bradsher, 'Union accuses China of illegal clean energy subsidies', *New York Times* (9 September 2010), www.nytimes.com/2010/09/10/business/energy-environment/10steel.html. For a commentary see S. Lincicome, 'USW to China: subsidies for me, but not for thee' (9 September 2010), http://lincicome.blogspot.it/2010/09/usw-to-china-green-subsidies-for-me-but.html.

There is, however, another angle which represents, in our view, the key legal question to be addressed. As we are about to explain, subsidies can be challenged under WTO subsidy laws only inasmuch as they are determined to be specific to certain enterprises or sectors. General subsidies, that is, subsidies available across the board, are not objectionable. Now, can we visualise a claim against electricity infrastructure? Is this infrastructure, which is provided (partly or fully) by the government, 'general' or 'specific'? More to the point: can we build a claim of specificity vis-à-vis the electricity sector? Or, should the infrastructure be viewed as the physical vehicle used to transmit and distribute electricity to everyone connected to the network? What if a significant infrastructure is built in a remote region, where the predominant, if not the only, customer is a company engaged in international trade? Again one can see that, at least at the level of speculation, it is possible to consider various factual scenarios. It appears, however, that these types of claims, like general subsidy claims related to infrastructure provision, would be difficult to win.[6]

V Subsidy for Energy Consumption

The last scenario is that in which subsidies are channelled not to the energy producer but to the consumers. The first important question in determining the applicability of subsidy laws is always that related to the existence of cross-border trade in the products or markets that could be affected by the subsidised demand. The second set of questions focuses on the type of subsidy at issue. In particular, is this subsidy broadly available (e.g. to all energy produced through clean sources) or is it only targeted at certain sectors? To put it another way, is it technology-neutral or is it designed to foster a certain, comparatively disadvantaged, green source? As already noted, the more targeted and specific subsidies are, the more likely it is that they may face problems under subsidy laws. We will return to this key issue when discussing the law and, in particular, the specificity requirement.

C Green Electricity Support: What Does the Law Say?

After a very brief overview of the regulatory framework emanating from the WTO Agreement on Subsidies and Countervailing Measures

[6] For two cases in which infrastructure subsidies claims were tested see *EC – Aircraft*, WT/DS316 and *US – Aircraft*, WT/DS353.

(ASCM),[7] this section attempts to show that WTO subsidy disciplines are both seriously unclear and unsound.

I The Subsidy Regulatory Framework: A Primer

It is apparent that WTO law does not object to subsidies in themselves. Being a trade discipline, subsidies are objected to only if the subsidy causes some form of negative trade effects, as defined in the laws. This crucial element has already been repeatedly highlighted in the previous section.

In terms of legal analysis, there are three successive steps to be taken.

First, subsidy disciplines can only apply if a given measure is first defined as a subsidy under the requirements of Article 1 of the ASCM. In particular, to be deemed a subsidy, a measure:

- must be granted by a government (or a public body which is equated to the government);
- must confer one of the given forms of financial contribution (or alternatively constitute any form of income or price support);[8]
- must be economically beneficial.

Second, only subsidies that are specific to certain firms or industries are covered by the disciplines.[9]

[7] A more detailed account can be found in L. Rubini, 'WTO subsidy laws: the international regulation of state aid' in H. C. H. Hoffmann and C. Micheau (eds.), *State Aid Law of the European Union* (Oxford: Oxford University Press, 2016), pp. 469–507.

[8] According to Article 1.1(a)(1) of the ASCM, 'there is a financial contribution by a government or any public body within the territory of a Member (referred to in this Agreement as "government"), that is, where: (i) a government practice involves a direct transfer of funds (e.g. grants, loans, and equity infusion), potential direct transfers of funds or liabilities (e.g. loan guarantees); (ii) government revenue that is otherwise due is foregone or not collected (e.g. fiscal incentives such as tax credits); (iii) a government provides goods or services other than general infrastructure, or purchases goods; (iv) a government makes payments to a funding mechanism, or entrusts or directs a private body to carry out one or more of the type of functions illustrated in (i) to (iii) above which would normally be vested in the government and the practice, in no real sense, differs from practices normally followed by governments'. While the case-law on the financial contribution requirement, in its various forms, is already rich, there is virtually no interpretation of the 'any form of income or price support' alternative, which may have important potential in the area of green energy support. It looks, in particular, very reasonable to conclude that certain measures such as feed-in tariffs or renewable portfolio standards may well amount to a form of 'price or income support'.

[9] Article 2 of the SCM Agreement outlines the sophisticated tests to establish specificity. A subsidy will be deemed specific if it is specific in law or *de facto*. Crucially, subsidies that

Third and finally, as already noted, specific subsidies are permitted unless they cause adverse effects in one of the forms defined by the ASCM (see Articles 5 and 6), or they are conditional on exports or on the use of local inputs (see Article 3).[10]

Each step of this legal analysis raises important legal issues, all of which contribute to determining the scope of autonomy of governments.

1 First Claim: Subsidy Laws Are Uncertain

Subsidy laws are notoriously uncertain. The regulation of public support for electricity is no exception, for two reasons in particular. First, the electricity sector and its market are complex. Second, at the time when subsidy laws were conceived, their framers certainly did not have trade in electricity in mind.

While not aiming at exhaustiveness, the following list sketches the main areas of uncertainty regarding subsidy laws that may have a specific impact on public support for electricity.

a) What amounts to a financial contribution? The various forms of financial contribution still raise many questions (e.g. can we have a cumulative application of more than one form of financial contribution?), but the most relevant in terms of support for electricity arguably focus on interpretation of the requirements when a third party is involved in the transfer of economic resources (see item (iv) of Article 1.1(a)(1) of the ASCM). This depends on the distinct regulatory nature of many support measures in this field, where the action of subsidisation involves both public and private agents.

b) What is the status of tax incentives? The language of the 'otherwise due' clause has already caused headaches among adjudicators.[11] But this is only the beginning. We have never had any dispute in which energy or environmental taxes have come under scrutiny.[12] Such a case may require an interesting analysis of the 'objectives' of these tax measures

are granted according to 'objective criteria and conditions' are not held to be specific. We will delve into the important, and still open, legal issues arising from these three tests below.

[10] If a subsidy is proved to be conditional on exports or the use of local inputs, there is no need to prove specificity.

[11] See, e.g., the *US – FSC* (WT/DS 108) litigation. For an extensive analysis see chapter 9 in Luca Rubini, *The Definition of Subsidy and State Aid – WTO and EC Law in Comparative Perspective* (Oxford: Oxford University Press, 2009).

[12] See the chapter by K. Holzer, I. Espa and T. Payosova in this volume.

and a search for general rules and exceptions. EU state aid law offers good examples of the difficulties involved in this type of enquiry.[13]

c) What is the status of 'regulatory' measures? The status of regulatory measures under subsidy laws is a recurring issue and is still the true 'elusive frontier' of subsidy laws, in much the same way as for EU state aid law. In the field of green energy subsidies one can think of the 2001 *PreussenElektra* decision, where the European Court of Justice famously held that the German FIT was not state aid. The *Canada – Renewable Energy/FIT* ruling in the WTO has provided some indications, but many issues remain open, and there is certainly no final statement of the law with respect to regulatory subsidies.

d) What does 'any form of income or price support' actually mean? One big issue is what role this language will play with respect to the scope of the definition of subsidy. The issue is particularly relevant with respect to regulatory measures which are pervasive in the green energy sector, like those at issue in the *Canada – Renewable Energy/FIT* case. We have good, and arguably sound, guidelines coming from the Panel Report in the *China – GOES* decision,[14] but the question of this provision's coverage with respect to regulatory measures has not yet been addressed. (In the *Canada – Renewable Energy* case, the Panel, seconded by the Appellate Body, did exercise judicial economy on the relevant claim made by the complaining parties.)

e) What does 'benefit' mean? Against the specific – and unclear – language of many of the other requirements of the definition of subsidy, as well as of the specificity and adverse effects tests (for these tests, see immediately below), the requirement that any government measure may raise concerns under WTO subsidy laws only if it confers a 'benefit' is clearly open-ended. The legal language naturally opens to arguments with respect to both the positive goals and effects of a subsidy. This is what happened in the *Canada – Renewable Energy/FIT* case. For the reasons discussed below, this interpretative approach is seriously wrong and potentially undermines the effectiveness of subsidy control as well as transparency.

f) Is the specificity test really 'specific'? Despite the length and sophistication of the legal language (and its three 'principles' under letters (a), (b) and (c) of Article 2.1 of the ASCM), one is often left to wonder whether

[13] For a review of this case-law, see L. Rubini, 'Ain't wastin' time no more', pp. 535–7.
[14] WTO Panel Report, *China – GOES*, WT/DS414/R, para. 7.86.

or not a given measure will be held to be specific.[15] In other words: against a clearly expansive approach to the interpretation of the test, one wonders when *any* measure will be held *not* to be specific.

The key interpretative issue – which is still open – is the relationship between *de facto* specificity (Article 2.1(c) of the ASCM) and the relevance of the cases where the subsidy is not specific because it is granted according to 'objective criteria or conditions' (Article 2.1 (b)).[16] In particular, does *de facto* specificity always prevail? What is the real relevance of the 'neutral and objective conditions'? Can these – despite the specific and suggestive language of the law – ever discard a finding of predominant benefit of the subsidy to certain firms or sectors only?[17] It is often noted that the answer to this last question may potentially open up interesting possibilities for policy autonomy.[18]

g) How difficult is it to prove adverse effects? If a subsidy is not automatically prohibited (because it directly impinges on exports or on localisation), in order to be actionable, it needs to cause adverse effects. There are three types of adverse effects under Article 5 of the ASCM: nullification and impairment of benefits, in the form of frustration of tariff concessions; injury; and the five forms of serious prejudice as outlined in Article 6. Now, all these 'trade-impact' standards are associated with their own difficulties and uncertainties, with the result that it may not be easy to prove adverse effects. As the analysis of the various factual scenarios in Section B showed, proving adverse effects in the electricity sector may be particularly challenging.

[15] As the seasoned lawyer knows, lengthy and sophisticated language is often not the path to legal clarity; quite the contrary.

[16] Which are defined as 'criteria or conditions which are neutral, which do not favour certain enterprises over others, and which are economic in nature and horizontal in application, such as the number of employees or size of enterprises'.

[17] The chapeau of subparagraph (c) reads: 'If, notwithstanding any appearance of non-specificity resulting from the application of the principles laid down in subparagraphs (a) and (b), there are reasons to believe that the subsidy may in fact be specific, other factors may be considered.'

[18] See, e.g., S. Bigdeli, 2011. 'Resurrecting the dead? The expired non-actionable subsidies and the lingering question of "green space"', *Manchester Journal of International Economic Law*, 8(2) (2011), 2–37, at 22–3. See also R. Howse, 'Securing policy space for clean energy under the SCM Agreement: alternative approaches' (2013), *E-15 'ThinkPiece'*, http://e15initiative.org/publications/securing-policy-space-for-clean-energy-under-the-scm-agreement-alternative-approaches/.

At the same time, the vagaries of these standards and the relevant evidence arguably mean no government can be reasonably certain that its subsidies will *not* be found to cause adverse effects. We therefore believe the statement that the system already offers policy autonomy, if governments abstain from causing adverse effects, is partly fallacious (we will return to this point below). Clearly, the regulatory framework, both in theory and in practice, involves significant legal uncertainty.[19] If one wants to offer a *definite and effective* degree of protection to certain desirable subsidies, this cannot rest on the lack of adverse effects, but needs positive recognition.

One final gloss on the real meaning of legal uncertainty may be useful. Some may argue that the ambiguity in the law is positive because it means flexibility for governments. Various readings of the law are possible, satisfying the preferences and conduct of different Members. This may be true, but only on one crucial negative condition: *to be positive, ambiguity must be permissive but not divisive.* If the legitimacy of different and *contrasting* readings of the law becomes a contentious issue, good flexibility turns into bad legal uncertainty. The next step is litigiousness and litigation.

2 Second Claim: Subsidy Laws Are Not Sound

What is referred to here is not a general critique of the economic soundness of WTO subsidy laws.[20] The claim is more specific and centres on whether WTO subsidy laws, as they currently stand, are in line with what the best policy practice tells us with respect to good green subsidies. This is measured in terms of the goals and effectiveness of these measures in relation to these goals, in particular as they relate to environmental concerns and security of supply.[21] In other words, certain features and prescriptions of the law may be clear. The question in this case is: if the

[19] This claim is fully developed in L. Rubini, 'Ain't wastin' time no more'.

[20] See A. O. Sykes, 'The questionable case for subsidies regulation: a comparative perspective', *Journal of Legal Analysis*, 2(2) (2010), 473–523.

[21] Although extremely common as a motive for support, it is difficult to consider job creation as a relevant benchmark for assessing subsidy laws, or for suggesting law reform. In other words, investment in the green energy sector does often generate jobs – and local ones – and this is the public discourse that invariably surrounds the support measure at issue. That being said, trade laws are concerned with negative trade spillovers and may take on board the pursuit of legitimate public policy goals which are essentially non-economic (e.g. health, safety, environment, public morals). Economic or socio-economic objectives may be relevant when they are horizontal, such as innovation promotion or regional development.

law is clear, is it also sound? Are these legal prescriptions in line with good policy indications and guidelines?

We return to these questions in the following section. Suffice it to say here that the question is only partly legal. It engages the 'factual' dimension of the issue of policy autonomy that we stressed at the beginning of the chapter.[22]

Two final glosses are necessary. First, when we ask whether the laws are 'in line with' the policy prescriptions, we do not only point towards the scenario in which blanket bans or prohibitions are legally enforceable. Arguably, the assessment of the laws would not be equally positive if the regulatory framework *de facto* renders certain support measures, which are deemed desirable, liable to be challenged. In other words, what matters is the actual misalignment between law and good policy, and the probability and scope of over-inclusion. In a nutshell, the issue is one of *degree of legal security.*

The second comment refers to the *coherency* of the system. Economics tells us that in some cases subsidies are the best intervention. They are less trade-distorting than, for example, tariffs, because they affect only one margin (the producer's) rather than two (the producer's and the consumer's). If this is the case, one would expect the law to reflect this difference. In the legal framework of the General Agreement on Tariffs and Trade (GATT), tariffs or quotas may eventually be justified under the general exceptions of Article XX, even if they cause trade distortions. This is not the case for subsidies, however. Pending the complex legal issue of the applicability of GATT Article XX to subsidies, if one looks at the ASCM, no exceptions for legitimate subsidies can be found (the old, and limited, shelter of 'non-actionable' subsidies expired at the end of 1999). Do we have double standards? Is this apparent difference in treatment correct?[23] We will return later to this claim of lack of coherency between the GATT and the ASCM.[24]

22 There is significant agreement among scholars that the current rules are not up to the job of facing the challenges of the twenty-first century. The most common example given is that of climate change subsidies. See note 42 for some references.

23 This point is raised in R. Howse, *Climate Change Mitigation Subsidies and the WTO Legal Framework: A Policy Analysis* (Winnipeg, Manitoba: International Institute for Sustainable Development, 2010).

24 We can anticipate the main point made below, that is, that there are various questions that need to be answered before concluding that one discipline is stricter or more generous than another, and the mere fact that one includes exceptions or justifications is certainly not conclusive. Other relevant questions thus concern the relevant default rule and the allocation of the burden of proof.

D Conceptualising Policy Autonomy in Regulation: Point of Balance and Trade-offs

The amount of policy space or autonomy provided is determined in two ways. It is first dictated by the scope of the definition of a subsidy. If a measure does not amount to a subsidy in the first place, it does not come under the scope of the discipline. Other rules, perhaps even stricter ones, may apply, but there is no issue under subsidy laws. If the application of these other rules and subsidy rules is cumulative, the fact that the latter are not applicable means we have, at least in principle, more policy space. Less regulation intuitively means more autonomy. If the above reasoning is correct, one can easily understand why many legal battles in WTO subsidy cases turn on the seemingly neutral and preliminary issue of the definition.

Policy autonomy is also, and most importantly, determined by the actual content of the disciplines that are applicable to a measure that has been classified as a subsidy. The rules may expressly state or, on construction, imply that certain measures are permissible, or, conversely, may prohibit them. We have seen the main tenets of current WTO subsidy laws.

Two concepts are key to assessing the autonomy that WTO subsidy rules allow to governments: the *point of balance* and the *idea of trade-off*.

I Point of balance

The first concept attempts to identify the *point of balance* of the disciplines. Let us use an example to explain this idea. Here, we are talking of subsidies that are aimed at promoting the generation and use of green electricity, and this is definitely linked to an environmental goal. The WTO legal regime in general, and WTO subsidy rules in particular, are trade disciplines. They are intended to foster trade liberalisation and remove obstacles to trade. Traditionally, however, trade laws have also recognised other-than-trade values and accepted that in certain cases there may be clashes between trade interests and non-trade interests. Crucially, for example, under the general exceptions of GATT Article XX, it is also accepted that, under certain conditions, non-trade values may trump trade liberalisation. In other words, in some cases, it is legal under WTO law to adopt measures that, despite in principle breaching even fundamental trade obligations (such as non-discrimination or the prohibition of quotas), also pursue legitimate objectives in a proportionate way and do not entail unjustified discrimination. Going back to the idea of *point of balance*, it is clear that, in the original regulatory framework of the GATT, in some cases this was tilted in favour of non-trade interests and values (and, to a limited extent, to the detriment of trade interests and values).

Now, the key question is: where is the point of balance of subsidy rules? Arguably, when the rules on non-actionability were in force, the ASCM and the GATT were partly aligned. The ethos (and balance) was the same. Both agreements object to some trade-distorting measures. However, in some cases, certain measures, even if trade-distorting, were justified and permitted. With the expiry of Articles 8 and 9 of the ASCM, the point of balance was tilted and is now clearly swinging towards the trade camp. Therefore, green or environmental subsidies are certainly permitted, but only if they do not cause adverse trade effects (as defined by the various standards of Article 5 of the ASCM). If they do, they are actionable. At least in principle, if the non-actionability provisions were still in force, or if – as some have argued – the general GATT Article XX exceptions were applicable to subsidies, we could, in some cases, have more policy space and autonomy, inasmuch as certain trade-distorting measures would be acceptable (and could not be subject to legal challenge).

Let us elaborate on the comparison between the GATT, the original outline of the ASCM (which we can call ASCM 1.0) and the current version of the ASCM (ASCM 2.0), always from the angle of the point of balance.

Under the GATT, if a measure discriminates against (or between) foreign product(s), or if it operates as an obstacle to market access, it is banned. This means that the threshold for intervention is low. Quoting a common statement, Article III and Article XI of the GATT (and also the non-violation nullification and impairment route under Article XXIII:1 of the GATT) protect 'expectations' and 'competitive conditions', not actual 'volumes of trade'. In a nutshell, there is no need to prove the actual trade impact produced by the measure. But, if the general standard of prohibition of the GATT is rigorous, at the same time there are, at least in principle, significant opportunities for justification represented by the general exceptions of GATT Article XX.[25] Both prohibitions and

[25] We say 'in principle' because *de facto* this defence has so far not been very successful in the GATT/WTO jurisprudence.

possibilities for justification need to be accounted for when considering the point of balance of the GATT.

Under the ASCM, only certain subsidies are prohibited. The proof of the key legal requirement of conditionality is demanding but, when it exists, the ban is a foregone conclusion. Again, there is no need to prove actual trade effects. Crucially, however, the scope of the ban is limited to export and local content requirement (LCR) subsidies, which, in the eyes of the negotiators, were undoubtedly characterised by discrimination and protectionism. Thus the ban does not touch on the arguably more important and broader group of domestic subsidies, the discipline of which really defines the point of balance of the ASCM.

In brief, as already explained, subsidies are not prohibited in WTO subsidy disciplines. Apart from the case of export and LCR subsidies, there are no prohibitions in the ASCM. Subsidies are simply actionable. And, as case-law shows, establishing the existence of adverse effects is not easy.[26] Crucially, it is worth recalling that the burden of proof is on the complainant.

Most important to our discussion on the positioning of the point of balance of the disciplines is that the original version of the ASCM – ASCM 1.0 – also included specific provisions making certain R&D, environmental and regional development subsidies non-actionable under WTO dispute settlement and countervailing duty actions. Hence – no prohibitions and limited justifications.

What is the meaning of this extra layer of protection? Arguably, the Uruguay Round negotiators wanted to be sure that certain (albeit limited) subsidies would be granted the maximum possible degree of legal security. In other words, they did not consider the absence of prohibitions sufficient to ensure that certain measures would be protected under subsidy laws. This also meant that, by granting the benefit of non-actionability, they accepted that certain subsidies could cause adverse effects and still be legal. The limit of protection was defined by the precise scope of the exceptions, determined by their terms and conditions and by a measure of proportionality, represented by the threshold of 'serious adverse effects'.

This brief analysis shows that the ASCM 1.0 was definitely more generous towards governments than the GATT: first, because the default rule on subsidies is a no-ban one, meaning that they are allowed unless the complainant proves adverse effects; second, because the point of

[26] There are only three disputes so far in which actionable subsidies claims were successful: *US – Cotton*, *EC – Large Civil Aircraft* and *US – Large Civil Aircraft*.

balance was, at least in terms of the idea underlying the structure of the disciplines, tipped in favour of certain good subsidies (despite the difficulty in proving that these subsidies were actionable, they were nonetheless expressly made safe). These categories were, however, limited in scope, which, in a sense, significantly diminished the practical importance of the idea.[27]

The comparison is more challenging if we consider the second, and current, version of the ASCM – ASCM 2.0. The intervening factor was the expiry of the non-actionability category. In a word, the full legal security which this granted (up to the limit of serious adverse effects) is now gone.

Which set of rules – the GATT or the ASCM 2.0 – is more generous towards government? The answer is not clear. On the one hand, the GATT still has more definite prohibitions than the ASCM, but these can, in principle, be justified if the defendant manages to rely on GATT Article XX. (It is, however, important to highlight that the GATT Article XX general exceptions are for the defendant to establish and, as litigation shows, this has not been easy.[28]) On the other hand, although the exceptions of non-actionability have gone, the default position is still that subsidies – whose simple legal existence is based on the proof of derogation/discrimination– are not per se prohibited.[29] Objection to them is based on the proof of adverse effects and it is for the claimant to prove that such effects exist.

This is what an analysis of the law tells us. This analysis raises another crucial question – which is not a legal one but, once again, demands a

[27] We have also not considered in our assessment the presumptions of adverse effects under Article 6.1 of the ASCM. These were ideally counterparts to the exceptions of the non-actionability provision. The reason why we do not consider them is that their normative innovation – the inversion of the burden of proof in favour of the complainant in certain cases – did not touch upon the extra value of legal security granted by the justifications of Article 8.

[28] *US – Shrimp* is still the only WTO case so far in which a GATT Article XX defence was successful.

[29] This is another line of comparison between, say, national treatment provisions and subsidy disciplines. The legal test of national treatment, which, if positive, triggers a prohibition, is based on discrimination between like domestic and foreign products. Arguably, the core of the notion of subsidy is based on the conclusion that the measure at issue derogates from the otherwise prevailing market or regulatory benchmark. The essence of the two tests is similar, if not identical. But – and here we come to the crucial point – proof of derogation or discrimination is not enough to object to a subsidy. It is necessary to prove a specific trade impact to have it banned. See Rubini, *The Definition of Subsidy and State Aid*, pp. 285–6.

'factual' analysis. What does the best policy practice tell us about 'good' green energy or environmental subsidies? Can we think of measures of support which, in order to achieve their desired goals, need to cause adverse effects? How probable is this? Do we have a real risk that some good green subsidies may also distort trade? Can governments considering support measures design or implement them in such a way that their negative effects are minimised, if not eliminated? How much allowance should the regulatory framework make for the complexity, and sometimes experimental nature, of certain policies?

In a word, where do Members want the point of balance of the disciplines to be? What type of practical and symbolic signals do they want to send with respect to good and bad subsidies? What type of legal security do they aim to grant? What type of flexibility should governments have, especially considering that the fight against climate change is a complex one and the success and effects of the policies may be uncertain?[30]

In conclusion, the question is not ultimately whether there is full coherence in the WTO system (although, in principle, one would expect this, unless there are specific reasons for divergences in treatment). The issue is whether the current point of balance of the subsidy laws is acceptable in the light of the economic, political and legal circumstances of the twenty-first century.

The fact that there have been few challenges to energy subsidies may be good evidence that Members have so far been happy with the status quo. Against mounting litigiousness and claims, and increased investment and competition, one wonders whether this scenario still reflects what is needed. This is – at the very least – a legitimate question to ask.

II *Trade-offs*

The second idea – which is partly implied in the concept of balance, and which explains the difficulty in determining the point of balance of subsidy disciplines, or indeed renegotiating it – is that of the trade-offs that are inherent in subsidisation.

Subsidies are equivocal because they may have various positive and negative effects at the same time. As we have noted elsewhere, underlying

[30] This, in essence, is what the WTO Appellate Body Report, *Brazil – Tyres*, WT/DS332/AB/R, para. 151, seems to be hinting at when it suggests that certain environmental measures, e.g. climate change measures, can be fully evaluated only with the benefit of time.

any policy decision and any legal compromise, there is a trade-off.[31] Economic distortions are thus accepted if, in view of the preferences and choices of the granting government, it is expected that the benefits will outweigh the disadvantages. There is no precision or inevitability in where the line is drawn (which reminds us of the concept of point of balance above). The main difficulty, however, comes when negative and positive effects are produced in different countries, since it is not easy to gain acceptance for transnational trade-offs. In these cases, the policy discourse clearly transcends the domestic and local level to reach the international and global one. Complex issues of settlement of conflicts of interest enter into play. This scenario may explain the stalemate in which Members find themselves when they even think of changing WTO rules.

The recently suggested linkage of law reform discourse to the idea of 'public goods', which is particularly relevant when one talks of common challenges such as the fight against climate change, may thus be a welcome move towards breaking this stalemate.[32]

E The Role and Limitations of the Jurisprudence

By now it should be obvious why subsidy laws are notoriously unclear. The ultimate reason for this is that Members have – and have always had – different views on subsidisation.[33] And these views change with time, and with the rise or fall of certain stakeholders. Ideologies also come and go. As repeatedly noted, technology develops, markets evolve and the political economy alters. For these reasons, fundamental features of subsidy laws – such as the definition itself of what constitutes a subsidy, or the conclusion that a certain subsidy is good or bad – have always been controversial.

Law can be clarified by and through jurisprudence. This is – in the end – courts' most important contribution to the administration of

[31] Rubini, 'Ain't wastin' time no more', p. 532. See also Luca Rubini, *The Definition of Subsidy and State Aid*, at pp. 54–6.

[32] Aaron Cosbey and Petros Mavroidis, 'A turquoise mess: green subsidies, blue industrial policy and renewable energy: the case for redrafting the subsidies agreement of the WTO', *Journal of International Economic Law*, 17(1) (2014), 11–47, at 45–6. See also Rubini, *The Definition of Subsidy*, p. 55.

[33] Gary Hufbauer and Joanna Shelton-Erb, *Subsidies in International Trade* (Cambridge, MA: MIT Press, 1984), noted early on the confrontation between the 'injury-only' and the 'anti-distortion' schools of thought.

justice. But there is a limit to this possibility. If WTO Panels and the Appellate Body can – and must – provide clarification on certain issues, it is also equally clear that there is a line that cannot be crossed. Courts are 'agents' and not 'principals'.[34] Unless the relevant legal system permits it, they cannot create law by filling gaps *and* making fundamental decisions on what is good and what is bad. This, according to the numerous and detailed commentaries to date, is the main problem with the 'benefit' interpretations of the Panel majority and the Appellate Body in the *Canada – Renewable Energy/FIT* disputes. The WTO dispute settlement body was called upon to decide whether Ontario's FIT system, which provides incentives to renewable energy producers and, critically, includes a local content requirement, was in line with WTO laws on national treatment and subsidies. After easily disposing of the local content requirement, which was patently in breach of the non-discrimination principle,[35] the Panel and the Appellate Body were both asked to establish whether the measure was also a prohibited subsidy. Determination of whether the FIT was a subsidy was the preliminary step in this analysis. In this respect, it was necessary to conclude that the incentive scheme was really providing a benefit to its recipients. Both courts rejected the claims, arguments and benchmarks put forward by the claimants, and concluded that they were not in a position to find for the existence of a benefit.

The Appellate Body's reasoning is particularly important, and has been the subject of much analysis and discussion. In a nutshell, the world trade tribunal introduced two innovations to its long-standing jurisprudence on the benefit analysis under Article 1 of the SCM Agreement. It first required that a full market definition be carried out.[36] The function of this exercise is to identify the universe within which the appropriate market benchmarks can be found. Crucially, the Appellate Body found that this analysis should not stop with the demand-side substitutability, but should extend to consider the supply side.[37] After concluding that the market at issue in the case was only that for renewable energy (as

[34] Unless the legal system itself requires them to create law. Think of what has traditionally been the role of the judge in common law systems.

[35] In another more recent case, the local content requirements enshrined in India's national solar programme were condemned under the non-discrimination principle. See *India – Solar Cells*, WT/DS456.

[36] See WTO Appellate Body Report, paras. 5.167–5.179.

[37] Ibid., para. 5.171 et seq.

opposed to the broader energy market),[38] the Appellate Body introduced its second innovation. It distinguished two scenarios: those in which governments (simply) create markets and those in which they intervene in already existing markets.[39] Importantly, the former is not 'in and of itself' a subsidy. It also transpires that, in the first scenario, the key requirement that a government's creative action needs to satisfy if it is to profit from this shelter is that the action of support is proportionate and not excessive. So, if we have a new market and the creative act is simply conducive to bringing this market into existence, WTO subsidy laws do not apply. It is clear that, through this interpretation, the Appellate Body managed to take certain forms of subsidisation in the clean energy sector outside subsidy control.

Our assessment of this decision is negative. In other writings we have severely criticised both the Panel (majority) and the Appellate Body.[40] The burning question was: how is it possible that what is almost a textbook example of a subsidy turned out not to be a subsidy at all?[41] More specifically, it is our view that their interpretation of the notion of benefit amounts to a fundamental misconstruction of the law. The result is a limited but unwarranted carve-out of certain types of subsidies from subsidy disciplines. The implications of this exception are troublesome, especially for transparency and subsidy control at large. If our reading is correct, it is desirable that future decisions limit the impact of this decision. In a recent analysis, we speculated about this and eventually concluded that it may not be possible.[42] Significantly, the vast majority of the various commentaries on the case to date are in agreement with the conclusion that the Appellate Body (and the Panel majority before it) went too far.[43]

[38] Appellate Body Report, para. 5.178. [39] Appellate Body Report, para. 5.188.

[40] Rubini, 'The good, the bad, and the ugly'; Aaron Cosbey and Luca Rubini, 'Does it FIT? An assessment of the effectiveness of renewable energy measures and of the implications of the Canada – Renewable Energy/FIT disputes' (2013), *E-15 'Think-Piece'*, http://e15initiative.org/publications/does-it-fit-an-assessment-of-the-effectiveness-of-renewable-energy-measures-and-of-the-implications-of-the-canada-renewable-energyfit-disputes/.

[41] See also the *amicus curiae* brief submitted to the Appellate Body on 13 March 2013.

[42] See L. Rubini, '"The wide and the narrow gate": benchmarking in the SCM Agreement after the *Canada – Renewable Energy/FIT* ruling', *World Trade Review*, 14(2) (2015), 211–37.

[43] See, e.g., A. Breckenridge, *FIT for Purpose? A Review of the Economics of a WTO Panel Ruling on Feed-in Tariffs and Local Content Requirements*; C. Briefing, Frontier Economics (March 2013); A. Breckenridge, *A Matter of Definition – Commentary of Aspects of the Appellate*

The *Canada – Renewable Energy/FIT* ruling is an excellent example of the difficulties highlighted in the previous paragraph with regard to the interpretation of subsidy laws. It also well exemplifies the difficulties of delegating fundamental decisions to dispute settlement.[44]

If the law is uncertain and unsound, and case-law cannot really help to improve the situation and tell us firmly what type of government intervention is good and what is bad, what should be done?

F The Only Real Option

This final section simply outlines what, in our view, is the only way forward: law reform. Our mandate – the definition of the current green space allowed by WTO subsidy laws with respect to support to electricity – has been fulfilled. Most crucially, space constraints do not allow us to develop this entire new line of analysis.

There is broad agreement among scholars that reform of subsidies rules is needed.[45] In terms of law reform, however, reference is normally

Body's Ruling on the Canada – Renewable Energy case in the WTO (Client Briefing, Frontier Economics, October 2013); L. Casier and T. Moerenhout, *WTO Members, Not the Appellate Body, Need to Clarify Boundaries in Renewable Energy Support*, IISD Commentary (2013); Cosbey and Mavroidis, 'A turquoise mess'; R. Pal, 2014. 'Has the Appellate Body's decision in *Canada – Renewable Energy/Canada – Feed-In Tariff Program* opened the door for production subsidies?', *Journal of International Economic Law*, 17(1) (2014), 125–37; S. Charnovitz and C. Fischer, 'Canada – Certain Measures Affecting the Renewable Energy Sector/Canada – Measures Relating to Feed-In Tariff Program', *World Trade Review*, 14(2) (2015), 177–210. For pieces which view the case-law development less negatively, or even positively, see, e.g., A. Kent and V. Jha, 'Keeping up with the changing climate: the WTO's evolutive approach in response to the trade and climate conundrum', *Journal of World Investment and Trade*, 15 (2014), 245–71; A. Lang, 'Governing "as if": global subsidies regulation and the benchmark problem', *LSE, Law, Society and Economy Working Papers* 12/2014 (2014); R. Howse, 'Securing policy space for clean energy under the SCM Agreement'.

44 For a comprehensive analysis of these issues see L. Rubini, 'The good, the bad and the ugly'.

45 See, e.g., R. Howse, 'Do the World Trade Organization disciplines on domestic subsidies make sense? The case for legalizing some subsidies', in K. Bagwell, G. Bermann and P. Mavroidis (eds.), *The Law and Economics of Contingent Protection in International Trade* (Cambridge: Cambridge University Press, 2010), p. 85; Rubini, 'Ain't wastin' time no more'; G. Horlick and P. Clarke, 'WTO subsidies disciplines during and after the crisis', *Journal of International Economic Law*, 13(3) (2010), 859; P. Aerni et al., 'Climate change, human rights and international economic law: exploring the linkages between human rights, trade and investment', *German Yearbook of International Law*, 53 (2010), p. 139; Bigdeli, 'Resurrecting the dead'; D. Steger, 'The Subsidies and Countervailing Measures Agreement: ahead of its time or time for reform?', *Journal of World Trade*, 44(4) (2010), 779. These calls for reform have been repeated following the *Canada – Renewable Energy/*

made to *new substantive disciplines* specifically permitting certain good subsidies.[46] While we agree with most of what is said and proposed, we adopt a broader notion of 'reform', one that is 'process-based'.

It is crucial to consider law reform for what it is, especially in the international arena: a process. This means that we should not think only, or immediately, about what substantive disciplines should emerge from reform. This is the outcome of the process. The suggestion made here is that the focus and energy should be directed towards how to start and how to design the process so that it creates the right conditions for possible future negotiations.

Right now, the conditions for new hard, binding law on subsidies do not seem to be present.[47] For this reason, it is probably unrealistic to expect meaningful deliberations by Member representatives on subsidy issues. This is particularly true since the Appellate Body invested decisions taken with the Committees with the value of binding law.[48] Against this background, one does not expect government representatives to make statements that could then be held against them in litigation.

There is another dimension to reform apart from deliberation, which is preliminary to actual negotiations and decisions – that of knowledge generation.

What is fundamental, if we accept that the current regulatory framework is not satisfactory and if we are serious about the need for change, is to kick-start the process for reform and frame it in the right way. Reforms – and their climax, negotiations, if that stage is ever to be reached – must be duly prepared. Discussions need to be started, and for these discussions to be meaningful and to advance knowledge and understanding, it is necessary to rely as much as possible on facts and data and their assessment. The ultimate goal of the exercise is to assess the subsidy practices of Members and evaluate their effects and effectiveness. This – it is hoped – could clarify the terms of the debate and

FIT litigation. See, e.g., Casier and Morenhout, 'WTO Members, not the Appellate Body, need to clarify boundaries in renewable energy support'; Cosbey and Mavroidis, 'A turquoise mess'; Rubini, 'The good, the bad, and the ugly'.

46 Reform could be achieved in various ways. R. Howse, 'Securing policy space for clean energy under the SCM Agreement', put forward the idea of an interpretative understanding.

47 It remains to be seen whether the recent developments (such as the Bali and Nairobi Ministerial packages and the green goods initiative) can be viewed as positive indications of a changing mood and appetite.

48 WTO Appellate Body Report, *US – Tuna II*, WT/DS381/AB/R, paras. 373–378.

facilitate diplomatic discussion and negotiations when the right political conditions are present. In this respect, the advice of experts should also be welcomed. The right venues for this process of knowledge generation should be carefully chosen.

Use could be made of the mechanisms that are already built into the ASCM, such as the Group of Experts or the creation of bodies 'subsidiary' to the Committee on Subsidies and Countervailing Measures,[49] or into the WTO at large, such as the Trade Policy Review Mechanism. This, again, is a matter of rendering operational devices negotiated and agreed on in the Uruguay Round.

In any event, the two key words should be pragmatism and imagination – the essence of the spirit of the GATT. What is sought is not a revolution but a well-thought-out evolution of the system and its disciplines. This is the narrative that should prevail and should reassure Members, and all relevant stakeholders, that they are in the driving seat, and that no back-door coup is planned.

Bibliography

Aerni, P. et al. (2010). Climate change, human rights and international economic law: exploring the linkages between human rights, trade and investment. *German Yearbook of International Law*, **53**, 139.

Bigdeli, S. (2011). Resurrecting the dead? The expired non-actionable subsidies and the lingering question of 'green space', *Manchester Journal of International Economic Law*, **8**(2), 2–37.

Bradsher, K. (2010). Union accuses China of illegal clean energy subsidies. *New York Times*, 9 September 2010, www.nytimes.com/2010/09/10/business/energy-environment/10steel.html.

Brandeis, L. (1916). The living law. *Illinois Law Review*, **10**, 470.

Breckenridge, A. (2013a). *FIT for purpose? A Review of the Economics of a WTO Panel Ruling on Feed-in Tariffs and Local Content Requirements*, Client Briefing, Frontier Economics, March 2013.

(2013b). *A Matter of Definition – Commentary of Aspects of the Appellate Body's Ruling on the Canada – Renewable Energy Case in the WTO*, Client Briefing, Frontier Economics, October 2013.

Casier, L. and Moerenhout, T. (2013). *WTO Members, Not the Appellate Body, Need to Clarify Boundaries in Renewable Energy Support*, IISD Commentary, www.iisd.org/pdf/2013/wto_members_renewable_energy_support.pdf.

[49] See Article 24.2 of the SCM Agreement.

Charnovitz, S. and Fischer, C. (2015). *Canada – Certain Measures Affecting the Renewable Energy Sector/Canada – Measures Relating to Feed-In Tariff Program*. *World Trade Review*, **14**(2), 177–210.

Cosbey, A. and Rubini, L. (2013). Does it FIT? An assessment of the effectiveness of renewable energy measures and of the implications of the Canada – Renewable Energy/FIT disputes. *E-15 Think-Piece*, http://e15initiative.org/publications/does-it-fit-an-assessment-of-the-effectiveness-of-renewable-energy-measures-and-of-the-implications-of-the-canada-renewable-energy fit-disputes/.

Cosbey, A. and Mavroidis, P. (2014). A turquoise mess: green subsidies, blue industrial policy and renewable energy: the case for redrafting the subsidies agreement of the WTO. *Journal of International Economic Law*, **17**(1), 11–47.

Holzer, K., Espa, I. and Payosova, T. (2017). Chapter 18, this volume.

Horlick, G. N. and Clarke, P. A. (2010). WTO subsidies disciplines during and after the crisis. *Journal of International Economic Law*, **13**(3), 859.

Howse, R. (2010a). *Climate Change Mitigation Subsidies and the WTO Legal Framework: A Policy Analysis*. Winnipeg, Manitoba: International Institute for Sustainable Development, www.iisd.org/pdf/2009/bali_2_copenhagen_subsidies_legal.pdf.

(2010b). Do the World Trade Organization disciplines on domestic subsidies make sense? The case for legalizing some subsidies. In K. Bagwell, G. Bermann and P. C. Mavroidis (eds.), *The Law and Economics of Contingent Protection in International Trade*. Cambridge: Cambridge University Press, p. 85.

(2013). Securing policy space for clean energy under the SCM Agreement: alternative approaches. *E-15 'Think-Piece'*, http://e15initiative.org/publications/securing-policy-space-for-clean-energy-under-the-scm-agreement-alternative-approaches/.

Hufbauer, G. and Shelton-Erb, J. (1984). *Subsidies in International Trade*. Cambridge, MA: MIT Press.

Kent, A. and Jha, V. (2014). Keeping up with the changing climate: the WTO's evolutive approach in response to the trade and climate conundrum. *Journal of World Investment and Trade*, **15**, 245–71.

Lang, A. (2014). Governing 'as if': global subsidies regulation and the benchmark problem. *LSE, Law, Society and Economy Working Papers* 12/2014, www.lse.ac.uk/collections/law/wps/WPS2014-12_Lang.pdf.

Lincicome, S. (2010). USW to China: Subsidies for me, but not for thee. 9 September, blog, http://lincicome.blogspot.ch/2010_09_01_archive.html

Rajib, P. (2014). Has the Appellate Body's decision in *Canada – Renewable Energy / Canada – Feed-In Tariff Program* opened the door for production subsidies? *Journal of International Economic Law*, **17**(1), 125–37.

Rubini, L. (2009). *The Definition of Subsidy and State Aid – WTO and EC Law in Comparative Perspective*. Oxford: Oxford University Press.

(2012). Ain't wastin' time no more: subsidies for renewable energy, the SCM Agreement, policy space and law reform. *Journal of International Economic Law*, **15**(2), 525–79.

Rubini, L. (2014). The good, the bad, and the ugly: lessons on methodology in legal analysis from the recent WTO litigation on renewable energy subsidies. *Journal of World Trade*, **48**(5), 895–938.

(2016). WTO subsidy laws: the international regulation of state aid. In Herwig C .H. Hoffmann and Claire Micheau (eds.), *State Aid Law of the European Union*. Oxford: Oxford University Press.

Steger, D. (2010). The Subsidies and Countervailing Measures Agreement: ahead of its time or time for reform? *Journal of World Trade*, **44**(4), 779–96.

Sykes, A. O. (2010). The questionable case for subsidies regulation: a comparative perspective. *Journal of Legal Analysis*, **2**(2), 473–523.

Vermulst, E. and Meng, M. (2017). Chapter 17, this volume.

Wu, M. and Salzman, J. (2014). The next generation of trade and environment conflicts: the rise of green industrial policy. *Northwestern University Law Review*, **108**, 401–74.

17

Dumping and Subsidy Issues in the Renewable Energy Sector

EDWIN VERMULST AND MADISON MENG

A Introduction

Since the beginning of the twenty-first century, governments worldwide have prioritised the growth of renewable energy technologies through shifts in policies and increases in incentives. The rapid development of renewables production, especially in industrialising economies such as China, has led to a progressively globalised supply system, and consequently a surge in trade issues involving renewable energy products. Against this background, this chapter seeks to examine certain renewable energy cases brought before the World Trade Organization (WTO) and at the national level. The focus is on trade defence cases concerning solar products from China and the relevant disputes at the WTO with respect to domestic measures related to renewable energy, including the pertinent findings of the Appellate Body (AB) in *Canada – Renewable Energy/Feed-in Tariff Program*, as well as the disputes over biofuels/biodiesel between the EU, Argentina and Indonesia at the WTO.

Although the focal points in most of the cases being assessed are the subsidies granted and the domestic or local content schemes instituted to promote an emerging sector, it should be noted that those developments are a global phenomenon. As renewable energy policies tend to rely on some kind of government support to stimulate the growth of domestic manufacturing industries, it is not surprising that they run into conflict with international trading rules.

Having set out the premise above, the rest of the chapter is structured as follows. Section B examines the domestic trade defence cases on solar products, namely the anti-dumping (AD) and/or countervailing duty (CVD) cases brought by the United States, the European Union (EU), India and Australia against solar products from China. Section C provides an analysis of the relevant findings of the AB in the *Canada – Renewable Energy/Feed-in Tariff Program* WTO disputes and also briefly covers the complaint filed by the United States against India at the WTO regarding the local content requirement of India's solar programme.[1] Section D presents the issues in the biofuels/biodiesel disputes involving the EU, Argentina and Indonesia at both the EU and the WTO levels. Section E concludes the chapter.

B The AD and CVD Investigations against Chinese Solar Products by the United States, the EU, India and Australia

Before discussing the specific cases raised by the United States, the EU, India and Australia against Chinese solar products, a general introduction to the AD and CVD methodologies they have used against China is provided.

I The AD and CVD Methodologies Used against China by the United States, the EU, India and Australia

While China is certainly a world-leading producer and exporter of solar modules, it is an easy target for AD and CVD cases, as it is considered a so-called 'non-market economy' (NME) by the EU, the United States and India, among others.

According to the domestic laws of the EU countries, the United States and India, since China is considered a NME, Chinese domestic producers' sales prices or costs are not compared with Chinese export prices, as would normally be required by the WTO Anti-Dumping Agreement (ADA), in order to assess the existence of dumping. Instead, the EU uses domestic sales prices or costs of producers of the like product in a third market economy or analogue country, while the United States establishes the dumping margin by comparing the export price to a constructed price in a surrogate country based on the factors of production methodology. These calculation methods make a finding of dumping relatively simple and the dumping margins are high.

India basically follows an analogue country methodology similar to that of the EU. In the recent solar case, India used itself as an appropriate third market economy country (coincidentally, India also happened to be the analogue country selected by the EU in its AD case).

[1] Request for Consultations, *India – Certain Measures Relating to Solar Cells and Solar Modules*, WT/DS456/1, circulated on 11 February 2013.

Australia, on the other hand, granted China market economy status in 2004. Accordingly, in its AD cases against China, Australia compares the export price with the domestic price of the like product in China to determine the dumping margin, although it sometimes makes cost adjustments.

Turning to the CVD cases against China, the EU and the United States generally resort to the establishment of external benchmarks for loans, inputs and land-use rights, among others, on account of the alleged distortion of these cost elements in China as a result of government intervention. However, the findings of distortions (and specificity) are typically based on the alleged non-cooperation of the Chinese government, which is overburdened with information requests with which it cannot reasonably comply. As a result, findings of subsidisation become a matter of course based on publicly available information that is often historical and referred to by the complainants.

Similar to the United States and the EU, Australia has used external benchmarks to calculate the benefit of certain inputs by reasoning that market distortions exist because of the Chinese government's involvement in the domestic market.[2]

It is worth noting that the United States and the EU, among others, have started CVD cases against China in recent years in anticipation of the fact that they would no longer be able to use the NME methodology for China in AD cases after 11 December 2016, pursuant to China's WTO Protocol of Accession.[3] However, there is no provision precluding the countries from utilising external benchmarks in CVD cases. Therefore, more CVD cases are being initiated against China, and this trend can be expected to continue.

Lastly, in the case of parallel AD and CVD proceedings in the EU, the effect of the simultaneous use of the analogue country methodology and the out-of-country benchmarks is limited to a certain extent through the lesser duty rule, which caps the level of the total duty to be imposed and

[2] See' Report to the Minister No. 148 on certain aluminium extrusions exported to Australia from China', Australian Customs and Border Protection Service, 15 April 2011. China's domestic prices for aluminium were considered inappropriate for the benchmark price since 'prices of primary aluminium supplied by SOEs [state-owned enterprises] are likely to have influenced domestic primary aluminium prices generally'. Therefore, the prices for primary aluminium on the London Metal Exchange, with adjustment for delivery and other costs, were used as the benchmark.

[3] See Accession of the People's Republic of China, WT/L/432, Article 15(a), 23 November 2001.

prevents the double-counting problem. In other words, assuming that the dumping margin is 50 per cent, the subsidy margin is 10 per cent and the injury margin is 30 per cent, in the EU the CVD would then be 10 per cent and the AD duty would be limited to 20 per cent. Through this methodology the EU avoids the double-counting issue which was challenged by China in the context of certain US trade defence measures on Chinese products, since the United States has not adopted the lesser duty rule and the AD duty is therefore always equal to the dumping margin found by the United States. However, in keeping with the ongoing modernisation of the trade defence instruments (TDIs), the European Commission has proposed the abolition of this WTO-plus element of the EU's TDI practice in CVD cases and in AD cases where 'structural raw material distortions' exist.[4]

Like the EU, India has adopted the lesser duty rule, which operates to restrict the duty to the dumping margin or the injury margin, whichever is lower, in AD cases. In contrast, Australia recently underwent a reform that included the removal of mandatory consideration of the lesser duty rule. Since January 2014, Australia has no longer been required to consider the lesser duty rule where certain conditions are met:

> in dumping cases, where there is a finding of a particular market situation for the goods in the country of export; in subsidy cases, where the country of export has not complied with World Trade Organization (WTO) obligations to notify subsidies; and in dumping and/or subsidy cases, where the Australian industry producing like goods consists of at least two small-medium enterprises – whether or not that industry consists of other enterprises.[5]

The deletion of mandatory consideration of the lesser duty rule means that higher duties are likely to be applied to exporting producers, since duties could be imposed up to the full dumping or countervailing margin, even though such high duties may provide the domestic industry with protection above and beyond what is necessary to prevent injury.

II The USA's Solar Cases

The original US AD and CVD cases took place against the background discussed above from 2011 to 2012. These cases covered 'solar cells,

[4] Commission publication, 'Commission proposes to modernise the EU's trade defence instruments', European Commission, 10 April 2013.

[5] Updates to anti-dumping system – January 2014, Australian Anti-dumping Commission, Anti-Dumping Notice No. 2013/108, 20 December 2013.

Table 17.1 *Original US AD case – final determination*[a]

Exporter/producers (*sampled companies)	Duty rate (per cent)
Wuxi Suntech companies*	29.14
Trina Solar companies*	18.32
Non-sampled cooperating companies	24.48
All others	249.96

[a] Amended final determination of sales at less than fair value and anti-dumping duty order of crystalline silicon photovoltaic cells, whether or not assembled into modules, from the People's Republic of China, International Trade Administration, Federal Register Vol. 77 No. 236, 7 December 2012.

Table 17.2 *Original US CVD case – final determination*[a]

Exporter/producers (*sampled companies)	Duty rate (per cent)
Wuxi Suntech companies*	14.78
Trina Solar companies*	15.97
All others	15.24

[a] Countervailing duty order of crystalline silicon photovoltaic cells, whether or not assembled into modules, from the People's Republic of China, Federal Register Vol. 77 No. 236, 7 December 2012.

whether or not assembled into modules or panels', originating in China. In the AD case, the 'all others' final anti-dumping duty reached almost 250 per cent. While the duties in the CVD case were not as high as the duties in the AD case, of the two sampled companies, about 71 per cent of Wuxi Suntech's and 66 per cent of Trina Solar's subsidy margins were accounted for by a finding that the producers might have benefitted from export credits. This finding was based on facts available since the US Department of Commerce concluded that 'it could not verify non-use of export buyer's credits during the verification of the questionnaire responses of the company respondents'.[6] For reference, the 2011–12 duty rates from the AD and CVD cases are provided in Tables 17.1 and 17.2.

[6] Issues and decision memorandum for the final determination in the countervailing duty investigation of crystalline silicon photovoltaic cells, whether or not assembled into

Table 17.3 *2014 US AD case – preliminary determination (China)*[a]

Exporter/producers (*sampled companies)	Duty rate (per cent)
Trina Solar companies*	26.33
Renesola/Jinko companies	58.87
Non-sampled cooperating companies	42.33
All others	165.04

[a] Preliminary affirmative determination of sales at less than fair value and postponement of final determination of certain crystalline silicon photovoltaic products from the People's Republic of China, International Trade Administration, Federal Register Vol. 79 No. 147, 31 July 2014.

In 2013, SolarWorld America alleged that Chinese producers were circumventing the original duties by using Taiwanese cells in their modules. This led to the initiation of new AD and CVD investigations in January 2014, which covered 'solar cells and modules or panels, whether or not assembled into other products but excluding the cells covered by the original measures'. In the AD case, 'solar cells and modules or panels' from China and Taiwan were investigated;[7] in the CVD case, the same products from China only were investigated.[8] The preliminary duties imposed by the United States on the Chinese producers are shown in Tables 17.3 and 17.4. In reaching these AD and CVD rates for the Chinese producers, the United States employed the NME methodologies discussed above. It is also relevant that, even though the preliminary duties in the CVD case were generally lower than the duties imposed in the AD case, about 43 per cent of Wuxi Suntech and 76 per cent of Trina Solar's countervailing margins were based on the preliminary finding that the producers purchased solar glass from public bodies, which was made on the basis of facts available (the preliminary duty rates are provided in Tables 17.3 and 17.4). The final determinations had not been published at the time of writing.

modules, from the People's Republic of China, International Trade Administration, 9 October 2012, 61.

7 Initiation of anti-dumping duty investigations of certain crystalline silicon photovoltaic products from the People's Republic of China and Taiwan, International Trade Administration, Federal Register Vol. 79 No. 19, 29 January 2014.

8 Initiation of countervailing duty investigation of certain crystalline silicon photovoltaic products from the People's Republic of China, Federal Register Vol. 79 No. 19, 29 January 2014.

Table 17.4 *2014 US AD case – preliminary determination (China)*[a]

Exporter/producers (*sampled companies)	Duty rate (per cent)
Wuxi Suntech companies*	35.21
Trina Solar companies*	18.56
All others	26.89

[a] Preliminary affirmative determination countervailing duty investigation of certain crystalline silicon photovoltaic products from the People's Republic of China, International Trade Administration, Federal Register Vol. 79 No. 111, 10 June 2014.

Table 17.5 *2014 US AD case – amended preliminary determination (Taiwan)*[a]

Exporter/producer (* sampled companies)	Duty rate (per cent)
Motech Industries, Inc.*	20.86
Gintech Energy Corporation*	27.59
All others	24.23

[a] Notice of amended preliminary determination of sales at less than fair value of certain crystalline silicon photovoltaic products from Taiwan, International Trade Administration, Federal Register Vol. 79 No. 163, 22 August 2014. The duties were amended to correct ministerial errors.

For the sake of completeness, the preliminary determinations for the Taiwanese exporters or producers in the AD case are presented in Table 17.5.

III The EU's Solar Dispute

Following the example from the original 2011–12 investigations in the US, the EU AD and CVD cases were spearheaded by SolarWorld AG, together with some other EU producers such as Solsonica, under the banner of EU ProSun. The EU AD and CVD cases initially included 'solar modules, cells, and wafers', but eventually 'wafers' were excluded from the product scope. Additionally, the duration of the EU measures has been limited – for the time being – to two years, in contrast to the US cases, which are limited to five years.

Table 17.6 *EU CVD case – definitive measure*[a]

Exporter/producer (*sampled companies)	Duty rate (per cent)
Wuxi Suntech companies*	4.9
Trina Solar companies*	3.5
Jiangxi LDK Solar companies*	11.5
Delsolar and related companies*	0
JingAo companies*	5.0
Jinko Solar companies*	5.0
Renesola companies*	4.6
Yingli companies*	6.3
Non-sampled cooperating companies	6.4
All others	11.5

[a] Council Implementing Regulation (EU) No 1238/2013 of 2 December 2013 imposing a definitive anti-dumping duty and collecting definitively the provisional duty imposed on imports of crystalline silicon photovoltaic modules and key components (i.e. cells) originating in or consigned from the People's Republic of China, OJ 2013 No. L 325/1, 5 December 2013.

As mentioned above, in the US cases, the original measures were imposed against 'solar cells originating in China whether or not assembled into modules'. The EU cases, however, covered 'solar modules and cells originating in or consigned from China'. This implies that modules assembled in China from non-Chinese cells are also covered by the measure, even though they would normally have, for example, Taiwanese origin. In addition, the European Commission has adopted product-specific origin rules for solar cells and modules. The combined operation of the two means that the loophole existing in the original US cases does not exist as regards the EU measures. The definitive AD and CVD rates are provided in Tables 17.6 and 17.7.

Importantly, the EU used external benchmarks to calculate the main subsidies found in the CVD case, which were loans, credit lines and the provision of land-use rights. Of the three subsidies, the determinations of loans and credit lines were based on the finding that the banks or financial institutions supplying the loans or credit lines to the exporting producers were public bodies as found on the basis of facts available, and the determination regarding the provision of land-use rights relied on a specificity finding that was also based on facts available. Moreover, in constructing the external benchmarks for the loans, the EU again used

Table 17.7 *EU AD case – definitive measure*[a]

Exporter/producer (*sampled companies)	Duty rate (per cent)
Wuxi Suntech companies*	41.4
Trina Solar companies*	44.7
Jiangxi LDK Solar companies*	46.7
Delsolar and related companies*	64.9
JingAo companies*	51.5
Jinzhou Yangguang companies*	27.3
Yingli companies*	35.5
Jinko Solar companies	41.2
Renesola companies	43.1
Non-sampled cooperating companies not subject to CVD	41.3
Non-sampled cooperating companies subject to CVD	36.2
All others	53.4

[a] Council Implementing Regulation (EU) No 1239/2013 of 2 December 2013 imposing a definitive countervailing duty on imports of crystalline silicon photovoltaic modules and key components (i.e. cells) originating in or consigned from the People's Republic of China, OJ 2013 No. L 325/66, 5 December 2013.

facts available by claiming a lack of cooperation by the government of China.[9] The external benchmarks used by the EU for loans in renminbi (RMB) are illustrated in Table 17.8, which shows that they were significantly higher than the domestic rates. Also, the benchmark interest rates for loans in US dollars (USD) and euros (EUR) were based on the simple average of corporate bonds which were issued in the investigation period, rated BB+, BB– or BB, and are either in USD or EUR as sourced from Bloomberg. As for credit lines, the credit line obtained by one of the sampled exporting producers from a bank whose headquarters are established in a financial jurisdiction other than China was used as the external benchmark. This benchmark was subject to an arrangement fee of 1.29 per cent and a commitment fee of 0.5 per cent. With respect to the provision of land-use rights, the benchmark land prices were established on the basis of the land prices in Taiwan, adjusted for inflation and gross domestic product (GDP) development, and were higher than domestic prices.

[9] Ibid., recital 197.

Table 17.8 *EU CVD case – external benchmark interest rates for loans in RMB*[a]

Duration of the loan	People's Bank of China rate (%)	Mark-up (%)	Benchmark (%)
Less than 6 months	6.08	3.447	9.527
6 months to 1 year	6.54	3.447	9.987
1–3 years	6.63	3.447	10.077
3–5 years	6.88	3.447	10.327
More than 5 years	7.03	3.447	10.477

[a] General Disclosure Document, Countervailing duty investigation on imports of crystalline silicon photovoltaic modules and key components (i.e. cells) originating in or consigned from the People's Republic of China, Annex 2, August 2013.

On 2 August 2013, the European Commission accepted an undertaking by a large number of Chinese exporting producers – more than 130 companies or groups of companies, jointly with the China Chamber of Commerce for Import and Export of Machinery and Electronic Products (CCCME) – in the context of the AD and CVD cases. This undertaking sets a minimum import price for solar panels and cells from the exporting producers that signed the undertaking. It also includes a quantity threshold, which is allocated among the various Chinese exporting producers by CCCME based on their market performance. Additionally, the undertaking is subject to strict documentary and reporting requirements. All cooperating Chinese exporting producers that did not sign the undertaking are subject to a combined AD and CVD rate of 47.7 per cent, and those exporting producers that did not cooperate in the investigations are subject to a combined duty rate of 64.9 per cent.

However, the Commission's acceptance of the undertaking was not the end of the road. First, the Commission Decision accepting the undertaking was challenged by five EU producers, led by SolarWorld AG, before the General Court of the EU (General Court) in late 2013, on the grounds of a manifest error of assessment of the undertaking price and violation of the rights of defence of EU ProSun, as well as a violation of EU competition law to the extent that the undertaking results in the acceptance and reinforcement of a horizontal price fixing arrangement.[10]

[10] Case T-507/13, *SolarWorld and Others* v. *Commission* [2013].

Moreover, following the publication of the AD and CVD definitive regulations in December 2013, SolarWorld AG and two other EU producers brought actions against the European Council on 28 February 2014 for the annulment of Article 3 in the AD definitive regulation and Article 2 in the CVD definitive regulation, which exempted Chinese exporting producers that are parties to the undertaking agreement from the payment of the duties.[11] The applicants also asked for these cases to be joined with the action regarding the Commission's decision accepting the undertaking.

Third, in June 2014, EU ProSun requested that the Commission investigate its claims that the Chinese producers breached the undertaking agreement by selling products at prices below the minimum import price.

Fourth, SolarWorld AG is reportedly preparing a petition for an anti-circumvention investigation based on allegations that Chinese producers are circumventing the duties on solar products by exporting to the EU through ports in Japan, Taiwan and Malaysia.

All of these examples show that the disputes between the EU and China regarding solar products remain active and are far from over.

IV India's Solar Case

As well as the United States and the EU, India has investigated solar products from China. On 23 November 2012, the Directorate General of Anti-Dumping and Allied Duties (DGAD) under the Ministry of Commerce and Industry of India initiated an AD investigation into 'solar cells, whether or not partially or fully assembled into modules or panels or on glass or other suitable substrates', imported not only from China, but also from Malaysia, Taiwan and the United States.[12] During the course of the proceedings, the CCCME offered a price undertaking with respect to solar cells exported from China to India, but the proposal was rejected, as the DGAD stated in the final findings that the offer was considered 'impractical to monitor'.[13]

[11] Case T-141/14, *SolarWorld and Others* v. *Council* [2014] EU:T:2014:281, para. 8; Case T-142/14, *SolarWorld and Others* v. *Council* [2014] EU:T:2014:283, para. 8.

[12] Initiation notification of anti-dumping investigation concerning imports of solar cells from Malaysia, China, Chinese Taipei and USA, Department of Commerce of the Ministry of Commerce and Industry of India, 23 November 2012.

[13] Final findings of anti-dumping investigation concerning imports of solar cells, whether or not assembled partially or fully in modules or panels or on glass or some other suitable substrates, originating in or exported from Malaysia, China, Chinese Taipei and USA, DGAD, 22 May 2014, para. 161.

On 22 May 2014, the DGAD presented its final findings and recommended duties ranging from 11 to 81 US cents per watt.[14] However, under India's AD regulation, the findings of the DGAD are given effect by the central government (in this case, the Ministry of Finance) through a customs notification in the Official Gazette within three months of the date of publication of the final findings.[15] Since no such notification was issued within the three-month period, it appears that India has decided not to impose AD duties on the imported solar products.[16] Press reports have cited opposition from the Ministry of New and Renewable Energy and the domestic industry association, that is, the Associated Chambers of Commerce and Industry of India, as the reason for the decision not to impose the duties, since both organisations have reportedly claimed that domestic manufacturers do not have adequate capacity to meet the needs of the market.[17]

V *Australia's Solar Case*

Australia initiated its own AD investigation into 'solar modules or panels' from China on 14 May 2014. 'Cells, wafers and other related products such as solar chargers' were excluded from the scope of the investigation.[18] However, in 2015 Australia terminated the investigation, without the imposition of measures.

[14] Ibid., Duty Table.

[15] Customs Tariff (Identification, Assessment and Collection of Anti-dumping Duty on Dumped Articles and for Determination of Injury) Rules of India, Article 18, 1 January 1995.

[16] See 'India not to impose anti-dumping duty on solar panels: Nirmala', *The Hindu*, 10 September 2014. The article states that the State Minister for Commerce and Industry, Nirmala Sitharaman, said at a press conference on the first 100 days of the Modi government that '[t]here was no notification. We allowed it to lapse.'

[17] See, e.g., Y. Malik, 'India rejects anti-dumping duties', *PV Magazine*, 26 August 2014.

[18] Initiation of an investigation into alleged dumping of certain crystalline silicon photovoltaic modules or panels exported from the People's Republic of China, Australian Anti-dumping Commission, Anti-dumping Notice No. 2014/38, 14 May 2014. In detail, the products excluded are described as 'Cells and wafers of the type used in PV modules or panels; solar chargers that consist of less than six cells, are portable and supply to devices or charge batteries; and PV products that are permanently integrated into electrical goods, where function of the electrical goods is other than power generation, and where these electrical goods consume the electricity generated by the integrated crystalline silicon photovoltaic cell(s)'.

C The WTO Renewable Energy Disputes

Taking a step further towards the more international dimension, it is noteworthy that in the *Canada – Renewable Energy/Feed-in Tariff Program* disputes filed by Japan and the EU in September 2010 and August 2011, respectively, the AB upheld the panel's finding that the Ontario Feed-in Tariff (FiT) Program, the FiT and micro FiT contracts were trade-related investment measures, and that the domestic content requirement violated the national treatment principle in Article III:4 of the GATT 1994 and therefore also the Agreement on Trade-Related Investment Measures (TRIMs).[19]

However, the AB declared moot the panel ruling on the non-existence of a subsidy. In particular as regards the existence of a financial contribution, the AB upheld the panel's finding that the FiT programme and contracts should be legally characterised as 'government purchases of goods' within the meaning of Article 1.1(a)(1)(iii) of the Agreement on Subsidies and Countervailing Measures (SCM Agreement) and not as direct or potential direct transfers of funds, as alleged by Japan.[20]

Concerning the existence of a benefit, while the Panel considered that none of the proposed benchmarks provided by Japan and the EU could be used to assess the existence of a benefit, the AB considered that the panel committed an error in not conducting the benefit analysis.[21] The AB, however, did not complete the analysis because of a lack of sufficient factual findings and uncontested evidence on the panel's record, although it did indicate that certain benchmarks could have been used to assess the existence of benefit. In other words, the AB did not rule out the existence of a subsidy.

In November 2012, China brought a complaint to the WTO against the EU, Italy and Greece regarding certain measures affecting the renewable energy-generation sector, including the domestic content elements in the FiT programmes of EU Member States, which included but were not limited to Italy and Greece.[22] China claimed that the measures were inconsistent with Articles I, III:1, III:4 and III:5 of the GATT 1994, Articles 3.1(b) and 3.2 of the SCM Agreement and Articles 2.1 and 2.2

19 Report to the Minister No. 148 on certain aluminium extrusions exported to Australia from China, para. 6.1.

20 Ibid., paras. 5.122–5.128.

21 Ibid., para. 5.219.

22 Request for Consultations, *European Union and Certain Member States – Certain Measures Affecting the Renewable Energy Generation Sector*, WT/DS452/1, circulated on 7 November 2012.

of the TRIMs.[23] The claims made by China and the nature of the measures challenged are similar to those that were decided by the Panel and the AB in the *Canada – Renewable Energy/Feed-in Tariff Program* disputes.

More recently, the United States filed a WTO complaint against India in which it challenged the domestic content requirement of India's solar energy programme. The United States alleged that the domestic content requirement discriminates against US solar products by requiring developers in the programme to use solar cells and panels manufactured in India, and that India's measures are inconsistent with Article III:4 of the GATT 1994, Article 2.1 of TRIMs and Articles 3.1(b), 3.2, 5(c), 6.3(a) and 6.3(c) of the SCM Agreement.[24] A panel has been established for the dispute.

D Biofuels/Biodiesel Disputes

The EU, Argentina and Indonesia have all engaged in extensive and prolonged disputes concerning biofuels.

On 17 August 2012, Argentina filed a WTO complaint against a Spanish ministerial order implementing the EU Renewable Energy Directive as regards annual biofuel targets.[25] Argentina challenged that computation of the mandatory biofuel targets only in relation to biodiesel produced in Spain or another EU Member State was discriminatory and acted as a *de facto* prohibition on imports from non-EU countries when computing compliance with the biofuels target, and therefore violated Articles III:1, III:4, III:5 and XI:1 of the GATT 1994 as well as Articles 2.1 and 2.2 of TRIMs.

On 15 May 2013, Argentina filed a second complaint with regard to the EU and certain Member States' measures concerning the sustainability criteria which need to be met for the purpose of compliance with the biofuels target and to obtain incentives, alleging violation of Articles III:1, III:4, III:5 and XI:1 of the GATT 1994, Articles 2.1 and 2.2 of TRIMs and Articles 5.1 and 5.2 of the TBT Agreement.[26] Additionally, Argentina

[23] Ibid. [24] Accession of the People's Republic of China, WT/L/432, Article 15(a).

[25] Request for Consultations, *European Union and a Member State – Certain Measures Concerning the Importation of Biodiesels*, WT/DS443/1, circulated on 23 August 2012.

[26] Request for Consultations, *European Union and Certain Member States – Certain Measures on the Importation and Marketing of Biodiesel and Measures Supporting the Biodiesel Industry*, WT/DS459/1, circulated on 23 May 2013.

challenged the national measures taken by Belgium, France, Italy, Poland and Spain concerning the various types of local content requirements for obtaining domestic support, for example, in terms of reduced consumption tax, alleging a violation of various provisions of the GATT, TRIMs and/or the SCM Agreement.

Neither of these two complaints by Argentina reached the panel stage. However, in the interim, on 29 August 2012, the EU initiated an AD case against Argentinian and Indonesian imports of biodiesel, followed by a CVD case on 10 November 2012, pursuant to complaints by the European Biodiesel Board.

In the CVD case, the complainants argued against the high export taxes imposed by Argentina and Indonesia on the main raw material for producing biodiesel – namely soya bean and soya bean oil in the case of Argentina, and palm oil in the case of Indonesia – which allegedly depressed the domestic prices of these raw materials on the one hand, and indirectly propelled increased exports of biodiesel on the other. The CVD case was terminated on 27 November 2013 following withdrawal of the complaint.[27]

In the AD case, the EU effectively acted against the export tax issue in the AD context by constructing the normal value and adjusting the domestic production costs for the alleged distortions in the raw material costs. The EU did this by comparing a constructed normal value that included the export tax on the raw material with an export price that excluded the export tax. Additionally, the EU determined that the biodiesel markets in the two countries were distorted as a result of government regulation, and on this count the profit margin for the constructed normal value was based on 'a reasonable amount of profit that a young and innovative capital intensive industry of this type under normal conditions of competition in a free and open market could achieve', which was established at the level of 15 per cent.[28] Consequently, the definitive AD measures published on 26 November 2013 imposed AD duties ranging from 8.8 per cent to 25.7 per cent.

[27] Commission Regulation (EU) No 1198/2013 of 25 November 2013 terminating the anti-subsidy proceeding concerning imports of biodiesel originating in Argentina and Indonesia and repealing Regulation (EU) No 330/2013 making such imports subject to registration, OJ 2013 No. L315/67, 26 November 2013.

[28] Commission Regulation (EU) No 490/2013 of 27 May 2013 imposing a provisional anti-dumping duty on imports of biodiesel originating in Argentina and Indonesia, OJ 2013 No. L 141/6, 28 May 2013, recital 46.

Following the publication of the definitive AD measures, Argentina filed a third complaint at the WTO on 19 December 2013, alleging that the provisional and definitive EU AD measures on the imports from Argentina as well as the underlying investigation, together with Article 2 (5) of the EU Basic AD Regulation as regards the adjustment or establishment of costs associated with the production and sale of products in the determination of dumping margins, are inconsistent with various provisions of the ADA, Article VI of the GATT 1994 and Article XVI:4 of the WTO Agreement.[29]

Before the panel was composed for the Argentinian case, on 10 June 2014 Indonesia also filed a complaint with regard to the EU's provisional and definitive AD measures on biodiesel imports from Indonesia and the underlying investigation, as well as Articles 2(5) and 2(6)(b) of the EU Basic Anti-Dumping Regulation (Basic AD Regulation).[30]

The two complaints contain some similar claims as regards Article 2 (5) of the EU Basic AD Regulation, the AD measures and the underlying investigation. Article 2(5) of the EU Basic AD Regulation was challenged by both Argentina and Indonesia as inconsistent 'as such' with Articles 2.2 and 2.2.1.1 of the ADA on the grounds that Article 2.2 does not permit the adjustment or establishment of the production costs with data not pertaining to the country of origin, and that Article 2.2.1.1 requires the costs to be calculated on the basis of records kept by the exporter or producer under investigation, provided that such records are in accordance with the generally accepted accounting principles of the exporting country and reasonably reflect the costs associated with the production and sale of the product under consideration. The AD measures and the underlying investigation are challenged by both countries under various provisions of Articles 2, 3, 6 and 9 of the ADA, Article XVI:4 of the WTO Agreement and Article VI of the GATT 1994.

Indonesia has brought several additional claims that are not covered by Argentina's complaint.

On 23 June 2014, a panel was composed for the Argentinian dispute, and Indonesia is currently in consultations with the EU. The Panel, supported by the AB later on, ruled against Argentina's as-such claim, but in favour of important as-applied claims brought by Argentina. Notably, it was found

[29] Request for Consultations, *European Union – Anti-Dumping Measures on Biodiesel from Argentina*, WT/DS473/1, circulated on 8 January 2014.

[30] Request for Consultations, *European Union – Anti-Dumping Measures on Biodiesel from Indonesia*, WT/DS480/1, circulated on 17 June 2014.

that the EU had violated Articles 2.2.1.1 and 2.2 of the ADA by not accepting the costs recorded in the accounting records of the Argentina producers and by using costs outside the country of origin.

E Conclusions

While the outcome of the cases discussed in this chapter is largely uncertain, the relevant dumping and CVD issues in this new era of renewable energy-focused trade disputes have, at the very least, brought the contestable methodologies of certain user countries such as the United States and the EU to the fore. Interested parties can expect a better understanding of the treatment of subsidies and other support measures in trade disputes as countries and entities continue to scrutinise and challenge the relevant legal issues both domestically and at the WTO.

That being said, we can already draw the following conclusions:

1. Methods for calculating NME dumping margins make findings of dumping a foregone conclusion.
2. In CVD investigations, specificity findings and recourse to external benchmarks based on 'facts available' make subsidy findings similarly automatic. As a result, it remains unclear whether China, the main target of AD and CVD cases in the PV sector, would be dumping or subsidising more than others in the sector if regular calculation methods were to be applied.
3. The PV trade remedy cases once again offer an illustration of the snowball effect of such proceedings.
4. Finally, and contrary to what some authorities like to believe, tit-for-tat trade actions are not limited to cases involving China, but occur equally in other trade conflicts between trading powers with more or less equal leverage (e.g. in dumping cases between the United States and the EU regarding grain-oriented flat-rolled electrical steel and sodium metal).

Bibliography

Accession of the People's Republic of China, WT/L/432, Article 15(a), 23 November 2001.

Amended final determination of sales at less than fair value and anti-dumping duty order of crystalline silicon photovoltaic cells, whether or not assembled into modules, from the People's Republic of China, International Trade Administration, Federal Register Vol. 77 No. 236, 7 December 2012.

Case T-141/14, *SolarWorld and Others v Council* [2014] EU:T:2014:281, para. 8.

Case T-142/14, *SolarWorld and Others v Council* [2014] EU:T:2014:283, para. 8.

Case T-507/13, *SolarWorld and Others v Commission* [2013].

Commission publication, *Commission Proposes to Modernise the EU's Trade Defence Instruments*, European Commission, 10 April 2013.

Commission Regulation (EU) No 1198/2013 of 25 November 2013 terminating the anti-subsidy proceeding concerning imports of biodiesel originating in Argentina and Indonesia and repealing Regulation (EU) No 330/2013 making such imports subject to registration, OJ 2013 No. L315/67, 26 November 2013.

Commission Regulation (EU) No 490/2013 of 27 May 2013 imposing a provisional anti-dumping duty on imports of biodiesel originating in Argentina and Indonesia, OJ 2013 No. 141/6, 28 May 2013, recital 46.

Council Implementing Regulation (EU) No 1238/2013 of 2 December 2013 imposing a definitive anti-dumping duty and collecting definitively the provisional duty imposed on imports of crystalline silicon photovoltaic modules and key components (i.e. cells) originating in or consigned from the People's Republic of China, OJ 2013 No. L 325/1, 5 December 2013.

Council Implementing Regulation (EU) No 1239/2013 of 2 December 2013 imposing a definitive countervailing duty on imports of crystalline silicon photovoltaic modules and key components (i.e. cells) originating in or consigned from the People's Republic of China, OJ 2013 No. L 325/66, 5 December 2013.

Countervailing duty order of crystalline silicon photovoltaic cells, whether or not assembled into modules, from the People's Republic of China, Federal Register Vol. 77 No. 236, 7 December 2012.

Customs Tariff (Identification, Assessment and Collection of Anti-dumping Duty on Dumped Articles and for Determination of Injury) Rules of India, Article 18, 1 January 1995.

Extension of time granted to issue the Statement of Essential Facts, Australian Anti-dumping Commission, Anti-dumping Notice No. 2014/77, 1 September 2014.

Final findings of anti-dumping investigation concerning imports of solar cells, whether or not assembled partially or fully in modules or panels or on glass or some other suitable substrates, originating in or exported from Malaysia, China, Chinese Taipei and USA, DGAD, 22 May 2014, para. 161 and Duty Table.

General disclosure document, countervailing duty investigation on imports of crystalline silicon photovoltaic modules and key components (i.e. cells) originating in or consigned from the People's Republic of China, Annex 2, August 2013.

India not to impose anti-dumping duty on solar panels: Nirmala, *The Hindu*, 10 September 2014.

Initiation notification of anti-dumping investigation concerning imports of solar cells from Malaysia, China, Chinese Taipei and USA, Department of Commerce of the Ministry of Commerce and Industry of India, 23 November 2012.

Initiation notification of countervailing duty investigation concerning imports of castings for wind operated electricity generators originating in or exported from China, Department of Commerce of the Ministry of Commerce and Industry of India, 29 May 2014.

Initiation of an investigation into alleged dumping of certain crystalline silicon photovoltaic modules or panels exported from the People's Republic of China, Australian Anti-dumping Commission, Anti-dumping Notice No. 2014/38, 14 May 2014.

Initiation of anti-dumping duty investigations of certain crystalline silicon photovoltaic products from the People's Republic of China and Taiwan, International Trade Administration, Federal Register Vol. 79 No. 19, 29 January 2014.

Initiation of countervailing duty investigation of certain crystalline silicon photovoltaic products from the People's Republic of China, Federal Register Vol. 79 No. 19, 29 January 2014.

Issues and decision memorandum for the final determination in the countervailing duty investigation of crystalline silicon photovoltaic cells, whether or not assembled into modules, from the People's Republic of China, International Trade Administration, 61, 9 October 2012.

Notice of amended preliminary determination of sales at less than fair value of certain crystalline silicon photovoltaic products from Taiwan, International Trade Administration, Federal Register Vol. 79 No. 163, 22 August 2014.

Preliminary affirmative determination countervailing duty investigation of certain crystalline silicon photovoltaic products from the People's Republic of China, International Trade Administration, Federal Register Vol. 79 No. 111, 10 June 2014.

Preliminary affirmative determination of sales at less than fair value and postponement of final determination of certain crystalline silicon photovoltaic products from the People's Republic of China, International Trade Administration, Federal Register Vol. 79 No. 147, 31 July 2014.

Report to the Minister No. 148 on certain aluminium extrusions exported to Australia from China, Australian customs and Border Protection Service, 15 April 2011.

Request for Consultations, *European Union and Certain Member States – Certain Measures Affecting the Renewable Energy Generation Sector*, WT/DS452/1, circulated on 7 November 2012.

Request for Consultations, *European Union and a Member State – Certain Measures Concerning the Importation of Biodiesels*, WT/DS443/1, circulated on 23 August 2012.

Request for Consultations, *European Union and Certain Member States – Certain Measures on the Importation and Marketing of Biodiesel and Measures Supporting the Biodiesel Industry*, WT/DS459/1, circulated on 23 May 2013.

Request for Consultations, *European Union – Anti-Dumping Measures on Biodiesel from Argentina*, WT/DS473/1, circulated on 8 January 2014.

Request for Consultations, *European Union – Anti-Dumping Measures on Biodiesel from Indonesia*, WT/DS480/1, circulated on 17 June 2014.

Request for Consultations, *India – Certain Measures Relating to Solar Cells and Solar Modules*, WT/DS456/1, circulated on 11 February 2013.

Semi-annual report to the WTO Committee on Subsidies and Countervailing Measures from India, G/SCM/N/212/IND, 6 September 2010.

Updates to Anti-Dumping system – January 2014, Australian Anti-Dumping Commission, Anti-Dumping Notice No. 2013/108, 20 December 2013.

WTO Appellate Body Report, *Canada – Certain Measures Affecting the Renewable Energy Generation Sector, Canada – Measures Relating to the Feed-In Tariff Program*, WT/DS412/AB/R, WT/DS426/AB/R, adopted on 24 May 2013.

Yogi, M. (2014), 'India rejects anti-dumping duties', *PV Magazine*, 26 August.

18

Promoting Green Electricity through Differentiated Electricity Tax Schemes

KATERYNA HOLZER, ILARIA ESPA AND TETYANA PAYOSOVA

A Introduction

The increased attention given to climate change on the international agenda has led governments around the world to resort to a wide set of policy instruments aimed at promoting the use of renewable energy (RE) sources. In particular, many governments have introduced measures to promote the use of electricity produced from renewable sources (green electricity).[1] Specific regulatory instruments aimed at supporting green electricity have been put in place for two main reasons. First, the electricity sector is both the largest single source of energy-related greenhouse gas (GHG) emissions and the sector carrying the biggest potential for emissions abatement.[2] Second, in many cases green electricity remains unable to compete with conventional electricity produced by fossil fuel-fired and nuclear power plants, owing to fundamental differences in cost structures and operating costs and characteristics.[3]

1 In this chapter, the term 'green electricity' is used to designate electricity generated from renewable types of energy. 'Grey electricity' means electricity generated from fossil fuels. Electricity generated from nuclear power does not fall within the meaning of either of the terms and will be dealt with separately.

2 The energy sector is currently responsible for two thirds of total greenhouse gas (GHG) emissions and fossil fuels still account for the greatest share of global power generation. International Energy Agency (IEA), *Redrawing the Energy Climate Map: World Energy Outlook Special Report* (2013), www.iea.org/publications/freepublications/publication/WEO_Special_Report_2013_Redrawing_the_Energy_Climate_Map.pdf (accessed 9 April 2015), pp. 15–32.

3 Green electricity is characterised by high capital investment costs and cannot rely on economies of scale, contrary to conventional electricity. In *Canada – FIT Program*, the WTO Appellate Body (AB) noted that because of these supply-side factors, green electricity markets can only be created through government regulation. See AB Report, *Canada – Measures Relating to the Feed-in Tariff Program (Canada – FIT Program)*, 6 May 2013, WT/DS426/AB/R, para. 5.175.

In the attempt to reduce the competitiveness 'gap' between green electricity and conventional sources, governments of both developed and developing countries have adopted RE support measures that lower the cost, raise the price and stimulate purchases of green electricity.[4] Along with market-based instruments aimed at incentivising investment in low-carbon technologies, states have invested heavily in the promotion of RE development through various forms of subsidies. These measures, however, have proven quite costly, and their compatibility with the World Trade Organization (WTO) under the Agreement on Subsidies and Countervailing Measures (ASCM) is often uncertain.

This chapter aims to promote an alternative approach based on taxing electricity at different rates depending on energy sources or its carbon footprint. Such an approach could facilitate the transition from direct subsidisation of RE to a more efficient steering system, which is less burdensome for public accounts.[5] Taxing electricity at higher rates if it is produced from fossil fuel and applying lower tax rates or granting exemptions to green electricity can create appropriate incentives for RE production without burdening public resources. Moreover, the revenues derived from the implementation of a differentiated electricity tax could also be used to support RE.[6]

B Rationale for a Differentiated Electricity Tax

I Taxing Electricity for the Promotion of Renewable Energy

In recent years, the set of governmental measures introduced to promote RE development and, in particular, the use of electricity produced from

[4] According to recent statistics, developing countries have surpassed developed countries with regard to the amount of financial investment in renewable energy sources. A. Ghosh and H. Gangania, *Governing Clean Energy Subsidies: What, Why, and How Legal?* (Geneva: ICTSD, 2012), p. 2, www.ictsd.org/downloads/2012/09/governing-clean-energy-subsidies-what-why-and-how-legal.pdf (accessed 9 April 2015).

[5] The idea of privileging fiscal measures on a differentiated basis rather than providing direct support schemes for the promotion of RE electricity development is on the rise in countries such as Switzerland. See Botschaft zum ersten Massnahmenpaket der Energiestrategie 2050 (Revision des Energierechts) und zur Volksinitiative 'Für den geordneten Ausstieg aus der Atomenergie (Atomausstiegsinitiative)', 4 September 2013, BBl 2013 7561, 7574–5.

[6] Ibid.

renewable sources has sharply increased. These measures generally aim at lowering the cost of green electricity, thus stimulating purchases by final consumers, or raising the price paid to producers.[7]

To promote green electricity, governments use both market and command-and-control measures. The former can either be price- or quota-based mechanisms. Many countries have, for example, introduced so-called RE quota obligations (ROs), that is, domestic schemes requiring energy generators, suppliers or consumers to include a given percentage of energy from renewable sources in their production, supply or consumption.[8] ROs help achieve national targets for the share of energy from renewable sources.[9] They are usually fulfilled with the help of renewable energy certificates (RECs), including 'green certificates' used in the European Union (EU) countries.[10]

As to the price-based instruments, the EU and a number of non-EU countries have introduced emissions trading schemes (ETSs) based on the cap-and-trade principle.[11] Under such schemes, a 'cap' is imposed on the total amount of CO_2 and/or other GHG emissions[12] that can be emitted each year by the power plants, factories and other companies covered by the system, and it is gradually reduced every year. Under this

[7] For an overview of the policy instruments implemented by both developed and developing countries for RE development since the 1970s, see World Bank, 'Design and performance of policy instruments to promote the development of renewable energy: emerging experience in selected developing countries', *Energy and Mining Sector Board Discussion Paper* No. 22 (2011), http://siteresources.worldbank.org/EXTENERGY2/Resources/DiscPaper22.pdf (accessed 10 April 2015).

[8] Renewable energy obligation schemes have been introduced by several EU Member States in pursuance of Article 2 (l) of the Directive 2009/28/EC of 23 April 2009 on the promotion of the use of energy from renewable sources (RES Directive), *Official Journal* (OJ) L 140 of 05/06/2009. Very often, these schemes are implemented by using 'green certificates'.

[9] See Article 5 of Directive 2009/28/EC.

[10] See Article 2 (k) of Directive 2009/28/EC.

[11] The EU emission trading scheme (ETS), which was created in 2005, is the largest international system for trading greenhouse gas emission allowances. It also operates in the member countries of the European Economic Area (Island, Norway and Lichtenstein). Other countries implementing a cap-and-trade system are Switzerland, Australia, New Zealand and Kazakhstan. There are also countries that implement ETSs at regional levels, for instance the United States (California and the Regional GHG Initiative), Canada (Quebec), Japan (Tokyo and Saitama) and China (ETSs have been introduced in six Chinese provinces). Furthermore, various countries have scheduled the launch of an ETS (e.g. South Korea) or are considering it (Mexico, Chile, Ukraine and Brazil). IEA, *Redrawing the Energy Climate Map*, p. 24.

[12] For instance, the EU emission trading scheme covers CO_2, nitrous oxide (N2O) and perfluorocarbons (PFCs).

cap, the covered companies receive or buy emission allowances,[13] which they can trade and whose price is determined on the market by the interaction between supply and demand. By putting a price on carbon, ETSs give companies the flexibility to choose how to reduce GHG emissions in the most cost-effective way while promoting investment in low-carbon technologies.[14] The electricity sector is normally covered by existing ETSs.[15]

The use of electricity generated from RE sources is also promoted through direct support schemes (i.e. subsidies) implemented by governments. Clean energy subsidies take various forms, from financial transfers (e.g. consumer subsidies) to direct price support schemes such as feed-in tariffs (FiTs); from preferential tax credits (e.g. investment tax credits, production tax credits or tax credits for consumption) to other regulatory and investment support measures aimed at lowering the cost of green electricity production or facilitating the distribution and supply of green electricity to consumers.[16] The use of subsidies, however, has recently come under closer scrutiny as a result of concerns about long-term efficiency and, at least in some cases, uncertainty regarding their WTO compatibility.[17]

In light of all the foregoing, many countries are currently making a regulatory shift in the electricity sector, from RE promotion systems to RE steering systems. In simple terms, this means substituting subsidies with excise taxes on electricity. If a tax is imposed equally on all types of electricity, it will have the desirable effect of reducing overall electricity consumption but will be useless in terms of promoting generation of

[13] Each allowance conventionally confers the right to emit one tonne of covered GHG emissions.

[14] For a detailed analysis of the functioning of the existing emission trading systems in the world, see A. Talberg and K. Swoboda, *Emission Trading Schemes Around the World*, Background Note for the Parliament of Australia (2013), http://parlinfo.aph.gov.au/parlInfo/download/library/prspub/2501441/upload_binary/2501441.pdf;fileType=application/pdf (accessed 10 April 2015).

[15] IEA, *Redrawing the Energy Climate Map*, p. 24.

[16] For a thorough description of the various forms of clean energy subsidies currently in place in leading RE countries see Ghosh and Gangania, *Governing Clean Energy Subsidies*, p. 20 ff.

[17] Among the different types of clean energy subsidies, the use of FiTs coupled with local content requirements (LCRs) has in particular been challenged before the WTO Appellate Body under the WTO Agreement on Subsidies and Countervailing Measures (ASCM). See AB Report, *Canada – FIT Program*; AB Report, *European Union and Certain Member States – Certain Measures Affecting the Renewable Energy Generation Sector, Request for Consultation by China*, 7 November 2012; AB Report, *India – Certain Measures Related to Solar Cells and Solar Modules*, WT/DS456/AB/R, 14 October 2016.

electricity from renewable sources. To promote investments in RE, tax rates need to be applied so as to stimulate the generation and consumption of green electricity and discourage the generation and consumption of grey electricity. The differentiated electricity tax system thus consists of tax rate reductions or exemptions for green electricity.

II The Use of Electricity Certificates for Differentiated Electricity Taxation

When assessing the compliance of a differentiated electricity tax with international trade rules, it is necessary to take into consideration the special characteristics of electricity as a product. The intangible nature of electricity and electricity trade's dependence on the availability of grids constrain the circulation of electricity in the market and limit trade in electricity among countries. Electricity has traditionally been traded within national jurisdictions or between neighbouring countries connected by grids. However, it is expected that international trade in electricity will increase in the future, as investments in modern technologies will enable the construction of international interconnected power networks.[18] Indeed, supplying Europe with electricity from North Africa through interconnected grids seems to be feasible.[19] A system of interconnected grids between the EU and Eurasia is already in place. This all means that the relevance of WTO rules applicable to international trade in electricity is steadily increasing.

The implementation of a differentiated electricity tax system needs also to take into consideration the homogeneity of electricity as a product. It is impossible to distinguish between green and grey electricity based on appearance and physical characteristics. The application of different tax rates to different types of electricity needs to rely on an electricity certification scheme. Different tax rates would be levied on electricity based on certificates proving the source of the electricity. Electricity certification schemes are already used by some countries, in most cases for purposes of electricity source disclosure. One example of such certification is the system of guarantees of origin (GOs) envisaged

[18] IEA, *Cross-Border Trade in Electricity and the Development of Renewable-Based Electric Power: Lessons from Europe* (2013), Annex 2 on Liberalisation of Electricity Markets, Competition and the Drivers of Cross-Border Trade in Electricity, p. 50.

[19] S. Chatzivasileiadis, D. Ernst and G. Andersson, 'The Global Grid', *Renewable Energy*, 57 (2013), 372–83.

by the EU Renewable Energy Directive.[20] The aim of the GOs is to inform consumers about electricity generated from RE. Each certificate is issued electronically for 1 MW/h and is valid for one year. GOs are issued on the request of any RE electricity generator, subject to a minimum capacity limit. Usually, GOs do not correspond to the physical flow of generated electricity. They are traded virtually as financial assets on the European Energy Exchange (EEX) and other electricity markets.

There are other types of certificates related to electricity besides GOs. For instance, RE obligation schemes, which impose on the EU electricity suppliers the requirement to supply a certain percentage of electricity produced from RE in order to achieve a mandatory target of 20 per cent of green energy in total EU energy consumption by 2020, are based on green certificates. Furthermore, various electricity labels have been introduced with the purpose of disclosing information to consumers on different aspects of electricity quality (e.g. Swiss 'naturemade', TÜV SÜD, etc.).[21] Green electricity labels are issued by a specific certification body at the request of electricity generators, provided they fulfil certain (sustainability) requirements under the given labelling scheme. Green electricity labels are attached to the physical flows of electricity produced by certified electricity generators. Green labels are different to green certificates in that the former certify the green origin of electricity per kWh whereas the latter provide information on the sustainability footprint of electricity plants, including their environmental impacts, social and economic criteria and various process criteria.

Finally, the implementation of a differentiated electricity tax could also be based on RECs specifically issued for the purposes of granting electricity tax exemptions. The use of such tax exemption certificates (TECs), which are part of the Climate Change Levy (CCL) applied to electricity and energy resources in the United Kingdom, is discussed below.

III Electricity Tax Practices in the EU

The application of taxes to electricity in EU countries is guided by the EU Directive on restructuring the Community framework for the taxation of

[20] See Directive 2009/28/EC of 23 April 2009 on the promotion of the use of energy from renewable sources (RES Directive), Article 15. OJ L 140 of 05/06/2009, p. 16.

[21] PriceWaterhouseCoopers and WWF (2009), *Green Electricity: Making a Difference. An International Survey of Renewable Electricity Labels*, www.repower.com/fileadmin/user_upload/re-all/02_Files_PDF-DOC-XLS/z_to_be_classified/Berichte_und_Studien/pwc_green_electricity_making_a_difference.pdf (accessed 10 April 2016).

energy products and electricity.[22] Article 15(1) of the Directive allows the application of an electricity tax at different rates depending on the source of the electricity, subject to EU non-discrimination and state aid rules. It also allows Member States to apply total or partial tax exemptions or reductions. Article 15(2) of the Directive further stipulates that the tax level can also be reduced by paying some or all of the amount of tax back to the producer of RE electricity.

As regards the application of electricity taxes to imports, individual EU Member States' practices vary. Finland used to apply an electricity tax at different tax rates based on electricity sources for domestic electricity and an average tax rate on imports of electricity. It justified this practice by claiming that the source of imported electricity could not be easily traced.[23] In the United Kingdom, electricity tax exemptions are available for domestic and imported electricity alike if electricity comes from renewable sources.[24] The Netherlands used to grant electricity tax exemptions to consumers without any restrictions on the origin of RE electricity.[25]

Electricity taxes featuring exemptions for RE electricity are currently in place in a number of EU Member States, including Denmark, Germany, Lithuania, Poland, Sweden and the United Kingdom. Tax exemptions are provided either to suppliers of all types of green electricity or only to suppliers of particular types of green electricity (e.g. only for wind electricity in Sweden or only for wind, hydropower and solar in Denmark), or depending on the size of renewable power facilities (e.g. only to plants with installed capacity of less than 5 MW in Poland).[26]

EU countries applying different electricity tax rates based on electricity sources resort to various types of RECs to trace the source of electricity. In Poland, the implementation of exemptions for renewable electricity is

[22] See EU Directive 2003/96/EC. OJ L 283 of 31/10/2003, p. 51.

[23] The European Court of Justice (ECJ) found these differences in the application of the tax to be in breach of EU non-discrimination rules. See See Case C-213/96, Outokumpu Oy, 1998 ECR 1-1777.

[24] Ofgem, Climate Change Levy (CCL) exemption, www.ofgem.gov.uk/environmental-programmes/climate-change-levy-ccl-exemption (accessed 20 January 2016).

[25] T. Winkel, M. Ragwitz, G. Resch and I. Konstantinaviciute, *Renewable Energy Policy Country Profiles 2011*, www.reshaping-res-policy.eu/downloads/RE-SHAPING_Renewable-Energy-Policy-Country-profiles-2011_FINAL_1.pdf (accessed 20 January 2016). See also K. Holzer and I. Espa, 'Greening electricity through taxing: an analysis of GATT constraints', *NCCR Working Paper No. 2015* (9 April 2015).

[26] J. M. Cansino, M. del P. Pablo-Romero, R. Román and R. Yñiguez, 'Tax incentives to promote green electricity: an overview of EU-27 countries', *Energy Policy*, 38(10) (2010), pp. 6000–8.

based on green certificates, or certificates of origin. The exemption from an excise duty is granted on submission to the customs office of a written statement by the electricity generator or supplier stating that the green certificates were confirmed by the Energy Regulatory Office as proof of the green origin of the electricity.[27] In Lithuania, electricity tax exemptions are provided for national suppliers of domestic and imported green electricity from different sources.[28] The exemptions are provided for under Article 46(2) of the Law on Excise Duty against submission of GOs, which are authorised by the transmission system operator, or any other evidence confirming that electricity was generated from RE sources.[29] In the United Kingdom, suppliers of domestic or foreign green electricity are exempted from the tax based on the tax exemptions certificates specifically introduced for these purposes. Electricity in the United Kingdom is taxed in accordance with the climate change levy (CCL) scheme, which applies an excise tax on electricity and fossil fuels in relation to climate change policy.[30] The CCL scheme provides exemptions for both domestic and imported green electricity. To receive exemptions, an electricity supplier must be in possession of renewables levy exemption certificates (LECs) obtained from an accredited green electricity producer in the United Kingdom or abroad. Certificates for domestic and foreign green electricity producers are issued by Great Britain's Office of Gas and Electricity Markets (Ofgem) or the Northern Ireland Authority for Utility Regulation through electronic registration.

C WTO Law Implications of a Differentiated Electricity Tax

I Status of Electricity in the WTO Legal System

Under WTO law, electricity qualifies as a good and falls under the provisions of the General Agreement on Tariffs and Trade (GATT).[31]

[27] Ibid., p. 6005.
[28] Law on Excise Duty of the Republic of Lithuania, 29/01/2004, No. IX-1987.
[29] Order of the Ministry of Economy on the 'Electrical Energy, Generated from Renewable Energy Sources, Guarantees of Origin Authorization Rules', 07/10/2005, No. 4–346.
[30] Ofgem, Climate Change Levy (CCL) exemption.
[31] Panel Report, *Canada – Measures Relating to the Feed-in Tariff Program (Canada – FIT Program)*, 19 December 2012, WT/DS426/R, para. 7.11, footnote 46. It should be noted, however, that electricity was classified as a good in the GATT era, when there was no legal framework for trade in services (this was established later by the WTO's General Agreement on Trade in Services [GATS]). There is currently discussion in the literature as to whether electricity should not be better dealt with as a service.

It is enclosed in the GATT Schedule of Concessions as an optional commitment under the heading HS 2716.00. Electricity is listed in subsection 27, 'Mineral fuels, mineral oils and products of their distillation; bituminous substances; mineral waxes', of section V, 'Mineral products'. Since the tariff commitment for electricity is optional, some WTO Members have left themselves a right to impose import duties on electricity at their own discretion. Yet, even these WTO members are obliged to observe all other provisions of GATT, including non-discrimination rules, when they trade electricity with other countries. This also applies to tax matters. First, countries have an obligation to provide tax treatment in accordance with the most favoured nation (MFN) principle to electricity imported or exported from/to all other WTO members under GATT Article I.[32] Second, they need to observe the national treatment (NT) principle in relation to taxing electricity under GATT Article III:2.[33]

Furthermore, electricity trade, and particularly operations with RECs, may fall under the disciplines for financial services and thus be subject to the provisions of the General Agreement on Trade in Services (GATS) and the Understanding on Commitments in Financial Services.[34] Countries' obligations under the GATS may also be relevant to the extent that electricity tax schemes may have an impact on service providers, particularly in relation to transportation and distribution of electricity. However, GATS issues are beyond the scope of this chapter.

II Relevant GATT Rules and Exceptions

1 Non-Discrimination Principles

Consideration of the compliance of a differentiated electricity tax with the MFN and NT obligations will have to be based on like product analysis, because the MFN and NT obligations only apply to trade in

[32] The MFN provision of GATT Article I requires that any benefit provided to a product imported from or exported to any WTO member must also be provided to a like product imported from or exported to all other WTO members.

[33] In general terms, the NT provision of GATT Article III prohibits a protectionist treatment for domestic products.

[34] See P. Delimatsis, 'Financial innovation and climate change: the case of renewable energy certificates and the role of the GATS', *World Trade Review*, 8(3) (2009), 439–60; P. Delimatsis and D. Mavromati, 'GATS, financial services and Trade in Renewable Energy Certificates (RECs) – just another market-based solution to cope with the tragedy of the commons?' in T. Cottier, O. Nartova and S. Z. Bigdeli (eds.), *International Trade Regulation and the Mitigation of Climate Change* (Cambridge: Cambridge University Press, 2009), pp. 231–58.

like products or products which are directly competitive or substitutable. The NT obligation prohibits the application of a tax on imports at a rate which is in excess of the tax rate applied to like domestic products. If the products qualify as like, even a small difference in the tax rate to the detriment of imports would lead to a finding of discrimination.[35]

In examining the compliance of a differentiated electricity tax with the non-discrimination principles under the GATT, the main question is whether electricity originating from RE can be viewed as different to or unlike electricity originating from fossil fuels.[36] Neither WTO jurisprudence nor the literature gives a clear answer to this question, which is part of an old debate on the accommodation of measures imposed on processes and production methods (PPMs) under the GATT regulatory framework.[37] Unlike EU law, which accepts trade measures based on production methods,[38] WTO law does not provide a clear answer as to the legal status of such measures, especially with regard to those PPMs that do not change the physical qualities of a product (i.e. non-product-related PPMs).[39]

When assessing whether products are like or different, WTO panels look at the competitive relationship between products and use four criteria that signal likeness of products: products' physical characteristics, products' end uses, consumer preferences and tariff classification.[40] Since electricity is electricity no matter whether it is generated from carbon-intensive or carbon-free energy sources, it falls under the same tariff code in the countries' schedules of concession and is intended for the same use as power in all cases. Accordingly, only a consumer's preference for electricity produced from renewable sources can render green and grey electricity 'unlike products'. An argument can be made that consumers in

[35] AB Report, *Japan – Taxes on Alcoholic Beverages (Japan – Alcoholic Beverages II)*, adopted 1 November 1996, WT/DS8/AB/R, WT/DS10/AB/R, WT/DS11/AB/R, p. 22.

[36] It should be noted that likeness of products is always assessed on a case-by-case basis.

[37] For more on this, see Ecoplan, WTI and University of Zurich (2013), *Border Tax Adjustments: Can Energy and Carbon Taxes Be Adjusted at the Border?* Final report prepared for the Swiss State Secretariat for Economic Affairs and the Swiss Federal Finance Administration, pp. 80–1.

[38] For instance, the ECJ did not find a PPM-based electricity tax applied in Finland in the 1990s to constitute a violation of EU law. See Case C-213/96, Outokumpu Oy, 1998 ECR 1-1777.

[39] C. R. Conrad, *Processes and Production Methods (PPMs) in WTO Law: Interfacing Trade and Social Goals* (New York: Cambridge University Press, 2011), pp. 275–81.

[40] See, e.g., AB Report, *European Communities – Measures Affecting Asbestos and Asbestos Containing Products (EC – Asbestos)*, adopted 5 April 2001, WT/DS135/AB/R, para. 101.

a particular market prefer green electricity to grey electricity under the availability of an electricity certification system that can facilitate their choice.[41] Uncertainty exists, however, as to whether consumer preferences for green electricity will be considered to be strong enough to prevail over other likeness criteria which are the same for grey and green electricity (physical qualities, end uses and the tariff code).

In *EC – Asbestos*, consumer preferences played a prominent role in the Appellate Body's (AB) finding of unlikeness of asbestos-containing and asbestos-free products. Without going into an examination of evidence, and based on the public awareness of hazards to human health from the use of asbestos, the AB made an assumption that consumers prefer to purchase products that do not contain asbestos. Thus, health risks associated with the use of asbestos-containing products rendered these products unlike those products that do not contain asbestos.[42] However, the consumption of fossil fuel electricity is different in this respect. The consumption of electricity generated from coal or nuclear energy does not create a direct association with health risks for consumers. The negative consequences of the use of coal-fired electricity are less tangible than the risks inflicted by the use of asbestos-containing products. Moreover, the realisation of consumer preferences for green electricity depends on the availability of a certification scheme established by the regulator in the electricity market. Without the use of certificates, consumers will not be able to distinguish between different types of electricity and to make their choice based on such a distinction.[43]

Consumers' heedfulness of the methods by which electricity is generated can also serve as evidence that 'green' electricity and 'grey' electricity are in a competitive relationship in the market, and as such are not completely identical (like) products but rather products that are directly competitive or substitutable. In *Canada – FIT Program*, the AB noted that directly competitive or substitutable products in the sense of GATT Article III:2, second sentence, are 'products that are in a competitive relationship. What constitutes a competitive relationship between products

[41] R. Howse, *World Trade Law and Renewable Energy: The Case of Non-Tariff Barriers* (Geneva: UNCTAD, 2009), p. 3.

[42] AB Report, *EC – Asbestos*, paras. 123–126.

[43] Certificates to facilitate consumption of RE electricity are not needed in the situation where RE electricity is generated by households and companies using solar PV panels. However, this constitutes only a small part of RE electricity consumption. A prevailing part of electricity is distributed to households and companies from the single electricity grid fed by a physically inseparable mix of 'green' and 'grey' electricity.

may require consideration of inputs and processes of production used to produce the product.'[44] Based on this argument, different types of electricity generated by different production methods can fall under the category of directly competitive or substitutable products.

Taxation of directly competitive or substitutable products is subject to a more lenient set of NT rules than those applicable to the category of like products.[45] As per GATT Article III:2, second sentence, the tax treatment of directly competitive or substitutable products must not be identical in order to satisfy the NT requirement. In contrast to the first sentence of GATT Article III:2, domestic and imported directly competitive or substitutable products are to be taxed similarly; certain variations in the amount of tax can thus be accommodated as long as they do not result in the protection of domestic production.[46] Although a differentiated electricity tax which is levied at higher rates on fossil fuel electricity and at lower rates on RE electricity with respect to all electricity sold in the market is origin-neutral *de jure*, discrimination can occur *de facto* if the proportion of fossil fuel electricity in electricity imports is significantly higher than the proportion of fossil fuel electricity in electricity generated domestically. This means that to meet the requirement of the second sentence of Article III:2, the amounts of imported and domestic electricity disadvantaged by a tax should be commensurate.

Finally, based on the recent WTO jurisprudence, it is still possible that electricity generated from RE sources will be found to be a different or unlike product from electricity generated from fossil fuels. In the *Canada – FIT Program* case, when assessing the compliance of Ontario's FiT scheme with the rules on subsidies under the ASCM, the AB found that electricity generated from solar photovoltaic (PV) and wind power technology and electricity generated from fossil fuels were sold in different markets. The markets were considered to be different because of the differences in the type of power (solar and wind plants generate peaking power, whereas coal-based and nuclear plants generate base-load power), the differences in contracts and the differences in consumers' sizes. Yet, the biggest difference was in the supply-side factors. According to the AB, 'supply-side factors suggest that wind-power and solar PV producers of electricity cannot compete with other electricity producers because of

[44] AB Report, *Canada – FIT Program*, para. 5.63.
[45] Alcoholic beverages (e.g. soju, whisky, brandy, gin, rum) are an example of products that were considered to be directly competitive or substitutable by WTO adjudicative bodies.
[46] See GATT Article III:2 and Ad Article III:2 read together.

differences in cost structures and operating costs and characteristics'.[47] The AB also noted that, unlike the market for conventional electricity, the market for green electricity would have not come into existence, due to high production costs and competitive disadvantages, if a government had not created it through regulation.[48] Thus, when determining the likeness of product markets in the context of the ASCM analysis, the AB looked not only at demand-side factors but also at supply-side conditions. If supply-side factors are taken into account in the analysis of likeness of different types of electricity under the GATT, green and fossil fuel electricity will be found to be different or unlike products. In that case, the differences in their tax rates will not raise issues of compliance with the NT obligation under the GATT. It remains to be seen whether WTO adjudicative bodies will be willing to incorporate the ASCM approach to likeness in the likeness analysis under the GATT.

To sum up, three different scenarios can be envisaged with respect to the analysis of likeness of 'green' and 'grey' electricity:

1. they may be found to be 'unlike products' and thus allowed to be treated differently in terms of taxation;
2. they may be found to be 'like products' and differentiated tax rates would thus be in violation of the NT rule;
3. they may qualify as 'directly competitive or substitutable products' and a breach of the NT rule could arise if a disproportionate tax burden is placed on electricity imports.

As follows from the *Shrimp – Turtle* jurisprudence, PPM measures that fail to meet obligations under the GATT – in our case, these would be scenarios (2) and (3) – may be justified under the general exceptions of GATT Article XX.[49] Whether the application of differentiated tax rates to electricity generated from different sources can be defended under the general exceptions is discussed in the next section.

2 Applicability of GATT Article XX Exceptions

Exceptions to GATT rules are available for measures taken in pursuit of one of the non-trade policy objectives specified in GATT Article XX. An important initial step in the analysis of whether a measure can be justified under GATT Article XX is the determination of its objective.

[47] AB Report, *Canada – FIT Program*, para. 5.174. [48] Ibid., para. 5.175.
[49] AB Report, *United States – Import Prohibition of Certain Shrimp and Shrimp Products (US – Shrimp)*, adopted 6 November 1998, WT/DS58/AB/R, para. 121.

A differentiated electricity tax can fall within the scope of Article XX if a link can be established with protection of the environment and/or public health. In the first case, a country introducing a differentiated electricity tax may seek justification under paragraph (g) of Article XX, which protects measures 'relating to a conservation of exhaustible natural resources'. In the second case, it may be able to justify a tax under paragraph (b) as a measure 'necessary to protect human, animal or plant life or health'.

In most cases, a differentiated electricity tax can qualify as an environmental or climate change-related measure falling under GATT Article XX (g). This may also be the case where the objective of a differentiated electricity tax is officially formulated as the promotion of production and consumption of RE. While this objective *prima facie* has a link to industrial policy, ultimately the industrial policy-related objective is driven by climate change concerns.[50] It could be argued that the promotion of green electricity aims to substitute the generation of electricity from fossil fuels (coal, oil and gas), which causes carbon emissions and consequently leads to climate change.

The question is, however, whether paragraph (g) can also be invoked to justify a higher tax rate on nuclear electricity, which is generally considered to be carbon-neutral.[51] While it is not associated with GHG emissions and the problem of climate change, electricity generated from nuclear power can still be connected with environmental problems. The environmental problems are caused by nuclear waste, which is stored underground.[52] The issue of nuclear waste storage is likely to be sufficient for arguing under paragraph (g). Moreover, a higher tax rate for nuclear electricity might be defended under paragraph (b) on the grounds that the operation of nuclear power plants presents risks to human life and health from possible accidents at nuclear power plants, such as those that

[50] For instance, a proposal for implementation of the second phase of Swiss Energy Strategy 2050, which considers differential electricity taxation as an option, is based on the assumption that renewable energy promotion and climate change mitigation goals are closely intertwined. See Verfassungsbestimmuing über ein Klima- und Energielenkungssystem. *Erläuternder Bericht zum Vorentwurf* (EFV, BFE, BAFU, 2015), pp. 16–17.

[51] J. Sathaye, O. Lucon and A. Rahman, 'Renewable energy in the context of sustainable energy', in O. Edenhofer, R. Pichs-Madruga, Y. Sokona, K. Seyboth, P. Matschoss, S. Kadner, T. Zwickel, P. Eickemeier, G. Hansen, S. Schlömer and C. von Stechow, *IPCC Special Report on Renewable Energy Sources and Climate Change Mitigation* (Cambridge: Cambridge University Press, 2011), pp. 732–3.

[52] Ibid., pp. 745–7.

happened in Chernobyl and Fukushima. Thus, justification of a differentiated electricity tax may require an invocation of several exception clauses at the same time, which is acceptable under WTO law, as a measure can consist of different elements that can be subject to justification under different exceptions.[53] In the *US – Gasoline* dispute, for instance, the panel scrutinised US standards for reformulated and conventional gasoline under three exceptions clauses.[54] Accordingly, justification of a higher tax rate for coal-based electricity compared to other types of electricity would have to be sought under Article XX (g), while justification of a higher tax rate for nuclear electricity would have to be sought under Article XX (b) and Article XX (g).

Once the link of a measure to the policy objective under a specific paragraph is established, the next step is analysis of the strength of this link. Paragraph (b) requires that a measure must be necessary for the achievement of the objective of health protection, while under paragraph (g) a measure must merely relate to the objective of conservation of exhaustible natural resources. In this respect, the link between a measure and a paragraph's objective is stronger under paragraph (b). Analysis of a differentiated electricity tax for nuclear electricity under paragraph (b) consists of the necessity test, which will look at whether the same objective could be met by alternative measures that are less trade-restrictive, whether these alternative measures could be reasonably available and whether they could equally guarantee the achievement of the objective.[55] Although alternative measures, such as a complete prohibition of sales of nuclear electricity or the use of labels (certificates) discouraging the consumption of nuclear electricity, could be found, they are unlikely to be considered as proper substitutes. A government cannot prohibit sales of nuclear electricity until the entire substitution of this type of electricity in the electricity supply is possible. Moreover, prohibition of sales is a more trade-restrictive alternative. As regards the use of labels, although these are less trade-restrictive measures, they

[53] AB Report, *European Communities – Regime for the Importation, Sale and Distribution of Bananas (EC – Bananas III)*, adopted 25 September 1997, WT/DS27/AB/R, para. 221.

[54] Panel Report, *United States – Standards for Reformulated and Conventional Gasoline (US – Gasoline)*, adopted 20 May 1996, WT/DS2/R.

[55] AB Report, *Korea – Measures Affecting Imports of Fresh, Chilled and Frozen Beef (Korea – Various Measures on Beef)*, adopted 10 January 2001, WT/DS161/AB/R, WT/DS169/AB/R, para. 164; Appellate Body Report, *Dominican Republic – Measures Affecting the Importation and Internal Sale of Cigarettes (Dominican Republic – Import and Sale of Cigarettes)*, adopted 19 May 2005, WT/DS302/AB/R, para. 70.

are unlikely to be found to be as efficient for achievement of the objectives as the use of a tax.

The 'relating to' link with the objective under paragraph (g) will be easier to establish. Paragraph (g), however, also contains a requirement that a measure must be taken in conjunction with the imposition of constraints on domestic production or consumption. It means that the environmental objective of a differentiated electricity tax needs to be consistently pursued in the internal market though national policies. In the context of a differentiated tax for nuclear electricity, the deployment of a strategy for the phasing-out of domestic production of nuclear electricity would be in line with this requirement.

The analysis of a differentiated electricity tax under Article XX can also raise the issue of extraterritorial application, that is, the impacts of such a tax on electricity-generation methods used by foreign producers. WTO case law does not exclude the possibility of justification of extraterritorial measures under GATT Article XX, especially if a link can be established between what happens in the exporting country and the risks inflicted on the importing country by the situation in the exporting country.[56] It should not be a problem to establish the territorial connection of risks in the case of a differentiated electricity tax. It could be argued that climate change has no territorial borders, and even if it is caused by the generation of electricity from fossil fuels in the territory of the exporting country, it still has effects on the climate in the importing country. Similarly, it could be argued that the environmental and health effects of nuclear plant accidents that may happen in the territories of exporting countries would also be felt in the territory of the importing country.

The accommodation of a differentiated electricity tax under paragraphs (g) or (b) of GATT Article XX will provide only a preliminary justification for the tax. In the second step, a differentiated electricity tax will have to be assessed for compliance with the conditions of the introductory paragraph (chapeau) of Article XX.[57] This stage will see the determination of whether or not a differentiated electricity tax is applied in a manner that constitutes arbitrary or unjustifiable discrimination between countries where the same conditions prevail, or that constitutes a disguised restriction on trade. To pass the test on arbitrary discrimination under the chapeau of Article XX, application of a differentiated electricity tax must take into account conditions prevailing in

[56] Ibid., para. 134. [57] AB Report, *US – Gasoline*, p. 22.

the countries from which electricity is imported. The meaning of 'conditions' here has a link with the objective of a paragraph or, more precisely, with the risks that a measure is aimed to mitigate.[58] Variations in the application of an electricity tax in relation to electricity that carries the same risks with respect to the environment or health would result in a finding of arbitrary discrimination preventing justification under GATT Article XX. However, as follows from the recent WTO case law, a measure's mode of application can somewhat deviate from the objective under a paragraph, if this is necessary to fulfil a country's obligations under an international agreement. The latter can apparently be untethered from the principal objective of a measure, that is, the objective of a paragraph of Article XX under which a measure is meant to be justified.[59] It is not entirely clear, however, how this recent addendum to the 'prohibition of arbitrary justification' standard's interpretation under the chapeau can be reconciled with the requirement that the discrimination has a link to the objective under a paragraph, which was advanced by WTO adjudicative bodies in past disputes.[60] Most probably, in the view of the AB, a measure pursuing multiple objectives must be not only the least discriminatory, but also the least inconsistent in relation to the principal legitimate objective.[61]

[58] AB Report, *Brazil – Measures Affecting Imports of Retreaded Tyres (Brazil – Retreaded Tyres)*, adopted 20 August 2009, WT/DS332/AB/R, para. 227. For interpretation of the prohibition of arbitrary discrimination under the Chapeau of Article XX by WTO adjudicative bodies, see K. Holzer, *Carbon-related Border Adjustment and WTO Law* (Cheltenham: Edward Elgar, 2014), pp. 167–75.

[59] In *EC – Seals*, the scrutiny of the EU regime for seals and seal-containing products under GATT Article XX focused, *inter alia*, on finding the right balance between the principal objective of the measure to address public moral concerns regarding seal-killing methods and the competing aim of accommodating the subsistence needs of Inuit communities acknowledged in the UN Declaration on the Rights of Indigenous Peoples and other international treaties protecting the rights of indigenous peoples, which the EU aimed to observe. See AB Report, *European Communities – Measures Prohibiting the Importation and Marketing of Seal Products (EC – Seal Products)*, adopted 22 May 2014, WT/DS400/AB/R, paras. 5.321–5.326.

[60] It should be noted that, in the end, both the panel and the AB in *EC – Seals* found the exception provided to Inuit (and the exception linked to the maritime management) to be incompatible with the principal objective of the seals regime, that is, the protection of public morals, and recommended adjusting the measure accordingly. For a detailed analysis of the *Seals* case, see T. Cottier, R. Liechti, I. Espa and T. Payosova (2015), 'The jurisprudence of the World Trade Organisation in 2014', *Schweizerische Zeitschrift für internationales und europäisches Recht*, 25(2) (2015), 239–64.

[61] L. Bartels, 'The chapeau of the general exceptions in the WTO GATT and GATS Agreements', *American Journal of International Law*, 109 (2015), 95–124.

That being said, there is nothing in the text of GATT Article XX and in the Article XX jurisprudence to suggest that an electricity tax imposed at different tax rates depending on the electricity sources would be difficult to justify under the general exceptions of the GATT. It is an important conclusion, taking into account that there will be a need for justification in the scenarios in which green and grey electricity are considered to be like or directly competitive or substitutable products and the differences in tax rates trigger a violation of GATT non-discrimination rules.

III Application of WTO Subsidies Disciplines to Differentiated Electricity Tax

A differentiated taxation scheme for electricity, especially where a lower tax rate is granted to particular electricity producers, may raise further compliance concerns under the WTO disciplines on subsidies.[62] WTO law addresses only those subsidies that, in the form of governmental intervention, distort international trade by giving 'an artificial competitive advantage to exporters or to import-competing industries'.[63]

WTO disciplines on subsidies – namely, the provisions of the ASCM –differentiate between three categories of subsidies: prohibited, actionable and non-actionable. Currently, only the first two categories exist.[64] The ASCM explicitly prohibits subsidies that are contingent (*de jure* or *de facto*) on export performance (export subsidies) or on the use of domestic goods over imported goods (import substitution

[62] The WTO disciplines on subsidies include the provisions of GATT Article XVI and the Agreement on Subsidies and Countervailing Measures. The provisions of the ASCM and GATT Article III contain cumulative obligations. The AB confirmed that, notwithstanding the fact that both agreements may deal with the same type of discriminatory measures, the WTO panel has to analyse the respective claims under both of the agreements, since the remedies provided for in case of violation are different. However, there is no clear-cut rule on the order of analysis: see AB Report, *Canada – FIT Program*, para. 5.5.

[63] WTO Secretariat, *World Trade Report 2006: Exploring the Links between Subsidies, Trade and the WTO*, www.wto.org/english/res_e/booksp_e/anrep_e/world_trade_report06_e.pdf (accessed 10 January 2016); see also Panel Report, *Brazil – Export Financing Programme for Aircraft (Brazil – Aircraft)*, adopted 20 August 1999, WT/DS46/R, as modified by AB Report WT/DS46/AB/R, DSR 1999:III, 1221, para. 7.26; Panel Report, *Canada – Measures Affecting the Export of Civilian Aircraft (Canada – Aircraft)*, adopted 20 August 1999, WT/DS70/R, upheld by AB Report WT/DS70/AB/R, DSR 1999:IV, 1443, para. 9.119.

[64] The so-called non-actionable subsidies envisaged in ASCM Article 8 were phased out in 1999, in line with ASCM Article 31.

subsidies).[65] All other subsidies will be considered as non-compliant with WTO law only if they cause adverse effects to the interests of another WTO member. To remedy the negative effects from imports of the subsidised products, WTO law provides for two different tracks of response. First, it gives the WTO members concerned the right to bring a claim to the WTO against the subsidies of another WTO member (a multilateral track). Second, WTO members can also resort to unilateral measures subject to strict procedural requirements set out in the ASCM and, if given conditions are met, apply countervailing duties. We provide an analysis of the differentiated taxation of electricity against this general background.

An electricity tax with different rates for green and grey electricity will fall within the scope of the ASCM only if it constitutes a subsidy, that is, either a financial contribution or an income or price support.[66] In addition, it should confer a benefit and comply with the specificity requirement.[67] Application of different tax rates to green and grey electricity is not linked *per se* to export performance or import substitution, and therefore would not fall within the category of prohibited subsidies.

With respect to the first element of analysis, a differentiated electricity tax will only constitute a financial contribution if a government, any public body or an entrusted private body (i) directly transfers funds, or (ii) provides fiscal incentives (government forgoes the revenue that is otherwise due), or (iii) purchases goods or provides goods or services apart from general infrastructure.[68]

The general scheme of differentiated tax on electricity would potentially fall under the second category of financial contribution, namely a government revenue (e.g. a tax) that is otherwise due, which is forgone. According to the WTO panel in *Canada – Autos*, the term 'government revenue' can be defined as '[t]he annual income of a government or State, from all sources, out of which public expenses are met'.[69] Furthermore, the AB has clarified on numerous occasions that 'the "foregoing" of revenue "otherwise due" implies that less revenue has been raised by the government than would have been raised in a different situation, that is,

[65] ASCM Article 3.1 (a) and (b). [66] ASCM Article 1(1)(a)–(b). [67] ASCM Article 2.
[68] ASCM Article 1(1)(a)(i)–(iii).
[69] Panel Report, *Canada – Certain Measures Affecting the Automotive Industry (Canada – Autos)*, adopted 19 June 2000, WT/DS139/R, WT/DS142/R, as modified by AB Report WT/DS139/AB/R, WT/DS142/AB/R, DSR 2000:VII, 3043, para. 10.159.

"otherwise". Moreover, the word "foregone" suggests that the government has given up an entitlement to raise revenue that it could "otherwise" have raised.'[70] In order to establish what is 'otherwise due', an analysis should be based on a 'defined, normative benchmark against which a comparison can be made between the revenue actually raised and the revenue that would have been raised "otherwise"'.[71] The basis of such comparison will always be the tax rules applied by the Member in question.[72] Thus, the benchmark will be different in each case depending on the tax rules applied by the WTO Member under consideration.[73] As confirmed by the recent jurisprudence, WTO members have the sovereign right to set the rate, structure and operation modalities of the domestic tax regime and can also adjust it, as tax systems are not static. In *US – Large Civil Aircraft*, the AB suggested that in order to determine whether the government revenue is due in a specific case the panel should compare the tax treatment of the alleged subsidy recipients, taking into consideration an objective reason behind the differential treatment. The comparison should be made with respect to tax treatment of comparable income of comparably situated taxpayers in light of the structure of the domestic tax regime and its organising principles. Finally, the panel should also take into consideration the background and reasons for differences between the challenged tax treatment and the benchmark tax treatment.[74] In the present case of differential taxation of green and grey electricity, the panel would have to compare the tax

[70] AB Report, *Canada – Certain Measures Affecting the Automotive Industry (Canada – Autos)*, adopted 19 June 2000, WT/DS139/AB/R, WT/DS142/AB/RDSR 2000:VI, 2985, para. 90; AB Report, *United States – Tax Treatment for 'Foreign Sales Corporations' (US – FSC)*, adopted 8 October 1999, WT/DS108/R, para. 90; AB Report, *United States – Measures Affecting Trade in Large Civil Aircraft (Second Complaint)* (*US – Large Civil Aircraft*), adopted 23 March 2012, WT/DS353/AB/R, para. 806.

[71] AB Report, US – FSC, para. 90.

[72] AB Report, *Canada – Autos*, para. 90; AB Report, *US – FSC*, para. 90; AB Report, *US – Large Civil Aircraft*, para. 806.

[73] AB Report, *US – FSC*, para. 88–89. According to the AB in *US – FSC*, paras. 8.18–8.37, 'examination as to whether there is revenue foregone that is 'otherwise due' must be based on actual substantive realities and not be restricted to pure formalism ... The key point is that the tax rules applied by the Member in question are the basis for the comparison': AB Report, *United States – Tax Treatment for 'Foreign Sales Corporations'. Recourse to Article 21.5 of the DSU by the European Communities (US – FSC (Article 21.5 - EC))*, adopted 14 January 2002, WT/DS108/AB/RW, paras. 86, 91–92. See also P. C. Mavroidis, G. A. Bermann and M. Wu, *The Law of the World Trade Organization* (St. Paul: West Publishers, 2010), p. 563.

[74] AB Report, *US – Large Civil Aircraft* (2nd Complaint), para. 831.

treatment of green and grey electricity and take into consideration the objectives pursued by the WTO Member through such differentiation (e.g. climate change mitigation).

In light of these requirements established in WTO jurisprudence, it seems to be better, in terms of WTO compliance, if the government concerned envisages the differential tax rates for different energy sources at the outset based on the environment-related objectives pursued (and not through a general electricity rate and tax exemptions for green electricity). Furthermore, for the analysis of differentiated electricity taxation, it might potentially be possible to consider an environmental objective as the rationale behind the tax.[75] Importantly, WTO law, unlike EU law, does not require the lower tax rate (or a tax exemption) to be proportionate in light of the objective pursued.

In a second step, if a differentiated tax on electricity amounts to a financial contribution in the form of government revenue forgone, the panel would have to determine whether this differentiation confers a benefit.[76] The benefit is conferred to a recipient, that is, a green electricity utility subject to a lower electricity tax rate, if it is better off in the marketplace with this financial contribution than without it.[77] Neither the GATT nor the ASCM define the term 'benefit'; nor do either provide a particular methodology to determine whether a benefit is conferred. However, the recent WTO jurisprudence provides quite a detailed explanation as to what constitutes a 'benefit' within the meaning of Article 1. The panel in *US – Large Civil Aircraft (2nd Complaint)* clarified that there is a benefit if a financial contribution was granted on terms more favourable than the market terms.[78] In order to determine a benchmark for this comparison, WTO adjudicating bodies often rely on ASCM Article 14[79] as the relevant context for ASCM Article 1.1(b). However, based on previous case law, this is only possible where the financial contribution, as determined under Article 1(1)(a)(i)–(iii), is also

[75] L. Rubini, *The Definition of Subsidy and State Aid: WTO and EC Law in Comparative Perspective* (Oxford: Oxford University Press, 2009), pp. 260–80.

[76] Panel Report, *European Communities - Countervailing Measures on Dynamic Random Access Memory Chips from Korea (EC - DRAMS)*, adopted 3 August 2005, WT/DS299/R DSR 2005:XVIII, 8671, para. 7.212; see also AB Report, *Canada – Aircraft*, para. 154.

[77] AB Report, *US – Large Civil Aircraft* (2nd Complaint), para. 873; Panel Report, *Canada – Aircraft*, para. 9.112.

[78] Panel Report, *US – Large Civil Aircraft* (2nd Complaint), para. 7.475.

[79] ASCM Article 14 provides for a calculation of a subsidy in terms of the benefit to the recipient.

referred to in ASCM Article 14(a)–(d) (e.g. loans or purchase of goods by the government). Thus, it is not clear whether the recent position of the panel and the AB with respect to the decisive role of government definition of the electricity market, namely the supply mix, would have any effect on determination of benefit from the revenue forgone.[80] Notably, the AB has previously stated that reduced tax payments do constitute a benefit.[81]

In a third step, if the panel finds a benefit conferred by the differentiated electricity tax, the financial contribution would have to meet the specificity requirement, in line with ASCM Article 2. As a differentiated electricity tax – specifically, its lower rate (or tax exemption) – will apply only to particular enterprises or groups of enterprises, that is, green electricity utilities, it will most probably be found to be industry-specific. Whereas the ASCM allows WTO Members to differentiate between enterprises, industries and geographical regions where differentiation is based on objective eligibility criteria, which are applied automatically,[82] the differentiated electricity tax is hardly likely to meet these conditions.

Finally, if the panel finds that the differentiated electricity tax constitutes an industry-specific subsidy, it will proceed with the analysis of adverse effects to the interests of other WTO Members. Given the special characteristics of the electricity trade, a differentiated electricity tax is not likely to cause an 'injury' through increased imports of green electricity from the subsidising WTO Member to the neighbouring WTO Member(s).[83] Similarly, it is unlikely that a differentiated electricity tax will lead to 'serious prejudice',[84] as it will not cause displacement of or impediment to imports of green electricity from abroad.

The compatibility of a differentiated electricity tax with the ASCM will depend to a large extent on its specific modalities. The probability of

[80] AB Report, *Canada – FIT Programme*, paras. 5.167–5.191; Panel Report, *Canada – FIT Programme*, paras 7.279–7.284.

[81] AB Report, *US – FSC (Article 21.5 – EC)*, para. 191; AB Report, *US – FSC*, para. 140.

[82] ASCM Article 2.1(b). Further, footnote 2 specifies that eligibility criteria for a financial contribution should be neutral (non-discriminatory), should not favour certain enterprises over others and should be 'economic in nature and horizontal in application'. It also gives examples of such criteria, namely size or number of employees of an enterprise. Negotiation history shows that earlier drafts of ASCM Article 2 referred also to other examples, e.g. incidences of pollution and health and safety standards, which at the very end were not included in the agreement. While it is not excluded that the panel may take into consideration negotiation history, the ordinary meaning of 'specificity' would prevail.

[83] Article 15 of the ASCM. [84] Article 6 of the ASCM.

adverse effects remains low, given the limited cross-border trade in electricity in most regions of the world and its dependence on the available infrastructure, but cannot be fully disregarded.

D Analysis of Various Tax Design Options

For countries with limited RE development potential, merely offering differentiated tax rates for electricity may not help to achieve the target of an increased share of production of green electricity. Policy-makers may therefore consider the possibility of combining a differentiated electricity tax with additional requirements, including those that determine eligibility of RECs for tax exemption purposes. Restrictions could be introduced, for instance, on the admissibility of RECs for the purpose of granting tax exemptions. These restrictions could be both quantitative and qualitative. A quantitative restriction could be put on foreign RECs on the grounds that they can be acquired at a lower price than domestic RECs.[85] In terms of qualitative restrictions, the admissibility of RECs could be conditioned on the actual attachment to the physical flow of electricity, or restricted to RECs originating from electricity installations that are also certified for their environmental footprint. These additional requirements may further complicate prospects of compliance of such a differentiated electricity tax with WTO law. Moreover, the implications of an electricity tax with regard to WTO law would be different should the tax be based on electricity's carbon footprint.

I A Differentiated Electricity Tax with Quantitative Restrictions on the Acceptability of Certificates

The implications of a quantitative limitation on foreign RECs eligible for tax exemptions would likely impinge on the obligation of general elimination of quantitative restrictions under GATT Article XI:1, which forbids both import 'prohibitions' and import 'restrictions ... whether effective through quotas, import ... licences or other measures'. Limiting the quantity of foreign RECs eligible for tax exemptions by imposing a certain fixed threshold could be found to be a measure constituting a 'restriction ... on importation' of green electricity. WTO jurisprudence has consistently interpreted Article XI:1 GATT in a broad manner,

[85] For instance, the price for GOs in the EU is presently four times lower than the price for GOs in Switzerland.

considering not only measures which may formally be considered quantitative restrictions (e.g. quotas), but also other measures constraining trade through reductions in the volume of imports (or exports), to fall within its scope of application. In *India – Autos*, the panel suggested that the term 'restrictions' includes all measures imposing a condition with a limiting effect.[86] In *Colombia – Ports of Entry*, the panel found that Article XI:1 would also cover 'measures which create uncertainties and affect investment plans, restrict market access for imports or make importation prohibitively costly'.[87] In *China – Raw Materials*, the panel further added that any measure with 'the very potential to limit trade ... constitute[s] a "restriction" within the meaning of Article XI:1 of the GATT 1994'.[88] Based on WTO case law, a quota on the number of foreign RECs eligible for tax exemptions is likely to be viewed as a measure having a limiting effect on importation within the meaning of GATT Article XI:1. This is because it would create uncertainty regarding foreign RECs' eligibility for the purposes of tax exemptions, while domestically produced electricity accompanied by domestic RECs would automatically be granted the exemption. Such a scenario would negatively affect the competitive opportunities of foreign green electricity compared to domestic green electricity.[89]

Furthermore, GATT Article XI:1 is not the only provision that may be relevant for the purpose of assessing the legal feasibility of quantitative limitations on foreign RECs eligible for tax exemptions. A violation of the NT rule under GATT Article III:4 may also arise to the extent that more favourable treatment would be granted to electricity suppliers submitting RECs of domestic origin, thus associated with green electricity produced

[86] Panel Report, *India – Measures Affecting the Automotive Sector*, circulated on 21 December 2001, WT/DS/146/R and WT/DS/175/R, para. 1.14.

[87] Panel Report, *Colombia – Indicative Prices and Restrictions on Ports of Entry*, adopted on 20 May 2009, WT/DS366/R, para. 7.240.

[88] Panel Report, *China – Measures Related to the Exportation of Various Raw Materials*, adopted on 22 February 2012, WT/DS394/R, WT/DS395/R and WT/DS398/R, para. 7.1081.

[89] This conclusion is supported by *Colombia – Ports of Entry*, where the Panel opined that the limiting effect on importation would not have to be proved based on the trade impacts of a measure so long as 'changes in trade volumes result not only from governmental policies, but also from other factors, and that, in most circumstances, it is not possible to determine whether a decline in imports following a change in policies is attributable to that change or to other factors'. Panel Report, *Colombia – Ports of Entry*, ft. 35, para. 7.254. Thus, there is no need to quantify the impact determined by the preference of domestic over foreign RE GOs on the volume of green electricity imports for the purpose of proving the violation of Article GATT XI:1.

domestically, rather than foreign RECs obtained from importing green electricity. This would modify the conditions of competition between imported and domestic green electricity in the country imposing the electricity tax, to the detriment of the importer.[90]

Moreover, designing an electricity tax scheme in such a way that the tax exemptions would be fully available to domestic green electricity upon submission of domestic RECs, while not granting them to imported green electricity accompanied by foreign RECs exceeding the quota, would likely reduce the chances of successfully defending the whole tax scheme under GATT Article XX. As already discussed, recourse to the GATT exceptions may be needed to justify the PPM nature of an electricity tax or the possible discriminatory effects on foreign green electricity compared to domestic green electricity. As a quota imposed on foreign RECs for the purposes of tax exemptions would ultimately discourage green electricity imports while stimulating the production of green electricity in the country imposing a tax, the question is whether the preference for domestic over foreign green electricity could be considered impartial. Pursuant to the chapeau of Article XX, a measure shall not be 'a means of arbitrary or unjustifiable discrimination between countries where the same conditions prevail, or a disguised restriction on international trade'. The country imposing an electricity tax with such requirements would need to prove that the 'conditions' prevailing domestically, on the one hand, and in the country from which green electricity is imported, on the other, are 'relevantly different'.[91] It seems unlikely that the importing country could rely on the existence of the tax scheme to argue that the domestic conditions are different from those prevailing in the exporting country. As mentioned above, because the quota on foreign RECs would result in discrimination between countries with the same conditions, the reasons for such discrimination should be related to the objective under

[90] A detailed analysis of the case law interpreting Article III:4 GATT and a thorough explanation of its applicability to a quantitative limitation imposed on GOs for the purposes of tax exemptions was provided in T. Cottier, I. Espa, S. Hirsbrunner, K. Holzer and T. Payosova, *Differential Taxation of Electricity: Assessing the Compatibility with WTO Law, EU Law and the Swiss-EEC Free Trade Agreement*, legal opinion commissioned by the Swiss Federal Finance Administration, the Swiss Federal Office of Energy and the Swiss State Secretariat for Economic Affairs (2014), www.efv.admin.ch/e/downloads/finanzpolitik_grundlagen/els/Differentiatial%20_Taxation_e.pdf?lang=de&msg-id=50122 (accessed 10 January 2016), p. 55.

[91] AB Report, *EC – Seal Products*, para. 5.299.

Article XX (g) or (b).[92] However, the reasons for discriminating against imported green electricity and in favour of domestic green electricity do not seem to relate directly to public health or environmental protection considerations, as these goals would be promoted just as effectively by measures incentivising the consumption of green electricity irrespective of its origin. For all these reasons, a quota put on foreign RECs eligible for tax exemptions would likely run afoul of GATT provisions.

Finally, restricting the availability of tax exemptions by having a quota for foreign green utilities may raise questions as to whether such a tax scheme could constitute a prohibited import substitution subsidy within the meaning of ASCM Article 3. In such a case no determination of adverse effect will be required.[93]

II A Differentiated Electricity Tax with Qualitative Restrictions on the Acceptability of Certificates

One could also envisage limiting the admissibility of RECs for tax exemptions based on certain qualitative criteria. The admissibility of RECs could, for instance, be conditioned on their actual attachment to the physical flows of electricity. This requirement seems to reflect the environmental goal of stimulating the use of electricity from renewable sources in the importing country better than the quantitative restrictions on RECs. This is because, from an environmental perspective, it is irrelevant whether the increase in the production of green electricity occurs in the territory of the importing country or elsewhere, owing to the global nature of climate change. Furthermore, it is an objective criterion that allows RECs to be used as a tool to extend a domestic tax to imports, while still treating domestic and foreign green electricity alike. Under this scenario, exclusion from tax exemptions would only happen with regard to certificates not linked to the importation of green electricity, while imports of green electricity would be treated the same as domestic green electricity. For this reason, this design option does not seem to entail any violations of relevant WTO provisions. Moreover, because such a criterion would be equally applicable to domestic and foreign RECs accompanying green electricity flows in a way similar to labels, no additional legal hurdles would arise out of the obligations under the Technical Barriers to Trade (TBT) Agreement.

[92] AB Report, *Brazil-Retreaded Tyres*, para. 227. [93] Article 3.2 of the ASCM.

Another qualitative criterion for the acceptability of RECs could consist in granting the tax exemptions upon the submission of RECs (e.g. GOs) coming from electricity installations that meet the requirements for green electricity labels, such as 'naturemade' labels issued in Switzerland. The idea behind this option is to restrict the availability of tax exemptions not simply to green electricity as such (i.e. CO_2-free electricity generated from renewable sources), but more specifically to green electricity produced in an ecologically sustainable manner.[94] This option could be implemented through GOs, as GOs contain information on 'the identity, location, type and capacity of the installation where green electricity was produced'.[95] Although in principle this criterion would be applicable to both domestic and foreign GOs, the choice of a national quality label such as 'naturemade' would likely entail *de facto* discrimination against imported green electricity. This is because national labels are not diffused in other countries. This means that GOs qualifying for tax exemptions will almost exclusively be those of domestic origin, whereas foreign GOs would rarely qualify for tax exemptions. Limiting the admissibility of GOs for the tax exemptions to those that correspond to a national electricity label has implications for trade in electricity similar to those arising out of the introduction of a formal quota of foreign RECs eligible for tax exemptions. Domestic GOs originating from certified installations would in fact be preferred over foreign GOs, even if the latter come from installations certified under another green electricity label based on ecological requirements similar to those assessed under the national label. The introduction of such a qualitative criterion is thus likely to entail a violation of Article III:4 and Article XI:1 GATT, while posing problems for justification under the relevant environmental exceptions available under the GATT. In order to avoid *de facto* discrimination against imported green electricity, a tax exemption scheme should thus be based on a label that is equally available to domestic and foreign suppliers of renewable electricity.

Qualitative restrictions on certificates' acceptability for the purposes of tax exemption would potentially entail a violation of WTO subsidies disciplines only if there were adverse effects as referred to above.

[94] Green electricity labels take into account a wider range of requirements concerning ecological, social, organisational and professional factors. See *Naturemade Swiss Quality Label: A Top Global Brand*, available at www.naturemade.ch/Dokumente/Kommunikation/PWC-Report-kurz-e.pdf (accessed 10 January 2016).

[95] See Article 15(6)(c) of Directive 2009/28/EC.

III A Tax Based on the CO_2 Footprint of Electricity

Instead of taxing electricity at different rates depending on the energy source used for its generation, policy-makers may consider taxing electricity at rates determined on the basis of its CO_2 footprint. A domestic CO_2 levy could then be extended to imported electricity as a border tax adjustment (BTA) measure.[96] WTO law issues arising out of a CO_2 levy on electricity are basically similar to those arising out of an electricity tax based on the sources of electricity. Similarly to the case of a tax levied on the sources of electricity, issues of inconsistency with the MFN and NT rules under the GATT might arise owing to the PPM nature of the levy, and thus recourse to Article XX GATT to defend the measure would probably be needed.[97] Moreover, to avoid a violation of non-discrimination rules, a CO_2 levy adjusted at the border must be levied on domestic power plants without exemptions. It should also be noted that a levy based on the CO_2 footprint of electricity cannot be imposed on carbon-neutral nuclear electricity.

Compared to a tax based on the source of electricity, it would currently be more difficult to impose a CO_2 levy in light of the practical issues related to the administration of the tax, including tracing electricity's carbon footprint and determining tax rates for imported electricity, the carbon footprint of which cannot be verified. The use of available RECs (e.g. GOs or green certificates) for the implementation of a CO_2 levy would only allow for tracing renewable sources of electricity. It would not provide the information on the CO_2 footprint of imported electricity on which a CO_2 levy would need to be based. Given this deficiency in existing electricity certification schemes, CO_2 levy rates for imported electricity cannot rely on the actual carbon footprint of imported electricity and need to be based on a benchmark level of emissions, that is, a 'constructed' carbon footprint of electricity.

Pursuant to the rules on the application of BTAs, a CO_2 levy applied to imported electricity should correspond to the same levy imposed on domestic electricity not only in terms of the tax burden but also in terms of the manner in which it is applied. Consequently, if a CO_2 levy on imported electricity is based on the average footprint of carbon emissions, domestic electricity should also be taxed on the basis of the average

[96] Ecoplan et al., *Border Tax Adjustments*, p. 84.

[97] For a comprehensive study on WTO compliance of border carbon adjustments, see Holzer, *Carbon-related Border Adjustment and WTO Law*.

CO_2 footprint. Different approaches taken with respect to the determination of electricity tax rates for domestic and imported electricity were found unacceptable in the *Outokumpu Oy* dispute settled by the European Court of Justice (ECJ).[98] While applying tax rates to domestic electricity that varied on the basis of electricity-generation methods, Finland applied to imported electricity a flat rate corresponding to the average rate. The ECJ concluded that the rules of the European Community precluded the imposition of a tax based on the application of different criteria for the calculation of the tax with regard to domestic and imported products.[99]

IV Tax Revenue Recycling Issues

The way in which tax revenues are used or recycled might also be a factor influencing the compliance of a differentiated electricity tax with WTO law. When examining a tax revenue recycling scheme's compliance with the law of the WTO, legal analysis focuses on whether a particular mode of allocation of the tax revenues subsidises national producers or consumers. To constitute an actionable subsidy, that is, a subsidy that can be challenged through the WTO dispute settlement procedures or counteracted by trading partners through countervailing import duties, a tax recycling system must fall under the WTO definition of a 'subsidy', be 'specific' and cause adverse effects within the meaning of the ASCM. To qualify as a subsidy, according to Article 1 ASCM, a measure must constitute a financial contribution by a government and confer a benefit. Tax revenue recycling can be a governmental financial contribution if tax rebates are forgone budget revenues or a direct transfer of funds. This would depend on the design of the tax rebate system. Yet, it does not seem to be reasonable to consider that electricity tax revenue allocation constitutes forgone government revenues in the situation where the 'normal' state of play in a country is not to tax grey electricity or carbon at all. It is more logical to consider tax rebates to be a redistribution of funds between private entities.[100] For similar reasons, it would be difficult to argue that rebates of such costs confer a benefit to the firms.[101] Instead

98 Case C-213/96, *Outokumpu Oy* [1998] ECR I-1777, para 34. 99 Ibid.

100 N. Shariff, 'Enhancing competitiveness and addressing carbon leakage: a value added based approach to emissions pricing system design', master's thesis, University of Bern (2012), p. 48.

101 Ibid.

of being viewed as a subsidy, '[i]n a country where the status quo is not to tax emissions at all ... the institution of a charge and rebate system should constitute ... a means of taxation that limits the cost impacts of the measure on its industry'.[102] And even if the tax revenue rebates to domestic electricity companies were found to constitute a subsidy, there is practically no way that this subsidy, that is, a specific subsidy causing adverse effects for the foreign electricity industry, would be considered actionable, especially in a situation in which foreign electricity producers bear no fossil fuel or carbon-related environmental costs. In light of all the foregoing, it is unlikely that electricity tax or CO_2 levy rebates would raise serious issues under WTO subsidy rules.

Nevertheless, when designing an electricity tax revenue recycling system, certain elements should be included to ensure its compliance with WTO law. First, the system should be administered in such a way as to demonstrate a clear connection between the tax and the rebate. This means that the tax is not deposited in the budget account but rather is clearly redistributed, with only a small portion being used to fund the administration of the tax system.[103] Second, the system should apply to all electricity installations so that they will automatically be subject to the tax and entitled to receive the rebate (or exemption). Third, tax revenues used even partially to support environmental and climate change mitigation and adaptation programmes may serve as evidence of the environmental rationale of the tax and an important indicator of its neutrality. By contrast, using revenues from a tax on imported electricity solely to support the development of domestic industries may impair justification of the tax under the environmental exceptions. Finally, it is important to note that WTO law does not prohibit the redistribution of tax revenues through a national tax reform. A government may, for instance, decide to use revenues from an electricity tax to lower corporate and income taxes.[104]

[102] Ibid. [103] Ibid., p. 50.

[104] In this respect, it is noteworthy that revenues from the UK Climate Change Levy are largely recycled back into industry through a 0.3 per cent reduction of the employer payment to national insurance contributions. Part of the revenues is further diverted to the Carbon Trust, an institution which fosters research and promotion of energy efficiency and renewable energy sources. R. Martin, U. J. Wagner and L. B. de Preux, 'The impacts of the Climate Change Levy on business: evidence from microdata', *Centre for Climate Change Economics and Policy Working Paper No. 7* (2009), http://eprints.lse.ac.uk/37676/1/The_impacts_of_climate_change_levy_on_business_evidence_from_microdata(lsero).pdf (accessed 20 November 2016).

E Conclusions

An electricity tax with different tax rates applied to different types of imported electricity against submission of RECs can be compliant with GATT rules, provided a number of requirements are met. Domestic and imported electricity of a particular type must be taxed equally and differences in taxation between electricity of different types must not exceed what is needed for the achievement of environmental or public health policy objectives. The compatibility of a differentiated electricity tax with WTO subsidies disciplines will depend to a large extent on its specific modalities, and primarily on the existence of adverse effects. The probability of such adverse effects in the current electricity market conditions remains low, but cannot be fully disregarded.

Introducing additional requirements and constraints for imported green electricity eligible for tax exemptions could, however, raise complications for the compliance of a differentiated electricity tax with WTO law. Limiting the number of foreign certificates eligible for tax exemptions would be likely to affect the volumes of green electricity imported, and thus trigger a violation of the NT obligation and potentially amount to an import substitution subsidy. However, restrictions on the eligibility of RECs might be defended under WTO law if they are based on qualitative criteria, such as the attachment of RECs to green electricity flows or to a green electricity label that is equally available to domestic and foreign suppliers of green electricity.

A differentiated electricity tax could also be based on the CO_2 footprint of electricity. A CO_2 levy on imported electricity may pass the test of compliance with WTO law if framed as an extension of a CO_2 levy on domestic electricity within the meaning of a non-discriminatory border tax adjustment scheme. A CO_2 levy on imported electricity will be consistent with WTO law only when the same levy is imposed without exemptions on domestic power plants. A CO_2 levy applied to imported electricity should correspond to the CO_2 levy imposed on domestic electricity not only in terms of the tax burden but also in terms of the manner in which it applies.

Assessing an electricity tax revenue recycling system's compliance with WTO law is not straightforward. However, the preliminary legal analysis shows that a partial redistribution of the electricity tax or CO_2 levy back to the industry would be unlikely to raise serious issues under WTO subsidy rules. To improve a tax recycling scheme's WTO compliance, tax revenues should, at least partially, be used to fund environmental and climate change mitigation and adaptation programmes. This would serve as evidence of the environmental rationale of the tax and an important indicator of its neutrality. Moreover, WTO law does not prohibit redistributing tax revenues through a national tax reform resulting in a decrease in other taxes.

Bibliography

Bartels, L. (2015). The chapeau of the general exceptions in the WTO GATT and GATS Agreements. *American Journal of International Law*, **109**, 95–124.

Cansino, J. M., Pablo-Romero, M., Román, R. and Yñiguez, R. (2010). Tax incentives to promote green electricity: an overview of EU-27 countries. *Energy Policy*, **38**(10), 6000–8.

Chatzivasileiadis, S., Ernst, D., and Andersson, G. (2013). The Global Grid. *Renewable Energy*, **57**, 372–83.

Conrad, C. R. (2011). *Processes and Production Methods (PPMs) in WTO Law: Interfacing Trade and Social Goals* (New York: Cambridge University Press).

Cottier, T., Espa, I., Hirsbrunner, S., Holzer, K., and Payosova, T. (2014). *Differential Taxation of Electricity: Assessing the Compatibility with WTO Law, EU Law and the Swiss-EEC Free Trade Agreement*, legal opinion commissioned by the Swiss Federal Finance Administration, the Swiss Federal Office of Energy and the Swiss State Secretariat for Economic Affairs.

Cottier, T., Liechti, R., Espa, I., and Payosova, T. (2015). The jurisprudence of the World Trade Organisation in 2014. *Schweizerische Zeitschrift für internationales und europäisches Recht*, **25**(2), 239–64.

Delimatsis, P. (2009). Financial innovation and climate change: the case of renewable energy certificates and the role of the GATS. *World Trade Review*, **8**(3), 439–60.

Delimatsis, P., and Mavromati, D. (2009). GATS, financial services and trade in Renewable Energy Certificates (RECs) – just another market-based solution to cope with the tragedy of the commons? In T. Cottier, O. Nartova and S. Z. Bigdeli (eds.), *International Trade Regulation and the Mitigation of Climate Change* (Cambridge: Cambridge University Press), pp. 231–58.

Ecoplan, WTI and University of Zurich (2013). *Border Tax Adjustments: Can Energy and Carbon Taxes Be Adjusted at the Border?* Final report prepared for the Swiss State Secretariat for Economic Affairs and the Swiss Federal Finance Administration.

Ghosh, A. and Gangania, H. (2012). *Governing Clean Energy Subsidies: What, Why, and How Legal?* Geneva: ICTSD.

Holzer, K. (2014). *Carbon-Related Border Adjustment and WTO Law.* Cheltenham: Edward Elgar.

Holzer, K. and Espa, I. (2015). Greening electricity through taxing: an analysis of GATT constraints. *NCCR Working Paper* No. 2015 (9 April).

Howse, R. (2009). *World Trade Law and Renewable Energy: The Case of Non-Tariff Barriers* (Geneva: UNCTAD).

IEA. (2013). *Cross-Border Trade in Electricity and the Development of Renewable-Based Electric Power: Lessons from Europe*, Annex 2 on Liberalisation of Electricity Markets, Competition and the Drivers of Cross-Border Trade in Electricity, www.oecd-ilibrary.org/trade/cross-border-trade-in-electricity-and-the-development-of-renewables-based-electric-power_5k4869cdwnzr-en (accessed 24 November 2016).

(2013). *Redrawing the Energy Climate Map: World Energy Outlook Special Report*, www.iea.org/publications/freepublications/publication/weo_special_report_2013_redrawing_the_energy_climate_map.pdf (accessed 24 November 2016).

Martin, R., Wagner, U. J., and de Preux, L. B. (2009). The impacts of the Climate Change Levy on business: Evidence from microdata. *Centre for Climate Change Economics and Policy Working Paper* No. 7.

Mavroidis, P., Bermann, G. A., and Wu, M. (2010), *The Law of the World Trade Organization* . St. Paul: West Publishers.

Naturemade Swiss Quality Label: A Top Global Brand, www.naturemade.ch/Dokumente/Kommunikation/PWC-Report-kurz-e.pdf (accessed 24 November 2016).

Ofgem. (2015). Climate Change Levy (CCL) exemption, www.ofgem.gov.uk/environmental-programmes/climate-change-levy-ccl-exemption (accessed 24 November 2016).

PriceWaterhouseCoopers & WWF. (2009). *Green Electricity: Making a Difference. An International Survey of Renewable Electricity Labels*, www.repower.com/fileadmin/user_upload/re-all/02_Files_PDF-DOC-XLS/z_to_be_classified/Berichte_und_Studien/pwc_green_electricity_making_a_difference.pdf (accessed 24 November 2016).

Rubini, L. (2009). *The Definition of Subsidy and State Aid: WTO and EC Law in Comparative Perspective.* Oxford: Oxford University Press.

Sathaye, J., Lucon, O. and Rahman, A. (2011). Renewable energy in the context of sustainable energy. In Ottmar Edenhofer, Ramon Pichs-Madruga, Youba Sokona, Kristin Seyboth, Patrick Matschoss, Susanna Kadner, Timm Zwickel, Patrick Eickemeier, Gerrit Hansen, Steffen Schlömer and Christoph von Stechow (eds.), *IPCC Special Report on Renewable Energy Sources and Climate Change Mitigation.* Cambridge: Cambridge University Press.

Shariff, N. (2012). Enhancing competitiveness and addressing carbon leakage: a value added based approach to emissions pricing system design. Master's thesis, University of Bern.

Talberg, A. and Swoboda, K. (2013). Emission Trading Schemes around the World, Background Note for the Parliament of Australia.

Winkel, T., Ragwitz, M., Resch, G. et al. (2011). *Renewable Energy Policy Country Profiles 2011*, www.reshaping-res-policy.eu/downloads/RE-SHAPING_Renewable-Energy-Policy-Country-profiles-2011_FINAL_1.pdf (accessed 20 January 2016).

World Bank. (2011). Design and performance of policy instruments to promote the development of renewable energy: emerging experience in selected developing countries. *Energy and Mining Sector Board Discussion Paper* No. 22.

Verfassungsbestimmuing über ein Klima- und Energielenkungssystem. Erläuternder Bericht zum Vorentwurf (EFV, BFE, BAFU, 2015).

19

Competition Policy, Monopolies and the Role of the State in Promoting Green Electricity

OLGA NARTOVA

A Introduction

Increasing concerns about the environment call for government involvement to handle market failures related to the emission of greenhouse gases and investment in green energies.[1] Although global policy frameworks such as the Kyoto Protocol have attained importance in the public debate, it is national policies that represent the primary tools to fight global warming and climate change.[2]

However, storage of renewable energies is problematic and they often require backup capacities from traditional fossil fuels, implying that states have a central role to play in promoting green technology.[3] This chapter explores general trends in the effect of competition policy, institutional monopolies and the role of governments in promoting green electricity.

I The Association between Environmental Policy and Competition Policy

Both competition policy and environmental policy are formulated by the state to rectify certain market failures, which often occur because of a lack of incentives to preserve the physical environment.[4] They may also result

[1] S. Borenstein, 'The trouble with electricity markets: understanding California's restructuring disaster', *Journal of Economic Perspectives*, 16(1) (2000), 191–211.

[2] P. D. Domah and M. G. Pollitt, 'The restructuring and privatisation of the regional electricity companies in England and Wales: a social cost benefit analysis', *Fiscal Studies*, 22(1) (2001), 107–46.

[3] Borenstein, 'The trouble with electricity markets'.

[4] Nordic Competition Authorities, *Competition Policy and Green Growth: Interactions and Challenges*, Report No. 1/2010 (2010); J. Kilponen and T. Santavirta, 'When do R&D subsidies boost innovation? Revisiting the inverted U-shape', *Bank of Finland Research Discussion Papers* 10 (2007).

from frequent incentives for organisations to participate in anti-competitive activities.[5] Therefore, competition and environmental policies share a common objective of promoting and safeguarding social welfare.

Promoting fair competition can potentially facilitate environmental policies that allow price indicators which show that environmental externalities are being transmitted efficiently.[6] In addition, competition can contribute to the reinforcement of environmental policies in that efficiency improvement and competition-triggered innovation efforts may be considered crucial mechanisms in promoting a successful environment-friendly policy.[7]

Despite the possible benefits of environmental policies, scholars have also argued that eco-friendly policies can harm competition. For instance, some environmental policies can contribute to increased market barriers, making it difficult for competitors to enter some market sectors.[8] For this reason, the Organisation for Economic Co-operation and Development (OECD) advocates that environmental governing agencies routinely assess the impact of competition on their environmental proposals. In such assessments, the national competition organisations can play a central role in pointing out the restrictive effects that policies might have on competition as a result of environmental regulations.[9]

Environmental remuneration may be presented as a justification for horizontal arrangements and practices otherwise considered proscribed under competition policy.[10] In this regard, however, a number of strict requirements need to be fulfilled. Most importantly, the measure under

[5] S. Scarpetta, P. Hemmings, T. Tressel and J. Woo, 'The role of policy and institutions for productivity and firm dynamics: evidence from micro and industry', *OECD Economics Department Working Paper* No. 329 (2002).

[6] R. Green, 'Electricity and markets', *Oxford Review of Economic Policy*, 21(1) (2005), 67–87.

[7] Domah and Pollitt, 'The restructuring and privatisation of the regional electricity companies in England and Wales'.

[8] Nordic Council, 'Benefits of green public procurement', *TemaNord* 953(51) (2009); S. Dröge and P. J. H. Schröder, 'How to turn an industry green: taxes vs subsidies', *Journal of Regulatory Economics*, 27(2) (2005), 177–202.

[9] P. Aghion, N. Bloom, R. Blundell, R. Griffith and P. Howitt, 'Competition and innovation: an inverted-u relationship', *The Quarterly Journal of Economics*, 120(2) (2005), 701–28.

[10] Domah and Pollitt, 'The restructuring and privatisation of the regional electricity companies in England and Wales'.

deliberation ought to be proportionate to its assertions. Moreover, the net economic benefits must be present in terms of declining environmental pressure emanating from the methods or practices concerned. In reference to the baseline where there is no action considered, the anticipated economic profits should outweigh the costs. Examples of these costs include compliance costs for business operations, costs of the effects of reduced competition and third-party effects.[11]

II The Environmental Impacts of Restrictive Practices

As noted earlier, environmental policies are beneficial in ensuring enhanced social welfare.[12] As stipulated in the competition policies, these benefits are represented by efficiency gains. As such, the environmental benefits achieved can be used in defence of horizontal arrangements or practices that may be perceived as restrictive under competition policies.[13] One may wonder how to turn this perception into a central part of an implementation procedure. According to Dröge and Schröder,[14] for organisations to turn the perception into an integral component of an implementation procedure, in line with the competition policy, it would be essential to perform two tasks. The first would be to evaluate the economic benefits to the environment, and the second to compare the benefits identified with the estimated social costs of the theoretically diminished competition stemming from implementing the methods or practices in question.

This raises the question of whether such restrictive practices can be defended. On their part, Aghion and colleagues express concern that it is not sufficient to defend restrictive policies because they bring unquantified and unspecified social benefits.[15] This implies that accepting such biased arguments can render competition policy and competition law ineffective, and, worse still, turn it into a general social policy framework. Green adds to this debate by categorically stating that if we can translate environmental impacts into specific cost savings or costs to society,[16] the requirements of proficiencies as spelled out in competition regulation

[11] Green, 'Electricity and markets'.

[12] Scarpetta et al., 'The role of policy and institutions for productivity and firm dynamics'.

[13] S. Borenstein and J. Bushnell, 'An empirical analysis of the potential for market power in California's electricity industry', *Journal of Industrial Economics* 47(3) (1999), 285–323.

[14] Dröge and Schröder, 'How to turn an industry green'.

[15] Aghion et al., 'Competition and innovation'.

[16] Green, 'Electricity and markets'.

are fully taken into account, while competition policies preserve their cutting edge.[17] Remarkably, it is clear in this case that policy-makers can ensure that both environmental and competition policy are compatible, in addition to increasing overall economic efficiency.

Incidentally, similar to the potential for a good environmental outcome associated with restrictive practices, it is possible to have negative environmental impacts linked with unrestrictive practices. This can also facilitate the achievement of competitive efficiencies that could demand the same balancing actions described previously.[18] Indeed, one can argue that increased competition, as opposed to declining competition, is likely to help in attaining environmental benefits, particularly where tradable permit systems have been put in place. This is because such provisions require effective competition to guarantee effective pricing of the permits.

III The Benefits of Competition

The primary objective of competition laws is to advance firms' competitive advantage and in this manner support efficient exploitation of environmental resources – that is, economic efficiency. Put simply, competition policy is often assumed to be an amalgamation of competition advocacy, competition laws and their enforcement.[19] If one can identify the exact benefits of competition and how they arise, it is possible to understand the importance of competition policies. Competition is necessary and desirable according to economic theory, since it increases consumers' welfare by supporting the efficiency of economic activities. Rivalry among individual organisations ensures that only highly innovative and efficient companies stay in the market.[20] In addition, competition is a prolonged dynamic process that works through the efficient exploitation of a firm's resources, ample incentives to innovate and free exit and entry from the market.

When the competitive edge is strong, companies are motivated to utilise resources efficiently. This is so because of the comparative gain

[17] Ibid.

[18] Domah and Pollitt, 'The restructuring and privatisation of the regional electricity companies in England and Wales'.

[19] Borenstein and Bushnell, 'An empirical analysis of the potential for market power in California's electricity industry'.

[20] G. J. Stigler, 'Competition' in J. Eatwell, M. Milgate and P. Newman, *The New Palgrave: A Dictionary of Economics* (London: Macmillan, 1987), pp. 1129–36.

as a result of increased efficiency in resource use, which is comparatively higher in markets that are more competitive than it is in non-competitive markets.[21] As Scarpetta and colleagues put it,[22] competition has been observed to reduce inefficient practices in organisations while contributing to increased production rates.[23]

In addition, it is clear that competition can only be enhanced when free market entry is the primary prerequisite. Firms that are unable to keep up with developments are more likely to exit the market if there is fair competition. The outcome is that more efficient businesses come in and gradually replace those that are less so, resulting in higher productivity for the whole economy.[24]

In terms of innovations, the desire to advance into the area of new technologies is stronger where there is vibrant competition. Increased innovativeness results in the creation of new products and services which benefit customers and other sectors of the economy. Nonetheless, the association between competition and innovation is complex.[25] Various practical studies have been undertaken to analyse this relationship. Generally, when the competitive environment is stronger, the level of innovation will be more advanced.[26]

The primary reason for innovation among companies is the need to survive by evading competition and staying ahead of their market rivals. In this regard, some scholars have recognised a possible relationship between competition and innovation, where increased competition leads to a significant increase in research and development investments by neck and neck-positioned firms, and the average technology level of the leading firm increases.[27] One can argue that, depending on the level of market competition, it is possible that practical findings on the source of innovation can be accredited to some institutional factors, such as regulations and laws in individual markets.

21 Kilponen and Santavirta, 'When do R&D subsidies boost innovation?'

22 N. Bloom and J. Van Reenen, 'Measuring and explaining management practices across firms and countries', *The Quarterly Journal of Economics* 72(4) (2007), 1351–408.

23 Borenstein, 'The trouble with electricity markets'.

24 Scarpetta et al., 'The role of policy and institutions for productivity and firm dynamics'.

25 A. Green and D. Mayes, 'Technical inefficiency in manufacturing industries', *The Economic Journal*, 101 (1991), 523–38.

26 R. Blundell, R. Griffith and J. van Reenen, 'Market share, market value and innovation in a panel of British manufacturing firms', *Review of Economic Studies*, 66 (1999), 529–54.

27 P. Aghion, S. Bechtold, L. Cassar and H. Herz, 'The causal effects of competition on innovation: experimental evidence', *NBER Working Paper* No. 19987 (2014).

Although there has been debate as to the quantification of how productivity and the economy are affected by effective competition, a number of extensive studies have reported that higher productivity is directly related to stronger competition.[28] Hence, there is an association between productivity and competition and innovation and competition, respectively. These interrelations may lend support to the perception that effective competition works to reinforce environmental policy because enhanced efficiency and increased innovation are considered critical aspects of a successful environmental policy.[29] This observation is based on findings by Scarpetta and colleagues,[30] who argued that increased costs for firms result in even more importance being placed on higher productivity and greater innovation. Guaranteeing reliable competition is very significant because it determines whether environmental policies succeed or fail.[31] This also acts to ensure that price indications showing environmental externalities are transmitted in an efficient manner. Such an effective competitive environment works to ensure market efficiency in tradable emission licences.[32]

IV Competition Assessment of Green Measures

So far, it is clear that in order to attain green growth, a number of green instruments must be implemented in various areas of the economy, although some may distort competition. Competition authorities have an important role to play in analysing and identifying regulations that have the potential to restrict or distort competition. This is essential in ensuring improved regulation of environmental quality.[33]

It is important to design and select green instruments that will allow competition to play a central role in market diversification and innovation of goods and services, thereby achieving environmental objectives

[28] P. Jamasb and M. Pollitt, 'Electricity market reform in the European Union: review of progress toward liberalization & integration', *The Energy Journal*, 26 (2005), 11–41; M. Grubb and K. Neuhoff, 'Allocation and competitiveness in the EU Emission Trading Scheme: policy overview', *Climate Policy*, 6 (2006), 7–30.

[29] Grubb and Neuhoff, 'Allocation and competitiveness in the EU Emission Trading Scheme'.

[30] Scarpetta et al., 'The role of policy and institutions for productivity and firm dynamics'.

[31] Blundell et al., 'Market share, market value and innovation in a panel of British manufacturing firms'.

[32] Kilponen and Santavirta, 'When do R&D subsidies boost innovation?'

[33] Nordic Council, 'Benefits of green public procurement'.

in a cost-efficient manner. In many cases, this necessitates restructuring green measures so as to reduce the possible harm to competition.[34] In cases of conflicting goals, competition authorities should ensure that an efficient and fair compromise is reached between environmental and competition policies.

To assess the effects that specific regulation of green measures might have on the concerned parties' interests, members can use the OECD Competition Assessment Toolkit, which provides relevant assessment criteria for those involved in competition. The OECD Toolkit stipulates that assessment of competition should be undertaken if a proposal has impacts that limit the range or number of suppliers, their incentives to compete or their ability to compete. In cases where distortions cannot be avoided, it would be crucial for the parties to select the option that is least distortive to the competition.

B The Role of State and Policy Instruments

I State Aid Control (Taxes and Subsidies)

Environmental policies often target the ratio of green to dirty output within a single industry. In order to achieve their objectives, policy-makers often subsidise green industry while taxing the dirty sector.[35] Where there is a monopolistic competition environment, the two policy instruments (subsidies and taxes) have different impacts on welfare benefits.[36] If there is going to be a stronger green policy (i.e. a high reduction in emissions), taxes should be dominant instruments. However, if moderate policy goals are to be attained, it is important to have superior subsidies. This reasoning is applicable where policy-makers may not have the required information about the impact that greenhouse gas emissions can have on the environment, in addition to making sure that they have been adequately informed. As such, the findings have substantial implications for policy formulations in line with state aid control.[37]

Some scholars suggest that subsidies can be a waste of taxpayers' money, although there are a number of reasons why state aid is

[34] Borenstein and Bushnell, 'An empirical analysis of the potential for market power in California's electricity industry'.
[35] Bloom and Van Reenen, 'Measuring and explaining management practices across firms and countries'.
[36] Green and Mayes, 'Technical inefficiency in manufacturing industries'.
[37] Dröge and Schröder, 'How to turn an industry green: taxes vs subsidies'.

controlled in the EU.[38] One reason is that subsidies can potentially affect the economy by distorting competition.[39] For instance, if the government subsidises one company but does not subsidise its competitors, the subsidised company is in a position to take larger shares in the market than it could have done if there were no subsidy from the government. So, the company that enjoys subsidies is likely to do well in the market, although it might be less efficient in that market than its competitors that do not receive government aid.[40]

In addition, when a company has access to government subsidies it might have less desire to compete on merit. As a result, such a company will have less incentive to improve its efficiency by investing in processes or product innovation and operations. If the market concerned is an expanding one, the competitors will not be competing in a fair environment in which they all enjoy a decent profit. In contrast, if the same company that enjoys subsidies is operating in a declining market, the subsidies would enable it to push more efficient competitors out of the market segment.[41]

Other dangers from such subsidies have also been noted. For instance, if the company that is given subsidies knows that it can depend on government bail-outs if the economy collapses, this company may be reluctant to put in place measures that enable it to become more efficient.[42] Such a company is likely to take excessive risks, knowing that profits are often privatised while losses are socialised. This situation is referred to as the problem of moral hazard.[43]

In most cases, state aid controls are directed at national companies rather than individual ones, and at broad input categories such as labour, capital and energy sectors. This is harmful, since subsidies may distort price signals and indicate that prices for input subsidies as estimated by economic actors may deviate from the actual cost-based prices. The outcome of this situation is a suboptimal combination of the production

[38] Scarpetta et al., 'The role of policy and institutions for productivity and firm dynamics'.

[39] Blundell et al., 'Market share, market value and innovation in a panel of British manufacturing firms'.

[40] P. L. Joskow and J. Tirole, 'Retail electricity competition', *Rand Journal of Economics*, 37(4) (2006), 799–815.

[41] Green and Mayes, 'Technical inefficiency in manufacturing industries'.

[42] P. A. Geroski, 'Innovation, technological opportunity, and market structure', *Oxford Economic Papers* 42(3) (1990), 586–602.

[43] Aghion et al., 'Competition and innovation'.

inputs.[44] In any case, subsidies and taxes have the potential to significantly distort the market competition, either positively or negatively, thereby giving a government an appropriate reason to introduce state aid control. In this regard, the primary concern is why this should be under the mandate of the EU instead of being performed by national governments. Perhaps this EU state aid control may be attributed to the need to protect the internal energy market sector.[45]

When Member States embarked on the creation of a standard green energy production process without any hurdles impeding trade, it became necessary to introduce stronger disciplines aimed at ensuring that trade and competition within the internal market would remain undistorted and unbiased.[46] The Member States understood that free and fair competition can be distorted not only by government interventions but also by abusive behaviour of private and public companies.[47] For this reason, various companies have been careful to put in place competitive policies to prevent individual governments from giving domestic companies unfair tax breaks or providing subsidies which give them an advantage over their foreign competitors from across the EU region.[48]

There are evidently potential problems and disadvantages associated with government control of aid in terms of the effects of taxes and aid on fair market competition. However, as the recent liberalisation of the electricity sector has shown, markets are not always well placed to function in an optimal manner; nor does unregulated and free energy sector competition always deliver the desired results.[49] At times, steps towards the realisation of green energy are likely to fail. That is, such government aid controls may produce inefficient outcomes.[50]

Various examples from the economic literature elaborate on these cases of market failures as a result of government aid control.[51] This is as common a challenge in the case of environmental conservation policies and promotion of green energy as it is in research and

44 Kilponen and Santavirta, 'When do R&D subsidies boost innovation?'

45 OECD, *Environmental Regulation and Competition*, DAF/COMP(2006)30 (2006), p. 19.

46 Ibid., p. 20. 47 Ibid., p. 26. 48 Aghion et al., 'Competition and innovation'.

49 Bloom and Van Reenen, 'Measuring and explaining management practices across firms and countries'; Blundell et al., 'Market share, market value and innovation in a panel of British manufacturing firms'.

50 OECD, *Green Growth Strategy Interim Report: Implementing Our Commitment For a Sustainable Future*, C/MIN(2010)5 (2010), p. 38.

51 Droge and Schroder, 'How to turn an industry green'.

development investment.[52] Moreover, markets may fail to deliver results which society considers fair or desirable.

The existence of market failures gives the government a reason to intervene actively in the market. There are a number of ways in which the government can do this, and the imposition of taxes and granting subsidies to companies can sometimes be an effective approach to handling such market failures.[53]

II Is State Aid Control Effective in the EU Member States?

The primary question is whether state aid control has been effective in the EU as far as the promotion of green energy is concerned. This may be hard to answer and not much work has been done to measure to the effectiveness of EU policies related to state aid control.[54] This implies that this is an area in which policy-makers need to make more effort. It is nevertheless possible to look at a few elements that are related to this concept. As elaborated above, the aim of the approach is to reduce inappropriate government aid while encouraging Member States to embrace reforms in their aid systems that will propel them towards horizontal aid that is centred on clearly identified market failures.[55]

The available information indicates that the EU state aid policy seems to have been quite successful with regard to this objective. Horizontal aid, in terms of the entire amount of aid granted by Member States, has been on the increase – from approximately 50 per cent in the early 1990s to more than 80 per cent in more recent years.[56] In addition, with respect to the secondary aim, which is limiting the total volume of state aid given in the EU, success rates increased in the 1990s and 2000s. Overall, the amount of aid granted to services and industries appears to have declined, going from more than 1 per cent of GDP back in the 1990s down to 0.5 per cent of GDP by 2008.[57] However, some scholars have questioned whether this decline reflects the discipline imposed by the EU state aid control or whether it is due to Member States' need to meet the required budgetary criteria for eurozone membership.[58]

[52] OECD, *Green Growth Strategy Interim Report*, pp. 21–2.
[53] Kilponen and Santavirta, 'When do R&D subsidies boost innovation?'
[54] OECD, *Green Growth Strategy Interim Report*, p. 27.
[55] OECD, *Environmental Regulation and Competition*, p. 36. [56] Ibid., p. 37.
[57] Kilponen and Santavirta, 'When do R&D subsidies boost innovation?'
[58] P. Milgrom, *Putting Auction Theory to Work* (Cambridge: Cambridge University Press, 2004).

Furthermore, the progress that had been made up to 2008 in limiting the amount of aid available in the EU was dealt a blow by the global economic crisis of 2008. In the wake of the 2008 financial crisis, total government aid increased from a low of 0.5 per cent of GDP recorded in 2007 to 2.5 per cent of GDP in 2008, before reaching 3.5 per cent in 2009. This increase has been associated with the negative impacts of the economic meltdown. If the crisis' effect, which was linked to aid in the banking sector, is removed from the figures recorded above, the total amount of aid is estimated to be around 0.5 per cent of GDP, as it was in 2007.[59] The financial crisis demonstrated that it would have been risky to introduce competitive policies that resulted in a total loss of state aid – which was in fact a largely unwelcome idea. If governments had been less willing to bail out banks during the 2008–9 financial crisis, there might have been a complete meltdown of the banking sector.[60]

The sharp increase in the amount of state aid assigned in the EU in response to the crisis has often been cited as proof of the weaknesses of the EU state aid control systems.[61] However, in a paper written for the American Antitrust Institute assessing the role of the EU state aid control systems, De Vito reports opposite results.[62] He notes that the European Commission facilitated the transparent and orderly release of aid with minimum anticompetitive results, and also indicates that the Commission placed meaningful limitations on bail-out measures with the aim of preventing financial institutions that received less aid from becoming unjustifiably disadvantaged. In his recommendation, policy-makers should draw upon the most successful aspects of the EU model and incorporate the competitive policies in future crisis situations.[63]

The bottom line is that the primary role of competition policy is to ensure there are no distortions, thus allowing proper market functioning. In the EU, markets are at risk of being distorted by certain companies' anti-competitive behaviour as a result of state interventions through

59 Bloom and Van Reenen, 'Measuring and explaining management practices across firms and countries'.

60 Nordic Council, 'Benefits of green public procurement'.

61 Lundberg et al., 2009. 'Assessment of green public procurement as a policy tool: cost-efficiency and competition considerations', *Umeå Economic Studies* No. 775.

62 J. M. Devito, 'The role of competition policy and competition enforcers in the EU response to the financial crisis', *AAI Working Paper No.11-01* (2001).

63 Ibid.

control of aid in terms of taxes and subsidies.[64] If the plans for securing green energy are to be realised, it might be reasonable to argue that state aid control should be integrated as a major instrument for enhancing fair market competition. In order to control the economy, there is a need for state aid control and for maintaining internal market integrity. In addition, this will ensure better utilisation of resources and public money by making sure that subsidies are adequately targeted to their market failures.[65]

III Emissions Trading System and the Significance of Effective Competition

The EU Emissions Trading System (EU ETS) is a cornerstone of the European Union's policy to combat climate change. It works on the 'cap and trade' principle, where a 'cap', or limit, is set on the total amount of certain greenhouse gas emissions and is reduced over time so that total emissions fall.[66] The price of carbon emissions in the ETS is determined by the number of permits issued.[67] These permits can be either grandfathered or auctioned, and can also determine the price of carbon emissions. This implies that for a given total quota, the price of the carbon emissions is dictated by the market in question.[68]

Therefore, the potential matters of concern that should be deliberated in this context include the auction, which is the primary allocation of the permits, and the subsequently issued trade permits.[69] The emission permits are supposed to be auctioned in such a way that incumbents are not accorded preferential status over newcomers.[70] A competitive and well-functioning market for greenhouse gas emission permits is important for the determination of suitable emissions pricing.

When prices are low – for example, as a result of collusive practices – in comparison with the equilibrium prices at competitive levels, this is

[64] Bloom and Van Reenen, 'Measuring and explaining management practices across firms and countries'.

[65] Nordic Council, 'Benefits of green public procurement'.

[66] EU ETS Handbook, http://ec.europa.eu/clima/publications/docs/ets_handbook_en.pdf (accessed 17 November 2015).

[67] Lundberg et al., 'Assessment of green public procurement as a policy tool', 14–16.

[68] Nordic Council, 'Benefits of green public procurement'.

[69] J. Moraga-Gonzalez and N. Padrón-Fumaro, 'Environmental policy in a green market', *Environmental and Resource Economics*, 22 (2002), 419–47.

[70] Jamasb and Pollitt, 'Electricity market reform in the European Union'.

likely to result in reduced incentives for firms to innovate.[71] Efficient operations in the area of trading permits and the auction market would be critical in this situation. Appropriate prices are necessary in the auction market given their role in generating government revenue, and should reflect the actual economic value of the emissions permits.[72]

If this objective is not achieved, it becomes less of an efficiency issue and more of a distribution concern. This is because prices are likely to adjust until they reach the actual market economic value through secondary market trading.[73] This is the case with the ETS market, given that the market concerned is a competitive one. When the objective of attaining a set market price at auction is achieved, its growth depends on how the auction is designed. Diverse designs can facilitate or deter collusion.[74] Where collusion is facilitated, the approach can undermine the policy aims of having in place correct pricing signals and competitive prices, as well as transfer of the commercial charges on emissions from the authority to the industry.

Experience with trading and auctioning emission licences has been gained both in the EU and in the United States. Where multiple licences and permits are issued for homogeneous goods, such as emission permits, various auction formats have been used. One example is the sealed bid auction, where potential bidders are required to submit their demand schedules. The demand schedules are added together by the auctioneer to produce an aggregate demand curve. The point at which the aggregate demand curve crosses the vertical supply curve is used to define the clearing price. Hence, the price paid can either be the clearing price or the bid price. The bid price is primarily referred to as the pay-your-bid price or discriminatory price, whereas the clearing price is known as union pricing.[75]

The auction formats have been utilised, for example, for selling treasury bills in various countries. Other auction formats include the simultaneous ascending auction that is used by the US Communications Commission and the clock auction used in the United Kingdom to sell emission licences. Paul Milgrom discusses these formats in his book *Putting Auction Theory to Work*,[76] in which he observes that buyers or

[71] Dröge and Schröder, 'How to turn an industry green'. [72] Ibid., p. 178.
[73] C. Wram, 'Electricity markets: should the rest of the world adopt the United Kingdom's reforms?' *Regulation*, 22 (2000), 48.
[74] Lundberg et al., 'Assessment of green public procurement as a policy tool', p. 7.
[75] Wram, 'Electricity markets', p. 9. [76] Milgrom, *Putting Auction Theory to Work*.

bidders tend to have a general incentive to reduce demand in an effort to keep prices as low as possible. Prices that are lower than their competition limits are thought to be unfavourable, and they apply when there is a fixed supply of goods.[77] It is true that the supply of goods for sale is mostly fixed when the mandated regulators sell emissions in an auction. Hence, it is important to design the emission permit auctions in such a way as to ensure that there is a reliable mechanism for setting competitive prices and thereby preventing collusive practices.[78] In most cases, the competition regulators have the important task of detecting collusive practices before and during the auction and providing a reliable ETS that is relatively competitive.

IV Green Public Procurement in a Competition Policy Perspective

Green public procurement (GPP) can actually influence competition and can speed up the development of markets for green goods.[79] If new requirements are introduced in public procurement policy, it is highly likely that this will result in barriers for companies wishing to enter the markets and in increased prices among the procurement firms.[80] Therefore, GPP can act as a barrier to potential new market entrants, and it is always important to take this aspect into consideration when evaluating GPP's impact on social welfare.[81]

If companies possess different production technologies, that is, are heterogeneous, it is also possible for competition levels to vary among them. This is because the environmental criteria used will attract certain companies while restricting the entry of others. This concept has been exemplified by Lundberg et al.,[82] who first assume that two types of companies exist: those companies that have not invested in environment-friendly technology, and those that have done so. Brännlund et al.[83] have postulated that when investing in an environment-friendly processes, there is a likelihood of a firm fetching both high bid prices and cost prices.

[77] Ibid., p. 294. [78] Dröge and Schröder, 'How to turn an industry green' p. 194.
[79] Wram, 'Electricity markets', p. 50.
[80] Lundberg et al., 'Assessment of green public procurement as a policy tool', p. 9.
[81] Jamasb and Pollitt, 'Electricity market reform in the European Union', p. 13.
[82] Lundberg et al., 'Assessment of green public procurement as a policy tool', p. 7.
[83] Nordic Council, 'Benefits of green public procurement'.

If one assumes that neither environmental criteria nor the above-mentioned factors are taken into consideration during the procurement process, it can be argued that companies that have made no investment in environment-friendly technology are likely to offer low bids compared to firms that have made such investment.[84] As a result, firms that have not invested in environment-friendly technology are more likely to win procurement contracts. Classically, companies with environment-friendly bids will be hampered in winning procurement contracts.[85]

In contrast, if the procurement entity incorporates environmental criteria into the procurement process, how will the picture change? It is highly likely that companies that have invested in environment-friendly technology will submit bids because they are confident that their investment will be highly valued. They will also be aware that they have a higher probability of securing the procurement contract.[86] On their part, companies that have not invested in environment-friendly technologies are likely either to stay out of the procurement process or to take the necessary steps to invest in such technology. The manner in which a company acts, that is, whether it invests or not, depends on the potential payoff expected from such an investment.[87] It goes without saying that if the expected profits far outweigh the cost of investment, it is possible for a firm to invest in the technology and then make its bid.

The above scenario implies that GPP can substantially affect the level of competition in three main ways. First, GPP can have a positive impact on competitiveness if the number of companies that have invested in environment-friendly technology so as to enter the auction and the market is higher than the number of companies that are moving out of the business because they are not willing to invest in environment-friendly technologies.[88] Second, GPP can have an adverse impact on market competition if the number of companies that are abandoning the market and that have not made any technological investments is greater than the number of companies that have made technological investments and are bidding for contracts.[89] Third, there will be no

[84] Wram, 'Electricity markets', p. 51.
[85] Dröge and Schröder, 'How to turn an industry green'.
[86] T. Valletti and A. Estache, 'The theory of access pricing: an overview for infrastructure regulators', paper prepared for the World Bank Institute (1998).
[87] Green, 'Electricity and markets'.
[88] Nordic Council, 'Benefits of green public procurement'.
[89] Valletti and Estache, 'The theory of access pricing'.

impact on market competition if the number of companies that exit the market is equal to the number of companies that enter the market.[90]

It is important to realise that the effect of product pricing and the result of competition are different factors. The overall result of using GPP is likely to be high prices even if there are more bids, because meeting the environmental criteria might require investments that result in additional costs for the companies submitting the bids.[91]

As well as ensuring improved environmental benefits, GPP is important as a driver of the development of green services and products and market development. If an environmental criterion is introduced in public procurement criteria, the procurement firms can motivate and influence companies to embrace the development of greener products.[92] The greener products that have been developed may go on to act as new market benchmarks that will attract more demand for green products. In this way, GPP can have a more dynamic impact in the market than policy-making, which often simply plays the role of removing products that do not fully meet the required standards for green products.[93]

V State-Owned Monopolies

For decades, energy markets were under government control and exposed to little competition, due to their specific economic and technical features.[94] The consensus is that retail and power transmission mostly constitute natural and historical monopolies. The general fragmentation of national markets led to restrictions on free trade in the energy business market being put in place. Thus, a completely state-owned energy market segment is typically vertically integrated, implying that a single actor is responsible for the entire process, including generation, delivery and retail.

State-owned monopolies restrict market competition through a number of barriers, regulations and non-transparent pricing systems, with the aim of maintaining market dominance. The key role of governments in this case is to ensure that consumers are protected, and this is

90 Jamasb and Pollitt, 'Electricity market reform in the European Union', p. 23.
91 Lundberg et al., 'Assessment of green public procurement as a policy tool', p. 10.
92 Ibid., p. 11.
93 S. Meritet, 'French perspectives in the emerging European Union energy policy', *Energy Policy* 25 (2007), 4767–71.
94 Valletti and Estache, 'The theory of access pricing'.

achieved by introducing restrictions and keeping the prices at bay. It is clear that a liberalised market allows greater competition, and in this way consumers have the freedom to choose between products, encouraging investments and transparent pricing. All these factors work to promote consumer welfare. Some good examples of liberalised energy markets in the EU include France, Germany and the United Kingdom. Understanding how they worked both before and after liberalisation is important in understanding the role of state-owned monopolies.

C Antitrust and Rules on Abuse of Dominant Position

The rules of the Treaty establishing the European Economic Community (EEC Treaty) which govern antitrust violations are found in Article 86.[95] Major significance is given to the concept of dominant position under Article 86, where the idea refers to the strength of the economic position enjoyed by an enterprise that empowers it to prevent effective competition.[96] Therefore, a firm in a dominant position can dictate business terms independent of its customers, its competitors and, ultimately, its consumers. Article 102 of the EEC Treaty is focused on preventing enterprises that hold a dominant position in a market from abusing their dominant positions to the detriment of consumers.[97]

Article 102 provides that any abuse of power by a dominant organisation within the common market should be stopped. Such activity is incompatible with the EU market because it can affect the normal business relations between Member States.[98] This may imply consideration of a number of practices, such as unfair imposition of purchase or selling prices, limiting production and applying contradictory conditions to comparable transactions to block market access.[99] It should be noted that a dominant position is an economic power possessed by a firm as a

[95] J. Barquin, L. Bergman, C. Crampes, R. Green, C. van Hirchhausen, F. Lévêque and S. Stoft, 'The acquisition of Endesa by Gas Natural: why the antitrust authorities are right to be cautious', *Electricity Journal*, 14(3) (2006), 62–8.

[96] M. Motta and P. Streel, 'Exploitative and exclusionary excessive prices in EU law', paper presented at 8th Annual European Union Competition Workshop, Florence, Italy, 6 June 2003.

[97] D. Newbery, 'The relationship between regulation and competition policy for network utilities', paper presented at Conference in Tribute to Jean-Jacques Laffont, Toulouse, France, 30 June 2005.

[98] Ibid.

[99] B. Allaz and J. Vila, 'Cournot competition, forward markets and efficiency', *Journal of Economic Theory*, 59 (2013), 1–16.

result of lack of competition, which the player can use to its own advantage to obstruct effective competition.[100]

The obvious example of exploitative abuse of a dominant position is the monopoly of pricing. The antitrust law in Article 86 (a) EEC Treaty seeks to prohibit a dominant player from unfairly setting selling prices.[101] Another prohibited practice under the antitrust law is production limitation by a dominant organisation, used to create an artificial shortage with the aim of increasing prices.[102] Article 86 (b) of the antitrust regulations seeks to deal with such exploitative abuses and stop a dominant enterprise from limiting its products, development or markets.[103] Therefore, any unfair practice that results in direct gain for a dominant firm at the expense of consumers is unlawful and unfair according to Article 86. Other prohibitions include tie-ins, as covered in Article 86 (d), because these impose upon buyers an obligation to purchase an item they do not want.[104] Similarly, discrimination is discouraged in Article 86 (c) because it may result in unlawful price maximisations that benefit a dominant firm and result in anticompetitive effects between the dominant firm and customers.[105]

I Unbundling and New Competitors Entering the Energy Market

In the first cases of energy market concentration, markets were national assets with few or no new entrants. Most energy markets were owned by governments, which made it difficult for competitors to join them, because these markets lacked liquidity. Vertical integration and supply of the energy services resulted in reduced trade in the markets.[106] The situation changed in the late 1990s, when the EU decided to change the concept of gas and electricity provision from a

[100] G. Immordino and M. Polo, 'Antitrust, legal standards and investment', *International Review of Law and Economics*, 40 (2014), 36–50.

[101] R. J. Fontenot and M. R. Hyman, 'The antitrust implications of relationship marketing', *Journal of Business Research*, 57(11) (2004), 1211–21.

[102] Allaz and Vila, 'Cournot competition, forward markets and efficiency'.

[103] Immordino and Polo, 'Antitrust, legal standards and investment'.

[104] Newbery, 'The relationship between regulation and competition policy for network utilities'.

[105] Immordino and Polo, 'Antitrust, legal standards and investment'.

[106] R. Leal-Arcas and A. Filis, 'The fragmented governance of the global energy economy: a legal-institutional analysis', *Journal of World Energy Law and Business*, 23(2) (2013), 1–58.

monopolistic to a competitive market one. This partially opened the energy markets to competition and large users were allowed to choose their suppliers.[107]

Article 11 (a) directs firms to comply with EU unbundling regulations as stipulated in Article 9 on fair competition and ownership, which eases the process of new market entry to the energy sector.[108] The primary intention of unbundling regulations is to end monopoly and increase competition by separating the grid from the power supply and to make it easier for new enterprises to join the market.[109] This is achieved by separating the production and supply of vertically linked management practices in monopolistic energy sectors, while encouraging horizontal alignments. Ownership unbundling has been proposed as the most appropriate way to achieve efficient unbundling. Hence, there are two main ways of implementing this measure to encourage new market entrants.[110] The first is by selling the supply business or the network business to a third party. The second is by splitting the shares in an integrated company into distinct shares for a network firm and for the production and supply business.[111] Therefore, successful unbundling relies on adopting non-discriminatory treatment to reduce pressure in the liberalised market while reducing transaction and capital costs for new entrants.[112]

II Disciplining Monopolies

Monopoly enterprises can contribute to consumer exploitation, for example by setting prices above the true marginal cost of supply. Therefore, it is important to discipline monopolies by setting rules that seek to control the operations of organisations that may have a market

[107] T. Cottier, O. Nartova and S. Matteotti-Berkutova, 'Third country relations in EU unbundling of natural gas markets: the "Gazprom Clause" of Directive 2009/73 EC and WTO law', *NCCR Trade Regulation Working Paper* No. 2010/06 (2010), pp. 1–16.

[108] Y. Selivanova. 'EU and Russia energy trade: impact of WTO, transit and integrated economic areas', First EU-Russian Energy Law Conference, University of Groningen (2013), pp. 41–52, www.rug.nl/research/groningen-centre-for-law-and-governance/onderzoekscentra/gcel/pictures/y.selivanova.pptx.

[109] Leal-Arcas and Filis, 'The fragmented governance of the global energy economy'.

[110] M. Cossy, 'Energy transport and transit in the WTO', *Global Challenges at the Intersection of Trade, Energy and the Environment Conference Draft* (30 September 2009), pp. 25–37.

[111] Motta and Streel, 'Exploitative and exclusionary excessive prices in EU law'.

[112] Cottier et al., 'Third country relations in EU unbundling of natural gas markets'.

monopoly in their own industries.[113] Disciplines on monopolies are designed to deal with the challenge of market failure in cases where markets are unable to achieve optimal resource allocation. As such, governments discipline monopolies by preventing operations that contravene the provisions of Article 86 and Article 102 on anti-competitive policies that are against the public interest.[114]

The Competition Commission is a body set up by the Competition Act of 1998 that came into effect on 1 April 1999. Its mandate is to report on referrals made by the Director General of Fair Trade, and to hear appeals against monopoly abuses and prohibitions under the Competition Act 1998.[115] Monopolies are also regulated under the European Competition Law under Articles 81 and 82 of the Treaty of Rome. The two provisions prevent certain market practices that are thought to be anti-competitive and can act against the consumers' interests.[116] Article 81 prohibits acts that can damage market competition, while Article 82 prohibits the abuse of dominant positions, thereby helping to discipline monopolies. The rules aim to achieve this by prohibiting monopolies in two important ways:[117] first, by enacting referrals to the Competition Commission, and second, by stipulating measures to be taken against anti-competitive agreements, such as fixing of selling and purchasing prices; limiting investment, technical development, and production; sharing supply and market sources; imposing selling prices or unfair purchasing; and applying different trading conditions.[118]

III *Market Access*

There are a number of antitrust regulations relevant to blockage and access to markets. Three possibilities are covered in the market access and antitrust regulations: cartels with boycotts; vertical agreements; and monopol-

113 E. Richard, R. Just and D. Pope, 'Second-best optimality of advertising when monopoly is sanctioned', *Journal of Economics and Business*, 64(6) (2012), 393–8.

114 R. S. Pindyck, 'On monopoly power in extractive resource markets', *Journal of Environmental Economics and Management*, 14(2) (2007), 128–42.

115 C.-H. Miao, 'Planned obsolescence and monopoly undersupply', *Information Economics and Policy*, 23(1) (2011), 51–8.

116 Richard et al., 'Second-best optimality of advertising when monopoly is sanctioned'.

117 E. Motchenkova, 'Determination of optimal penalties for antitrust violations in a dynamic setting', *European Journal of Operational Research*, 189(1) (2008), 269–91.

118 Miao, 'Planned obsolescence and monopoly undersupply'.

istic exclusions and discrimination.[119] A cartel with a boycott engages in illegal conduct contrary to antitrust and unfair competition laws. When the problem of extraterritorial access to a market is mooted by an agreement, there is a risk that disadvantaged nations and firms will face unfair market competition and restricted market access.[120] In the second category, vertical restraints may be detrimental to competitors as they prohibit anticompetitive foreclosures or unreasonable foreclosures, although the laws are usually applied differently.[121] The United States has abandoned this restrictive form of market access and firms have the freedom to impose vertical restraints unilaterally to encourage market access.

Legal suits under competition law have challenged vertical restraints aimed at limiting output and harming consumers. Such restraints can block competitors from joining the market.[122] In many EU Member States, the practices of firms with a market share of more than 40 or 50 per cent are often regulated to prevent them excluding smaller firms from accessing the market and supplying goods and providing services.[123] Thus, the problems faced by enterprises entering a new market in which there are vertical restraints should be addressed by effective horizontal rules. In the third category of monopolistic exclusions, most dominant firms seek to exclude their rivals by monopolising the market. This conduct is illegal under EU antitrust laws.[124] A significant gap exists between the United States' strong principle on whether firms can refuse a deal under narrowly defined conditions and the EU's broader rules on dominant firms' duties not to discriminate, unfairly exclude or refuse to deal with a new market entrant.[125]

IV EON Electricity Cases about Divestiture of Significant Network Access

The E.ON AG commitment decision was adopted in November 2008 by the Commission with regard to suspected infringement of EU

[119] Leal-Arcas and Filis, 'The fragmented governance of the global energy economy'.

[120] Ibid. [121] Barquin et al., 'The Acquisition of Endesa by Gas Natural'.

[122] Ma, T.-C., 'Competition authority independence, antitrust effectiveness, and institutions', *International Review of Law and Economics*, 30(3) (2010), 226–35.

[123] Cottier et al., 'Third country relations in EU unbundling of natural gas markets'.

[124] H. Grabowski and M. Kyle, 'Competition authority independence, antitrust effectiveness, and institutions', *International Review of Law and Economics*, 30(3) (2010), 226–35.

[125] A. Kleit, 'Defining electricity markets: an arbitrage cost approach', *Resource and Energy Economics*, 23(3) (2001), 259–70.

competition laws involving the German electricity markets. The concerns in the German case were removed by E.ON's proposal to divest some generation capability. A second case was about the electricity market in Germany and concerns that E.ON may have undertaken an approach as a transmission system operator (TSO) that favoured its own affiliated suppliers, and as a result raised the costs borne by the final consumers. The problems in this case were resolved by E.ON's proposal to divest its German electricity transmission systems.[126]

The E.ON cases have resulted in liberalisation of the EU electricity market sector, which has led to competition that allows consumers to reap the benefits of efficiency and choice. Capacity deterrence and withdrawal of investments can deny customers benefits to such a level, and for so many decades, that one may question the benefits of liberalised markets in the EU energy sector.[127] It is therefore part of the Commission's mandate to track down such cases and prevent them. This also applies to firms that distort competition due to vertical integration between suppliers and networks. In 2008, E.ON had already commenced its removal of power plants. Buyers of the plant's assets had to be approved by the Commission in accordance with several criteria, specifically being independent from E.ON and having suitable financial strength and competence, as well as the absence of any prima facie concerns about competition that may arise.[128] Preparations for sale of the transmission services and other practices in E.ON concerns are subject to the Commission taking any action necessary to stop unfair energy practices while ensuring that consumers reap the benefits of liberalised markets.[129]

V Can US Antitrust Laws Be Extended to OPEC Countries?

The United States has shown that non-discriminatory and vigorous enforcement of antitrust laws can contribute to the promotion of international marketplaces characterised by an open and competitive business environment.[130] Nonetheless, antitrust laws are, at best, a supporting

126 C.-G. Giorgio, 'How competitive are EU electricity markets? An assessment of ETS Phase II', *Energy Policy*, 73 (2014), 278–97.

127 A. B. Haney and M. G. Pollitt, 'International benchmarking of electricity transmission by regulators: a contrast between theory and practice?' *Energy Policy*, 62 (2013), 267–81.

128 Cossy, 'Energy transport and transit in the WTO'.

129 Ma, 'Competition authority independence, antitrust effectiveness, and institutions'.

130 Motta and Streel, 'Exploitative and exclusionary excessive prices in EU law'.

measure in the construction of a liberal multilateral trading organisation, and are unable to promote any single nation's exports. Therefore, the US antitrust laws can play a significant role in promoting the competitive global marketplace as well as developing a wave of competition law around the world; as such, they should be extended to OPEC countries.[131]

Antitrust is perceived as a relatively untested weapon that can be powerful against unfair competition, given the resources of the Antitrust Division and the Federal Trade Commission, who may investigate anticompetitive conduct that affects the United States through the use of Civil Investigative Demands, grand juries and the discovery process set out in the Federal Rules of Civil Procedure. This could force the disclosure of information, expose foreign conduct to judgement by judges and juries and help obtain injunctive, broad criminal and treble damage relief. Hence, the US model is an appropriate and successful approach to the application of antitrust laws that can be extended to OPEC countries.[132] Effective competition can be tested by looking at how a country takes action – criminal or civil – against its own national firms, and not just against foreign firms. If the US antitrust regulations are extended to OPEC, the United States should encourage other nations to realise the goal of making domestic competition law fair for businesses.

D Conclusion: Future Role of the State

Mitigating market failures has become a central activity for governments trying to ensure a competitive and liberal market. In planning for future green energy development, the EU has a lot to learn from the developments in its Member States' competitive and environmental policies. These include government incentives formulated to promote the deployment and development of renewable energy resources. As such, states also have a central role to play in ensuring that there is an effective competition policy, as well as in eliminating government monopolies and thereby promoting the development of green electricity.

If EU Member States are to implement renewable energy systems, governments must reform their energy regimes. Members should observe

[131] S. W. Waller, 'Can U.S. antitrust laws open international markets?' *Northwestern Journal of International Law & Business*, 20(2) (2000), 23–32.

[132] Newbery, 'The relationship between regulation and competition policy for network utilities'.

the Kyoto Treaty and other international climate change measures to reduce the use of fossil fuels in electricity generation, thereby reducing the emission of greenhouse gases. If the EU is to turn towards greener electricity, government incentives and environmental laws are needed to help it effect this transition. States should evaluate a number of areas to ensure effective development of green electricity. These include handling market failures by enacting appropriate horizontal agreements and practices in their environmental and competition policies. Governments should also evaluate the environmental impacts of their internal restrictive practices and how they affect competition in the energy sector.

It is possible that green markets could be disrupted by antitrust regulations, and governments should assess the extent to which business practices in the green markets can violate competition laws meant to create fair market practices. Importantly, the issue of state aid control should be evaluated to ensure liberal markets in which all members can compete equally. This will reduce state monopolies and allow for fair competition, in addition to encouraging emission trading systems.

Bibliography

Aghion, P., Bloom, N., Blundell, R., Griffith, R. and Howitt, P. (2005). Competition and innovation: an inverted-u relationship. *The Quarterly Journal of Economics*, **56**(8), 701–28.

Allaz, B. and Vila, J. (2013). Cournot competition, forward markets and efficiency. *Journal of Economic Theory*, **59**, 1–16.

Barquin, J., Bergman, L., Crampes, C., Green, R., van Hirchhausen, C., Lévêque, F. and Stoft, S. (2006). The acquisition of Endesa by Gas Natural: why the antitrust authorities are right to be cautious. *The Electricity Journal*, **14**(3), 62–8.

Betz, R., Rogge, K. and Schleich, J. (2006). EU emissions trading: an early analysis of national allocation plans for 2008–2012. *Climate Policy*, 6(4), 361–94.

Bloom, N. and Van Reenen, J. (2007). Measuring and explaining management practices across firms and countries. *The Quarterly Journal of Economics*, **72**(4), 1351–408.

Blundell, R., Griffith, R. and van Reenen, J. (1999). Market share, market value and innovation in a panel of British manufacturing firms. *Review of Economic Studies*, **66**, 529–54.

Borenstein, S. (2000). The trouble with electricity markets: understanding California's restructuring disaster. *Journal of Economic Perspectives*, **16**(1), 191–211.

Borenstein, S., and Bushnell, J. (1999). An empirical analysis of the potential for market power in California's electricity industry. *Journal of Industrial Economics*, **47**(3), 285–323.

Cossy, M. (2009). Energy transport and transit in the WTO. *Global Challenges at the Intersection of Trade, Energy and the Environment Conference Draft, Geneva, 30 September 2009*, pp. 25–37.

Cottier, T., Matteotti, B. and Nartova, O. (2010). Third country relations in EU unbundling of natural gas markets: the 'Gazprom Clause' of Directive 2009/73 EC and WTO Law 1–16. *Working Paper No 2010/06.*

Domah, P. D. and Pollitt, M. G. (2001). The restructuring and privatisation of the regional electricity companies in England and Wales: a social cost benefit analysis. *Fiscal Studies*, **22**(1), 107–46.

Dröge, S. and Schröder, P. J. H. (2005). How to turn an industry green: taxes vs subsidies. *Journal of Regulatory Economics*, **27**(2), 177–202.

Fontenot, R. J., and Hyman, M. R. (2004). The antitrust implications of relationship marketing. *Journal of Business Research*, **57**(11), 1211–221.

Geroski, P. A. (1990). Innovation, technological opportunity, and market structure. *Oxford Economic Papers*, **42**(3), 586–602.

Giorgio, C.-G. (2014). How competitive are EU electricity markets? An assessment of ETS Phase II. *Energy Policy*, **73**, 278–97.

Grabowski, H. and Kyle, M. (2010). Competition authority independence, antitrust effectiveness, and institutions. *International Review of Law and Economics*, **30**(3), 226–35.

Green, A, and Mayes, D. (1991). Technical inefficiency in manufacturing industries. *The Economic Journal*, **101**(406), 523–38.

Green, R. (2005) Oxford review of economic policy. *Electricity and Markets*, **21**(1), 67–87.

Grubb, M. and Neuhoff, K. (2006). Allocation and competitiveness in the EU Emission Trading Scheme: policy overview. *Climate Policy*, **6**, 7–30.

Haney, A. B. and Pollitt, M. G. (2013). International benchmarking of electricity transmission by regulators: a contrast between theory and practice? *Energy Policy*, **62**, 267–81.

Immordino, G. and Polo, M. (2014). Antitrust, legal standards and investment. *International Review of Law and Economics*, **40**, 36–50.

Jamasb, P., and Pollit, M. (2005). Electricity market reform in the European Union: review of progress toward liberalization and integration. *The Energy Journal*, **26**, 11–41.

Joskow, P. L. and Tirole, J. (2000). Transmission rights and market power on electric power networks. *Rand Journal of Economics*, **31**(3), 450–87.

(2005). Merchant transmission investment. *Journal of Industrial Economics*, **53**(2), 233–64.

(2006). Retail electricity competition. *Rand Journal of Economics*, **37**(4), 799–815.

Joskow, P. L. and Tirole J. (2007). Reliability and competitive electricity markets. *Rand Journal of Economics*, **38**(1), 60–84.

Kilponen, J., and Santavirta, T. (2007). When do R&D subsidies boost innovation? Revisiting the inverted U-shape. *Bank of Finland Research Discussion Papers*, **10**, 45–65.

Kleit, A. (2001). Defining electricity markets: an arbitrage cost approach. *Resource and Energy Economics*, **23**(3), 259–70.

Leal-Arcas, R. and Filis, A. (2013). The fragmented governance of the global energy economy: a legal-institutional analysis. *Journal of World Energy Law and Business*, **23**(2), 1–58.

Littlechild, S. C. (2003). Wholesale spot market passthrough. *Journal of Regulatory Economics*, **23**(1), 61–91.

Lizardo, J. and de Araujo, R. H. (2009). The case of Brazil: reform by trial and error. In F. Sioshansi and W. Pfaffenberger (eds.), *Electricity Market Reform: An International Perspective*. London: Elsevier, pp. 565–95.

Lundberg, S., Marklund, P.-O. and Brännlund, R. (2009). *Assessment of Green Public Procurement as a Policy Tool: Cost-efficiency and Competition Considerations*. Umeå Economic Studies No. 775.

Ma, T.-C. (2010). Competition authority independence, antitrust effectiveness, and institutions. *International Review of Law and Economics*, **30**(3), 226–35.

Mansur, Erin, T. (2007). Upstream competition and vertical integration in electricity markets. *Journal of Law and Economics*, **50**(1), 125–56.

Miao, C.-H. (2011). Planned obsolescence and monopoly undersupply. *Information Economics and Policy*, **23**(1), 51–8.

Milgrom, P. (2004). *Putting Auction Theory to Work*. Cambridge: Cambridge University Press.

Moraga-Gonzalez, J. and Padrón-Fumaro, N. (2002). Environmental policy in a green market. *Environmental and Resource Economics*, **22**, 419–47.

Motchenkova, E. (2008). Determination of optimal penalties for antitrust violations in a dynamic setting. *European Journal of Operational Research*, **189**(1), 269–91.

Motta, M. and Streel, P. (2003). 'Exploitative and exclusionary excessive prices in EU Law', paper presented at 8th Annual European Union Competition Workshop, Florence, Italy.

Newbery, D. (2005). 'The relationship between regulation and competition policy for network utilities', paper presented at Conference in Tribute to Jean-Jacques Laffont. Toulouse, France.

Nordic Council. (2009). 'Benefits of green public procurement'. *TemaNord*, **953**: 51.

OECD. (2006). *Environmental Regulation and Competition*, DAF/COMP(2006) 30: 19.

(2009). *A Proposal for Developing a Green Growth Strategy*, (2009)147/REV1.

(2010). *Green Growth Strategy Interim Report: Implementing Our Commitment for a Sustainable Future*, C/MIN(2010)5: 38.

Pindyck, R. S. (2007). On monopoly power in extractive resource markets. *Journal of Environmental Economics and Management*, **14**(2), 128–42.

Richard, E., Just, R. and Pope, D. (2012). Second-best optimality of advertising when monopoly is sanctioned. *Journal of Economics and Business*, **64**(6), 393–8.

Scarpetta, S., Hemmings, P., Tressel, T. and Woo, J. (2002). The role of policy and institutions for productivity and firm dynamics: evidence from micro and industry. *OECD Economics Department Working Paper* No. 329,.

Selivanova, Y. (2013). EU and Russia energy trade: impact of WTO, transit and integrated economic areas. The first EU–Russian Energy Law Conference, University of Groningen, pp. 41–52, www.rug.nl/research/groningen-centre-for-law-and-governance/onderzoekscentra/gcel/pictures/y.selivanova.pptx.

Stigler, G. J. (1987). Competition. In Eatwell, J., Milgate J. and Newman, P. (eds.), *The New Palgrave: A Dictionary of Economics*, vol. l. London: Macmillan, pp 1129–36.

Waller, S. W. (2000). Can US antitrust laws open international markets? *Northwestern Journal of International Law & Business*, **20**(2), 23–32.

20

Transfer of Technology and a Global Clean Energy Grid

FREDERICK M. ABBOTT

The idea of a global power grid linked to renewable resources is credited to Buckminster Fuller and dates back to the 1970s.[1] This chapter revisits that utopian concept. In the 1970s, the primary concerns regarding energy generation were the potential depletion of global resources and insecurity of supply. Today, climate change is a major driver of interest in green energy generation and distribution, while insecurity of supply appears to remain a significant concern.

Cross-border electricity transmission is already a substantial feature of various markets, including those in the European Union,[2] and also takes place between Canada and the United States.[3] And, certainly,

[1] Global Energy Network Institute, www.geni.org/: 'In 1969, Dr. R Buckminster Fuller developed a World Game™ simulation to explore global resources and trends.
It posed the critical question, "how do we make the world work for 100% of humanity in the shortest possible time through spontaneous cooperation without ecological damage or disadvantage to anyone?" In other words, how do we provide a decent living standard for everyone in an environmentally sustainable way? How does this change happen spontaneously around the globe?
The premier strategy was the interconnection of electric power grids around the world with an emphasis on tapping abundant renewable energy resources. Electricity provides the foundation of our modern society, and power grids act as freeways that deliver the electricity to power our homes, businesses and industry.
During the 20th Century, transmission lines delivered electricity from large, centralized plants that burned fossil fuels and nuclear energy. Today, developers of wind, solar and geothermal resources in remote locations are plugging into these expanding grid systems – decentralizing power production while increasing grid integration.'

[2] See, e.g., http://ec.europa.eu/energy/en/topics/markets-and-consumerss/wholesale-market/electricity-network-codes; www.entsoe.eu/Pages/default.aspx.

[3] See, e.g., www.electricity.ca/industry-issues/economic/canada-u.s.-affairs.php:
'Smart Grid: The movement towards the "Smart Grid" is a leading example of the technological transformation underway in the North American electricity sector. Utilities across Canada and the U.S. are taking advantage of the enormous potential of Smart Grid technologies to improve their overall efficiency, better manage system performance and increase consumer control over electricity use. With these benefits come challenges,

governments are interested in regulating this trade to promote the transition to clean energy technologies.

The prospects for creation of a global clean electricity grid (GCEG) must address two major elements: first, clean electricity generation, and second, the establishment of a global grid through and across which electricity may be shared. A number of significant studies have looked at the potential for technology transfer to facilitate clean energy generation, including identification of potential patent obstacles.[4] Some of those studies have taken into account the technologies that may be important to the creation of 'smart grids' for energy distribution, including identification of intellectual property (IP) issues involving computer software.[5] This

though. With Smart Grid components set to become a standard feature of new electrical infrastructure – including cross-border transmission infrastructure – Canada-U.S. coordination on the development of interoperability standards governing Smart Grid technologies will be critical to ensuring such technologies can be deployed securely, reliably and expediently.'

Electricity trade between the United States and Mexico is far more modest. See US Energy Information Administration, *Mexico Week: U.S.–Mexico Electricity Trade is Small, with Tight Regional Focus* (17 May 2013), www.eia.gov/todayinenergy/detail.cfm?id=11311.

4 See generally United Nations Environment Programme (UNEP), the European Patent Office (EPO) and the International Centre for Trade and Sustainable Development (ICTSD), *Patents and Clean Energy: Bridging the Gap between Evidence and Policy, Final Report* (2010), www.unep.ch/etb/events/UNEP%20EPO%20ICTSD%20Event%2030%20Sept%202010%20Brussels/Study%20Patents%20and%20clean%20energy_15.9.10.pdf. The results of this trilateral study are not particularly surprising, as its authors observe. In general, the study found that patents on 'clean energy technologies' or 'CETs' are predominantly owned by US, Japanese, German, French and South Korean companies, although in a few areas Chinese and Indian companies are increasing their patenting activity. Although academic institutions actively pursue out-licensing of their technologies, this is not a significant activity for private sector enterprises, although they are open to the possibility. Some of the constraints on out-licensing to developing countries involve finding suitable partner/licensees within those countries. While the potential inhibiting characteristics of patents and other IPRs are recognized, other factors are important, including access to large-scale capital needed to build energy generation facilities.

5 See, e.g., United Nations Department of Economic and Social Affairs (UN DESA), *Climate Change: Technology Development and Technology Transfer* (2008), https://sustainabledevelopment.un.org/content/documents/1465back_paper.pdf; F. M. Abbott, 'Innovation and technology transfer to address climate change: lessons from the global debate on intellectual property and public health', *ICTSD Programme on IPRs and Sustainable Development, Issue Paper No. 24 (2009)*, pp. 4–8, http://ssrn.com/abstract=1433579; F. M. Abbott, 'Innovation as imperative to sustainable development: meeting the new challenges', PowerPoint presentation, Citenel V Congress on Technological Innovation in Electrical Energy, Belem, Brazil (2009), slide 16, http://frederickabbott.com/presentations.

author is not aware of any research studies specifically concerning technology transfer issues in connection with the creation of a GCEG. That is, no studies so far appear to have answered the questions of what aspects of technology transfer may be relevant to global cross-border trade in 'green' electricity, whether there are potential obstacles to such trade created by the global system designed to protect rights in IP and what might be practical mechanisms for facilitating technology transfer.

A Defining Transfer of Technology

'Transfer of technology' embodies a broad range of activities and actors. The term is sometimes used as shorthand for 'low-cost provision of technology from developed to developing countries', but transfer of technology more broadly defined is a ubiquitous feature of international economic activity.[6] Multinational enterprises routinely exchange technology with their affiliates across borders, including through transmissions of data, shipments of materials and equipment, movement of employees, licensing of patents, provision of trade secret information, and so forth. Virtual research and development (R&D) centres have become a fairly common feature of multinational business, with researchers in various countries collaborating on projects. Enterprises in developed and developing countries frequently engage in collaborative projects involving transfers of technology that may flow in both directions.

The Internet makes information regarding many technologies accessible to scientists and researchers, wherever they are. Universities in Organisation for Economic Co-operation and Development (OECD) countries routinely accept and provide educational opportunities for students from developing countries, some of whom remain within the OECD, but many of whom return to their home countries.

'Open source movement' is relevant to technology transfer, as making technology publicly available is one way of diffusing it, and this may be relevant to the effective creation of a global power grid.[7] Cross-border

[6] See F. M. Abbott, *Trends in Local Production of Medicines and Related Technology Transfer* (Geneva: World Health Organization, 2011) (hereinafter *Local Production*), www.who.int/phi/publications/Trends_in_Local_Production_of_Medicines.pdf.

[7] See, e.g., the Elon Musk model of open innovation:
'We believe that Tesla, other companies making electric cars, and the world would all benefit from a common, rapidly-evolving technology platform.

'hacking' of proprietary data is also a form of technology transfer, even if accomplished through illicit means.

The above demonstrates that the term 'technology transfer' can be used in different ways. It may be used as a general shorthand for exchanges of technical information, personnel and materials in the private and public sectors, and it may be used in a developmental context, that is, referring to the movement of technology (and its various forms) from more technologically advanced countries or areas to those that are less advanced.

B Ownership and Control of Technology

Although circumstances are changing, commercialisable new technologies are still predominantly owned by large multinational enterprises that are mainly based in the OECD countries.[8] These technologies are typically available on the global market through facilities owned or controlled by these enterprises, and may not be shared with 'locally owned' enterprises in developing countries. Thus, the provision of products and services and the generation of returns on investment may predominantly benefit these multinational enterprises. The questions of whether and how technologies may be transferred to promote development and diffusion of green energy resources may be framed as questions of who owns and controls new technologies, and how the public is ultimately provided with access to the benefits of such technologies. The developmental question may be reframed, at least in part, as one of control over diffusion.

A related question is whether the diffusion of new technologies to locally owned enterprises is more beneficial to the population of a developing or developed country than external ownership and control.[9] There are arguments that favour local ownership and control of technologies. These include the possibilities for enhanced cooperation among

Technology leadership is not defined by patents, which history has repeatedly shown to be small protection indeed against a determined competitor, but rather by the ability of a company to attract and motivate the world's most talented engineers. We believe that applying the open source philosophy to our patents will strengthen rather than diminish Tesla's position in this regard.' E. Musk, CEO, *All Our Patent Are Belong to You* (12 June 2014), www.teslamotors.com/blog/all-our-patent-are-belong-you.
See also Institute of Electrical and Electronics Engineers (IEEE) Task Force on Open Source Software for Power Systems, http://ewh.ieee.org/cmte/psace/CAMS_taskforce/.

[8] See UNEP et al., *Patents and Clean Energy*. [9] Cf. Abbott, *Local Production*.

business and research/academic communities, adaptations of technologies to the local environment, generation of employment opportunities and reinvestment of profits in the local community, among others. It is evident that government policy-makers in OECD and developing countries encourage the development of 'national champions' and local control, and they do this for a reason that probably consists of more than just 'national pride'.[10]

C Access to the Benefits of Technology

A fundamental issue with the creation of a GCEG is the provision of affordable access to the ultimate users of the electricity product or service. Putting aside the issue of who develops, owns and controls the GCEG resources, there must be mechanisms in place to provide assurance that the product or service is reasonably accessible to the public in countries with significantly divergent income levels. Most countries control the price of electricity charged to the consumer through some form of regulation, whether the electricity is generated and distributed by the public or the private sector. Some challenging pricing issues may be raised by a GCEG. If production resources are shared, how will prices be determined? Traditionally, the owners of 'monopolies' created by IP rights have enjoyed pricing power that is not always ameliorated by regulation.

D The Characteristics of the Green Energy Market

Regarding the characteristics of the green electricity market as it relates to transfer of technology issues, a few contextual observations may be worthwhile.

First, 'electricity' as such is a fungible commodity (or service) and its utility is not determined or affected by its generation source.[11] A nuclear power generation facility, coal-fired plant, hydroelectric dam, wind turbine, solar cell installation and bioelectric generation facility all produce

[10] See, e.g., D. Neven, *Ownership, Performance and National Champions* (2008), http://ec.europa.eu/dgs/competition/economist/neven_ownership_performance_en.pdf.

[11] See J. H. Barton, 'Intellectual property and access to clean energy technologies in developing countries: an analysis of solar photovoltaic, biofuels and wind technologies, trade and sustainable energy series', *Issue Paper* No. 2, International Centre for Trade and Sustainable Development (2007).

the same product or service. The external cost to the environment based on these different forms of electricity generation may be very different, but it is an important aspect of the electricity market that there are substitutable production technologies, even if some substitutes are imperfect from an environmental standpoint.

Second, the basic sources for large-scale electricity generation are presumed to be largely 'known'. We do not know the unknown, and cannot wholly discount the possibility that 'gravitational waves from outer space', for example, will become a new source of earthbound electricity. But most of the new technologies for generating electricity, at least for the foreseeable future, are likely to represent incremental innovation and/or variations on known themes.[12] One of the consequences of this presumption is that it is not likely in the electricity generation sector that a specific invention and/or patent will substantially inhibit overall growth in global generation capacity, even if restrictions on diffusion affect the specific type of energy generation. Furthermore, there may be differences at the margins of efficiency, but wholly transformative innovation from a cost perspective does not appear likely in the near to medium term.

E Factors Relevant to the 'Grid'

As to the technologies that may be important for creating an international power grid, and the posited creation of a CGEG,[13] it should be noted that computer software and digital technologies will play an important role, and that patents may not be the only form of IP relevant in this area. While patents are used to protect computer software (with variations between national or regional jurisdictions), software is also and perhaps predominantly protected by copyright, and in certain instances trade secret. Moreover, questions arise around 'network effects', for example, and potential widespread adoption of standardised computer networking systems that may be important to a shared grid.

[12] The author does not wish to overstate this point. There are new technologies such as hydrogen fusion that, in theory, could produce vast amounts of energy at reasonable cost, and such technologies are 'being worked on'. But hydrogen fusion technology remains based on known science. The implementation of that science into commercially useful sources of energy would be incremental.

[13] See, e.g., www.ge.com/globalimpact2013/#/ecomagination/ge-ventures-energy; www.reliableplant.com/Read/25533/GE-power-grid-technology; www.ge.com/about-us/openinnovation; http://w3.siemens.com/smartgrid/global/en/pages/default.aspx.

Technical standards are likely to play an important role in the integration of systems and networks.[14]

The territorial nature of the international IP system raises some intriguing issues with respect to the potential emergence of a global power grid. We might begin with the question of exhaustion of IP rights. If electricity is created using a patented product or process, does the IP right holder have an interest in the electricity as such? The answer to that question is probably 'no', because electricity itself is not patentable as a natural phenomenon. On the other hand, Article 28.2 of the Agreement on Trade-Related Aspects of Intellectual Property Rights (TRIPS Agreement) requires that process patent holders be entitled to exclude the importation and sale of products directly produced by patented processes, so it may be that a process patent holder would have downstream interests in electricity. History is full of interesting assertions of rights in IP, and it is at least worth flagging the exhaustion issue, even if we can only agree to dismiss it.

Let us assume that a group of countries agrees on the creation of a shared electricity grid. There is a substantial possibility that the group will agree on equipment in order, for example, to establish interoperability and make possible the sharing of parts and repairs. Will patents play a role? The grid will require substantial investment in computer software, probably uniquely tailored to the specific countries and grid. How will that software be developed? Who will own the copyright? Who will have the right to modify the software?

Although the prospect may seem counterintuitive, it may be that the technologies involved in the establishment of cross-border trade in electricity, and a more extensive global electricity grid, will be more subject to concentration than the underlying energy-generation assets. A perusal of industry providers turns up some common names: ABB, Alstom, General Electric, Siemens and a few others. While electricity may be fungible, the software and other elements of the digital network may not be. While the relevance of smart grid technologies to green energy generation and the role of IP has previously been noted,[15] the empirical studies conducted to date with respect to potential technology

[14] See, e.g., IEE & the Smart Grid, http://smartgrid.ieee.org/ieee-smart-grid. 'In order for the Smart Grid to be successful, there needs to be a set of well-established standards in place that all industries and organizations involved can utilize': Dr. W. Charlton Adams, Jr., Past President IEEE Standards Association.

[15] See Abbott, 'Innovation as imperative'.

bottlenecks in the development of green energy resources have not focused on the 'grid' aspect.

Will all countries have the financial resources to participate in the grid if there are substantial royalties associated with operating on the network? Will private actors control the resources that are critical to the functioning of the network? Should the grid be based on an open source platform?

F Open Source, Cybersecurity and the Grid

From the standpoint of a shared global environment and economic development more generally, it would intuitively appear that grid technology (e.g. software and other digital elements) should be based on open source platforms. A host of different actors will need to modify the grid to address local electricity needs and specific characteristics of the market, and to integrate with specific equipment. An open source platform would allow a larger group of technologists to understand how the system operates, and contribute to its progressive development. Software based on open source platforms is used in a number of critical financial infrastructure environments (e.g. the Red Hat software running trades for the New York Stock Exchange).

Yet an open source platform is immediately confronted with issues of security and raises the questions of who would have access to the code underlying the platform, who would be allowed to modify it and under what circumstances this would be the case. The current global security environment suggests it would not be prudent to allow relatively open access to the technology underlying an energy grid that may meet the electricity needs of a substantial number of countries. To the contrary, the potential risks from a security standpoint may argue in favour of a heavily guarded proprietary/secret programme to which very few individuals have unrestricted access.[16]

G The Development Side

A significant issue in addressing climate change and the effect of greenhouse gas emission through energy generation involves the extent to

[16] Recall, for example, that Israeli and/or US software engineers were able to substantially inhibit Iranian nuclear centrifuges through the secret manipulation of computer code: see, e.g., W. J. Broad, J. Markoff and D. E. Sanger, 'Israeli test on worm called crucial in Iran nuclear delay', *The New York Times* (15 January 2011), www.nytimes.com/2011/01/16/world/middleeast/16stuxnet.html?pagewanted=all&_r=0.

which countries owning technology assets should provide access to them to public or private enterprises in countries that do not own such assets. Transfer of technology has been a major talking point at the United Nations Framework Convention on Climate Change (UNFCCC); however, very little has happened at the UNFCCC in terms of development of concrete programmes, or even theoretical approaches. There appear to be a number of reasons for this.

One, not surprisingly, is that the private sector enterprises investing substantial financial resources in developing new technologies to prevent or mitigate climate change are interested in profiting from their investments, not in making them available at a low cost to competitors and potential competitors.[17] General Electric and Siemens are certainly interested in the development and implementation of a GCEG, but they may not be so interested in assisting Chinese or Indian competitors. The views of the private sector are naturally taken into account by government delegations negotiating in the UNFCCC.

Second, it is sometimes forgotten that governments are seldom the owners of the technologies that policy-makers and non-governmental organisations (NGOs) suggest should be transferred. Of course, there are exceptions. The US Department of Energy, for example, works on the development of innovative technologies and may (in some cases) own the resulting patents or other IP (or share rights). But, assuming that technology is privately owned, governments are generally rather reluctant to intervene or appropriate private technology rights (e.g. patents) for the purposes of making technology available to third parties, including third parties that may reside in foreign countries.[18] Thus, a strategy that says 'Developed Country A should transfer clean energy technology to Developing Country B' may not be taking into account that this is not a choice that government officials in Country A can make without incurring substantial political or economic costs.

[17] See pronouncements by General Electric executives distinguishing the energy market from the pharmaceuticals market, e.g. 'Paying dividends: IP and green energy', *World Intellectual Property Review* (1 October 2010), www.worldipreview.com/article/paying-dividends-ip-and-green-energy; see also 'Letter from the Alliance for Clean Technology Innovation, to the Australian Productivity Commission' (28 September 2012), www.pc.gov.au/inquiries/completed/patents/submissions/submissions-test/submission-counter/sub009-patents.pdf.

[18] See, e.g., K. Stech, 'Energy department fights to keep solar patents in U.S. hands', *Wall Street Journal* (10 October 2011), http://blogs.wsj.com/bankruptcy/2011/10/10/energy-department-fights-to-keep-solar-patents-in-u-s-hands/.

One way to fill gaps in technology transfer is for developing countries to create exceptions in their IP laws, or to issue compulsory licences, to make such technology available to domestic enterprises. However, in the public health field such action has proven to come at a relatively high political cost from an international standpoint. If there is no reasonable alternative, exceptions and/or compulsory licensing are options that governments must look into, but the creation of a GCEG is by definition going to require collaboration and cooperation, so this would not be a 'first best' solution.

One practical solution for creating a system of global technology sharing is some agreement on mechanisms for establishing joint venture partnerships that takes into account the interests of private sector technology suppliers and private or public sector technology recipients, as well as the public interest in affordable access to the resulting energy products or services. Given the scale of potential global demand for electricity, it seems likely that there is enough potential profit in a global mechanism to satisfy the interests of all or almost all participants.

A part of that mechanism might involve the creation of a global patent or broader technology pool to which private enterprise and government research laboratories would agree to transfer relevant technologies, and from which grid operators throughout the global system could draw. Undoubtedly, establishing such a mechanism would be complicated, and would necessitate the development of an incentive that would cause private industry to continue investing in innovation. It has been noted that patent pooling endeavours run the risk of disincentivising private innovation because potential contributors may elect to free-ride on the work of others.[19] This reinforces the observation concerning the necessity of working through an appropriate incentive structure, which could also include prizes. Working on the side of possible solutions, the potential global demand for electricity based on clean technologies may be sufficiently great that a reasonable royalty would provide a large return on investment.

The Medicines Patent Pool (MPP), while addressing different subject matter and a different context, does provide a model and some hope that a mechanism could be devised whereby technology and expertise could be shared. Clearly the circumstances are very different, and this is not to suggest that a GCEG could or should work on the MPP model. Rather, it

[19] See J. H. Barton, *IP and Climate Technology* (19 October 2007), unpublished draft in author's files.

is saying that fifteen years ago, the effective implementation of the MPP might have been dismissed as an unrealistic dream. Yet, it has become a reality that is contributing substantially to addressing the HIV–AIDS epidemic.

Subsidies are a commonly used mechanism for incentivising innovation that might otherwise not occur owing to a lack of adequate market incentives. If a global innovation financing facility directed towards establishing a GCEG can be established, that could be used in association with a pooling arrangement.

If the international community is genuinely interested in creating a GCEG, the first steps would involve convening meetings of interested stakeholders. In principle, this could somehow be done under the auspices of the UNFCCC. The following might be considered.

- First, it would be desirable to bring together energy regulators from a wide geographical area to set ground rules for the discussions to follow. From the outset, it would be necessary to address competition law issues during cooperative discussions among the major industry stakeholders whose participation may otherwise raise serious competition law concerns.
- Another meeting would involve the major grid technology companies, such as GE and Siemens, to hear their views about what might be possible from a technological standpoint. It would be important to invite stakeholders representing the widest possible geographical area.
- Consumer groups would be another essential element, as any plan would ultimately need the support of the eventual users of the grid.
- Fiscal and tax authorities would need to be involved, as arrangements for a GCEG would have significant budgetary and tax implications.

H A Building Block Approach

The foregoing discussion views the potential establishment of a GCEG from a global perspective. Most real-world, large-scale projects do not start out as multilateral endeavours, but rather as bilateral public and private arrangements that may be extended geographically over time. In the same way, transfers of technology directed towards establishing a GCEG might well begin at a more modest level through bilateral and regional electricity-sharing agreements and programmes.

21

The Role of the Fair and Equitable Treatment Standard: Regulatory Coherence for Trade and Investment in Renewable Energy

SOFYA MATTEOTTI AND TETYANA PAYOSOVA

A Introduction

Foreign investment is vital for the diffusion of renewable energy technologies across countries. It also helps to develop domestic generation capacities for electricity from renewable energy sources (RES). In order to meet international targets for reduction of greenhouse gas emissions and national targets for the share of RES in the total energy mix, states resort to a number of policies to incentivise investment in the sector.[1] General challenges for development of the renewable energy sector include the high up-front capital costs of expensive components and, in most cases, the necessary infrastructure investment. However, cost reductions for some technologies, notably solar and wind power, have been reported recently. Solar photovoltaic (PV) modules, for instance, cost 75 per cent less today than they did in 2009, while wind turbine prices have declined by almost a third over the same period.[2] These two technologies are predicted to remain the leading ones for renewable energy investment up until 2020, with roughly USD 140 billion investment in each per year. Investment in less mature technologies – biomass, geothermal and tidal, wave and ocean energy – is expected to increase more rapidly, albeit starting from much lower levels.[3] Europe and North America, as the first movers, have been responsible

[1] For an overview of state policies to promote investment in renewable energy see B. S. Gentry and J. J. Ronk, *International Investment Agreements and Investments in Renewable Energy*, Yale Environmental Protection Law Clinic (2003), 59–64.

[2] IRENA, *REthinking Energy: Renewable Energy and Climate Change* (November 2015), p. 5, www.irena.org/rethinking/default2015.aspx.

[3] Ibid., 19.

for the majority of past investments, but trends indicate a rise in investments in developing markets in Africa, Asia and Latin America.[4]

Support schemes for renewable energy create incentives that aim to stimulate investment in its development. Low-carbon investors build their business cases on the basis of these policies and promises of support. Their investment relies upon the faithful implementation of support commitments made by host states, with an expectation that they will receive public support in accordance with the schemes in existence at the time of investment.[5] According to the International Energy Agency's 2010 report,[6] most of these investors' decisions in the renewable energy sector are significantly influenced by the availability of support measures in the host state. Apart from the direct cost disadvantage in comparison to conventional energy sources, renewable energy investors are often exposed to regulatory and political risks while trying to recover their investment.[7] A recent DIA-CORE assessment suggests that risks related to the changing framework of the support schemes and policy designs are perceived by investors as the most crucial.[8] National support for RES may be granted by a green certificates system (or white certificates), feed-in tariffs (FITs) or premiums or by granting investment aid or fiscal advantages to producers of electricity that use renewable energy sources. FITs are among the most widespread policy tools used to ensure the attractiveness of investment in renewable energy and its competitiveness as compared to conventional energy. In terms of cost efficiency, focusing support only on the most mature and most competitive RES technologies seems a rational approach at a first glance. However, in order to achieve large-scale RES deployment, a diverse mix of generation technologies, installation size and capacity and geographical location is preferable.

4 Ibid., 18.

5 A. Boute, 'Combating climate change through investment arbitration changes', *Fordham International Law Journal*, 35(3) (2011), 613–64.

6 IEA, *World Energy Outlook* (2010), p. 293, www.worldenergyoutlook.org/media/weo2010.pdf.

7 See, e.g., Gentry and Ronk, *International Investment Agreements and Investments in Renewable Energy*, 58–9.

8 See P. Noothout, D. de Jager, L. Tesnière, S. van Rooijen, N. Karypidis, R. Brückmann, F. Jirouš, B. Breitschopf, D. Angelopoulos, H. Doukas, I. Konstantinavičiūtė and G. Resch, 'Investment risks for renewable energy projects – the case of onshore wind power', DIA-CORE Policy Brief (21 May 2015). DIA-CORE is a programme launched in April 2013 under the auspices of the Intelligent Energy Europe programme. It primarily deals with assessment of the existing policy mechanisms and aims to establish a fruitful dialogue between stakeholders. For more information, please see http://diacore.eu.

As suggested in a recent study by the Council of European Energy Regulators (CEER), this can only be triggered through scheme design based on a mix of different support levels.[9]

A considerable number of investment cases have been initiated in recent years as a result of changes to the national FIT regimes in a number of European countries. In light of this policy environment, the question arises as to whether the current position in international economic law (both trade and investment) with regard to support schemes in the renewable energy sector is coherent. In this study we also aim to analyse whether international investment law grants foreign investors a sufficient level of protection with respect to their investments in the renewable energy sector. First, we will briefly address the principle of fair and equitable treatment (FET) in international investment law and its role in securing a favourable investment climate, as it has the potential to serve as a safety net for investors in cases of sudden regulatory reforms. Then, we will address the most recent developments in investment dispute settlement related to renewable energy. Next, we will offer an overview on whether WTO law can facilitate protection from changes to national regulatory frameworks that are unfavourable to investment. We conclude with an example of the upcoming legal reforms in the EU aimed at achieving the binding targets for renewable energy by 2030 through opening up the national support schemes, and the impact they might have on investors' legitimate expectations. Based on the FET jurisprudence, we provide a number of policy recommendations which governments may use when implementing new support regimes or amending the existing ones in the renewable energy sector.

B Fair and Equitable Treatment in the International Investment Agreements

I The Scope of the FET Concept

The concept of 'fair and equitable treatment' and the related notion of 'full protection and security' is largely a post-Second World War development. Reference to these terms is found in the 1948 Havana Charter and a number of draft conventions that followed, including the 1967 draft OECD Convention on the Protection of Foreign Property, the draft

[9] Council of European Energy Regulators (CEER), *Key Support Elements of RES in Europe: Moving Towards Market Integration*, Report C15-SDE-49-03, 26 January 2016, p. 13.

United Nations Code of Conduct on Transnational Corporations and a number of other international drafting initiatives.[10]

FET is acknowledged as one of the most commonly used standards in investment agreements.[11] A detailed analysis of 335 bilateral investment treaties (BITs) that was conducted in the early 1990s showed that only twenty-eight BITs did not include a reference to FET. From the 1990s onwards, the standard has been included in the majority of BITs.[12] Thus, the model BITs adopted by such countries as Canada, the United States, Germany, the United Kingdom and France; sectoral agreements such as the Energy Charter Treaty (ECT); and regional trade agreements such as the North American Free Trade Agreement (NAFTA) and many others all incorporate the FET clause. The remarkable feature of recent state practice is that the FET clause has also been included in the BITs entered into between the developing countries.[13]

In contrast to other standards of treatment, namely most-favoured-nation and national treatment, which are typically subject to varying lists of exceptions that aim to protect, inter alia, certain sectors or enterprises in which the host state has a particular national interest,[14] the vast majority of investment agreements do not contain indications of any exceptions or derogations, limiting the applicability of the FET standard.[15]

Klaeger claims that FET appears to be the most vaguely formulated investment protection standard.[16] The scope of the FET standard differs from treaty to treaty and remains controversial. According to some scholars, the obligation to grant foreign investment 'fair and equitable treatment' is not different from the obligation to treat investment in

[10] On the historic development of the FET principle, see Martins Paparinskis, *The International Minimum Standard and Fair and Equitable Treatment* (Oxford: Oxford University Press, 2013).

[11] R. Klaeger, *Fair and Equitable Treatment in International Investment Law* (Cambridge: Cambridge University Press, 2011), p. 9, citing R. Dolzer and M. Stevens, *Bilateral Investment Treaties* (The Hague: Martinus Nijhoff Publishers, 1995), p. 58.

[12] M. Malik, 'Fair and equitable treatment', *IISD Best Practices Series, Bulletin no. 3* (2009), 1.

[13] P. Dumberry, *The FET Standard: A Guide to NAFTA Case Law on Article 1105* (The Netherlands: Wolters Kluwer, 2013), 35.

[14] R. Dolzer and M. Stevens, *Bilateral Investment Treaties* (The Hague: Martinus Nijhoff Publishers, 1995), pp. 71–6.

[15] Klaeger, *Fair and Equitable Treatment in International Investment Law*, p. 43, noting that only some US, Indian and Canadian BITs contain certain public policy exceptions, or so-called non-precluded measures clauses, that limit the applicability of FET.

[16] Ibid., 3.

accordance with the international minimum standard. According to other scholars, however, 'fair and equitable treatment' means something different from the international minimum standard. According to this view, the term 'fair and equitable treatment' should be given its plain meaning. This results in a case-by-case application of a test based on equity in order to determine whether the standard has been infringed.[17]

The reference to customary international law raises additional challenges of interpretation as to what that standard is and how state action should be judged in relation to that standard. To put it another way, if the FET concept refers to a minimum standard recognised under customary international law, is that standard frozen in time, or has it evolved to take into account changing notions of acceptable conduct on the part of governments? Some clarification is provided by arbitration decisions under the ECT and NAFTA, as well as International Centre for Settlement of Investment Disputes (ICSID) awards under various investment protection agreements.

Dolzer and Schreuer, discussing FET in international investment law, identify from the arbitral case law five general areas in which the standard may operate:

(i) transparency, stability and the protection of legitimate expectations;
(ii) compliance with contractual obligations;
(iii) procedural propriety and due process;
(iv) good faith;
(v) freedom from coercion and harassment.[18]

1 FET Standard under NAFTA

NAFTA Article 1105 includes a provision on the minimum standard of treatment which states:

> Each Party shall accord to investments of investors of another Party treatment in *accordance with international law, including fair and equitable treatment and full protection and security.* (emphasis added)

It should be mentioned at the outset that the FET standard clause under Article 1105 must be analysed under specific parameters that do not exist with regard to most other investment treaties. This is due to the language

[17] UNCTAD, *Bilateral Investment Treaties, 1995–2006: Trends in Investment Rulemaking* (New York and Geneva: UNCTAD, United Nations, 2007), p. 28.

[18] Dolzer and Stevens, *Principles of International Investment Law*, pp. 133–47.

of the treaty. According to Article 1105 NAFTA, under 'international law', FET has to be accorded to investors. The vast majority of BITs with a FET clause make no reference to 'international law'. The Notes of Interpretation of Certain Chapter 11 Provisions clarify that the 'FET to be accorded under international law' constitutes a reference to the minimum standard of treatment under customary law and does not require anything beyond it.[19]

NAFTA case law suggests that while FET under NAFTA falls within the umbrella clause on the minimum standard of treatment, it is still not an 'empty box'. The NAFTA case-law further clarifies that only the prohibition of arbitrary conduct and denial of justice and the obligation of due process are unambiguously stand-alone elements of the FET obligation. Importantly, in the NAFTA context, no tribunal has ever interpreted the FET standard as an absolute obligation for the host state to ensure stable legal and business conditions.[20]

2 FET Standard under the ECT

The ECT also contains a FET standard, which is applicable to our second case study below – namely, the withdrawal of the FIT in Spain – and has also been invoked in a number of other recent investment cases on renewable energy. The FET provisions in the ECT are modelled on Chapter 11 of the NAFTA, with more expansive wording in Article 10, which provides:

> Each Contracting Party shall, in accordance with the provisions of this Treaty, encourage and create stable, equitable, favourable and transparent conditions for Investors of other Contracting Parties to make Investments in its Area. *Such conditions shall include a commitment to accord at all times to Investments of Investors of other Contracting Parties fair and equitable treatment.* Such Investments shall also enjoy the most *constant protection and security* and no Contracting Party shall in any way impair by unreasonable or discriminatory measures their management, maintenance, use, enjoyment or disposal. In no case shall such Investments be accorded treatment less favourable than that required by international law, including treaty obligations. Each Contracting Party shall observe any obligations it has entered into with an Investor or an Investment of an Investor of any other Contracting Party. (emphasis added)

[19] NAFTA Free Trade Commission, 'Notes of interpretation of certain chapter 11 provisions', 31 July 2001, www.sice.oas.org/tpd/nafta/Commission/CH11understanding_e.asp.

[20] Dumberry, *The FET Standard*, p. 163.

In this respect, commentators have noted a lack of clarity regarding how all of the obligations mentioned relate to each other and to FET.[21] Are they a precondition to find a FET violation? Are they self-standing obligations?

In *Petrobart* v. *The Kyrgyz Republic* the Tribunal took note of various specific elements in Article 10(1) of the Treaty, but highlighted that this paragraph in its entirety is intended to ensure fair and equitable treatment of investments. Thus, FET under the ECT is treated as an overarching principle that embraces all the other standards mentioned in that Article.[22]

The ICSID tribunal in *Plama Consortium* v. *Bulgaria* concluded as follows with respect to the FET obligation under ECT:

> The stability of the legal framework has been identified as 'an emerging standard of fair and equitable treatment in international law'. However, the State maintains its legitimate right to regulate, and this right should also be considered when assessing the compliance with the standard of fair and equitable treatment.[23]

Following this line of argumentation, some commentators suggest that the most relevant context in which to interpret FET under the ECT is that of the introductory sentence of Article 10(1): 'Each Contracting party shall, in accordance with the provisions of this treaty, encourage and create stable, equitable, favourable and transparent conditions for Investors of other Contracting Parties to make investments in its Area.' Such an interpretation might suggest that FET under ECT entails the encouragement and creation of stable, favourable (to investors) and transparent conditions.[24]

As has been demonstrated, the fragmented nature of international investment law may have negative impacts on legal security both for an investor and for a host state. The exact scope of many of the provisions is too vague and is subject to (subjective) interpretation by an arbitral tribunal on a case-by-case basis. Thus, in most cases, it is unclear whether FET is an objective standard based on public international law (i.e. which

[21] See T. Roe and M. Happold, *Settlement of Investment Disputes under the ECT* (Cambridge: Cambridge University Press, 2011), p. 112.

[22] *Petrobrat Limited* v. *The Kyrgyz Republic*, SCC Case No. 126/2003, Arbitral Award, 29 March 2005, paras. 77–78.

[23] *Plama Consortium Limited* v. *Republic of Bulgaria*, ICSID Case No. ARB/03/24, Arbitral Award, 27 August 2008, para. 177.

[24] See Roe et al., *Settlement of Investment Disputes under the ECT*, p. 115.

also incorporates WTO rights and obligations, in which case a breach of a WTO obligation that results in harm to a foreign investor could amount to a denial of FET) or a more pragmatic, flexible and subjective standard based on the circumstances and context of the particular treaty.[25] Despite the lack of a coherent approach, the FET standard could provide important guarantees of protection against changes by the state to the arrangements it has made to attract low-carbon investments. According to the *World Investment Report* issued by the United Nations Conference on Trade and Development (UNCTAD) in 2010, the FET standard 'could [potentially] be used to challenge the refusal of expected government support [and] the dismantling of market-creating mechanisms'.[26]

C FET Standard as Applied to Renewable Energy Sector Investment Disputes

I Recent Trends in Renewable Energy Investment Dispute Settlement

The years 2012–14 were notable for the large number of investment cases filed in the renewable energy sector. In fact, nearly a quarter of all arbitrations initiated in 2013 involved challenges to regulatory actions by Czech Republic and Spain that affected the renewable energy sector.[27] Most of them were submitted to the arbitral tribunals based on the ECT investment provisions; however, several cases are also subject to NAFTA investment rules. Table 21.1 offers an overview of investment arbitration cases in the renewable energy sector.[28]

The common denominator for all the cases listed in Table 21.1 (based on the notices of arbitration accessible to the public) is that all of them included a claim related to a change in the regulatory framework for renewable energy which allegedly violated the FET standard (in some cases, investors also argued that such regulatory changes constitute a form of indirect expropriation).

[25] Ibid., p. 118.
[26] UNCTAD, *World Investment Report 2010: Investing in a Low-Carbon Economy* (New York and Geneva: United Nations, 2010), p. 137.
[27] UNCTAD, 'Recent developments in investor-state dispute settlement (ISDS)', *IIA Issue Note* no. 1 (April 2014).
[28] Please note that the list is not complete. For up-to-date information on the cases, see UNCTAD Investment Policy Hub, Investment Dispute Settlement Navigator, http://investmentpolicyhub.unctad.org/ISDS/FilterByEconomicSector.

Table 21.1 *Overview of investment arbitration cases in the renewable energy sector*

Title of the case	Disputed measures	International investment agreement	Arbitration rules	Date of initiation	Status
Mesa Power Group, LLC v. *Government of Canada*	Introduction of the Ontario FIT programme	NAFTA	UNCITRAL, PCA Case No. 2012–17	2012	Award 24.03.2016
Windstream Energy LLC (U.S.A.) v. *The Government of Canada*	Same as above	NAFTA	UNCITRAL	2013	Award 27.09.2016
Antaris Solar GmbH (Germany) and Dr. Michael Göde (Germany) v. *The Czech Republic*	Levy on 'solar' electricity	BIT Germany–Czech and Slovak Republic; ECT	UNCITRAL	2013	Pending
The PV Investors v. *Spain*	Revocation of FIT for solar PV plants	ECT	Ad hoc UNCITRAL Arbitration Rules	2011	Pending
Charanne (the Netherlands) and Construction Investments (Luxembourg) v. *Spain*	7 per cent tax on power generators' revenues, reduction of RES subsidies	ECT	Arbitration Institute of the SCC	2013	Award, 21.01.2016
Isolux Infrastructure Netherlands B.V. v. *Spain*	Same as above	ECT	Arbitration Institute of the SCC	2013	Award 13.07.2016

CSP Equity Investment S.à.r.l. v. *Spain*	Regulatory reforms in the renewable energy sector	ECT	Arbitration Institute of the SCC	2013	Pending
Eiser Infrastructure Limited and Energía Solar Luxembourg S.à r.l. v. *Kingdom of Spain*	7 per cent tax on power generators' revenues, reduction of RES subsidies	ECT	ICSID Case No. ARB/13/36	2013	Pending
Masdar Solar & Wind Cooperatief U.A. v. *Kingdom of Spain*	Same as above	ECT	ICSID Case No. ARB/14/1	2013	Pending
RREEF Infrastructure (G.P.) Limited and RREEF Pan-European Infrastructure Two Lux S.à r.l. v. *Kingdom of Spain*	Same as above	ECT	ICSID Case No. ARB/13/30	2013	Pending
Antin Infrastructure Services Luxembourg S.à.r.l. and Antin Energia Termosolar B.V. v. *Kingdom of Spain*	Same as above	ECT	ICSID Case No. ARB/13/31	2013	Pending
NextEra Energy Global Holdings B.V. and NextEra Energy Spain Holdings B.V. v. *Kingdom of Spain*	Same as above	ECT	ICSID Case No. ARB/14/11	2014	Pending
InfraRed Environmental Infrastructure GP Limited and others v. *Kingdom of Spain*	Same as above	ECT	ICSID Case No. ARB/14/12	2014	Pending

Table 21.1 (*cont.*)

Title of the case	Disputed measures	International investment agreement	Arbitration rules	Date of initiation	Status
RENERGY S.à r.l. v. *Kingdom of Spain*	Same as above	ECT	ICSID Case No. ARB/14/18	2014	Pending
Blusun v. *Italy*	Cut of tariff incentives for some solar power projects	ECT	ICSID Case No. ARB/14/3	2014	Pending
Belenergia v. *Italy*	Same as above	ECT	ICSID Case No. ARB/15/40	2015	Pending
Eskosol v. *Italy*	Same as above	ECT	ICSID Case No. ARB/15/50	2015	Pending
Greentech v. *Italy*	Same as above	ECT	SCC	2015	Pending
Silver Ridge Power BV v. *Italian Republic*	Same as above	ECT	ICSID Case No. ARB/15/37	2015	Pending
EVN v. *Bulgaria*	Pricing of electricity and changes to the public compensation regime for RES	ECT, Austria–Bulgaria BIT	ICSID Case No. ARB/13/17	2013	Pending
ENERGO-PRO v. *Bulgaria*	Same as above	ECT, Bulgaria–Czech BIT	ICSID Case No. ABR/15/19	2015	Pending

The overarching questions to ask with respect to these cases are: what policy space – from the perspective of international investment law – does the state have for modifying measures such as FITs or for terminating support schemes in the renewable energy sector, and what is the appropriate balance with respect to the rights and legitimate expectations and interests of investors? We further raise the question of whether such balance could potentially be achieved through application of the FET standard.

To answer these questions, we turn to a short analysis on the substance of three selected cases. One of these was still pending at the time of writing and two had been recently resolved: they are *Mesa Power Group, LLC* v. *Government of Canada*, *PV Investors* v. *Spain* and *Charanne B.V. and Construction Investments S.A.R.L.* v. *Spain*.

1 *Mesa Power Group, LLC* v. *Government of Canada*

In *Mesa Power Group, LLC* v. *Government of Canada*, the claim arose due to regulatory changes with respect to renewable energy in Ontario, namely the introduction of a FIT programme in 2009.[29] Mesa is an American company which owns four wind farm investments in Ontario (all unlimited liability companies, registered in Alberta) producing electricity from RES. According to the FIT programme of 2009, Mesa was required to enter into contractual relations with the Ontario Power Authority (OPA) in order to be able to sell electricity to the grid. In assessing the wind projects that applied for the FIT, the OPA primarily favoured 'shovel-ready' projects.[30] In addition, in September 2009, the OPA incorporated provisions on certain levels of local inputs (local content requirements) that had to be complied with as a precondition for the FIT. Moreover, Mesa had concerns regarding the impartiality of the ranking methodology to qualify for the FIT, which was allegedly non-transparent and affected by a 'secret' governmental-level deal with Korean investors. Finally, in June 2011, new rules for the FIT programme were issued. Projects from selected geographical areas were artificially

[29] It should be noted that the same FIT program linked to local content requirements was also challenged in the WTO. See Appellate Body Report, *Canada – Certain Measures Affecting the Renewable Energy Generation Sector*, WT/DS412/AB/R/WT/DS426/AB/R, adopted 6 May 2013. Notably, this trade dispute focused primarily on the effect of Ontario's measures on the competitive opportunities in a given market, and to a lesser extent on protection of specific investors' interests.

[30] *Mesa Power Group LLC* v. *Government of Canada*, Notice of Arbitration from 4 October 2011, paras. 13–18.

assigned a higher priority ranking and could build long transmission lines to interconnect with another region, thus overcoming the FIT ranking. As a consequence, all four of Mesa's companies lost priority ranking and did not receive contracts under the FIT programme.[31]

Mesa claimed that by introducing and implementing the FIT programme, Canada had violated a number of its obligations under NAFTA Chapter 11, namely obligations regarding national treatment, most favoured nation treatment, standard of treatment, international law standards of treatment, performance requirements and certain provisions for state enterprises. Moreover, according to Mesa, the FIT programme prevented the development of its four wind power projects, which resulted in considerable damages.

With respect to the FET standard according to Article 1105 NAFTA, Mesa claimed, among other things, that the government of Ontario engaged in unfair, arbitrary and discriminatory actions, including unannounced changes to the regulatory framework which undermined Mesa's reasonable and legitimate expectations. It also claimed that there had been arbitrary application of the ranking methodology and non-transparent implementation of the FIT.[32] Canada in turn argued that the threshold of the FET standard should not be set too low, as it is supposed to protect investors only against unfair or manifestly arbitrary actions by the host state.[33] To support its position, Canada further referred to *Glamis Gold* v. *United States*, where the tribunal found that violation of the FET standard under NAFTA would require 'an act that is sufficiently egregious and shocking, a gross denial of justice, manifest arbitrariness, a complete lack of due process, evident discrimination or a manifest lack of reasons'.[34] In the same case, the tribunal clarified that a mere breach of a contract or not 'living up to expectations cannot be sufficient to find a breach of Article 1105 of the NAFTA'.[35] Canada also argued that two other recent cases, *Cargill* v. *United Mexican States*[36]

[31] Ibid., paras. 43–50. [32] Ibid., paras. 61–63.

[33] *Mesa Power Group LLC* v. *Government of Canada*, Hearing Transcript Day 6, 31 October 2014.

[34] *Glamis Gold* v. *United States of America*, Award, 8 June 2009, para. 616.

[35] Ibid., para. 620.

[36] *Cargill Inc.* v. *United Mexican States, ICSID Case No. ARB(AF)/05/2.* With respect to the high threshold of the FET violation, the tribunal stated: 'To determine whether an action fails to meet the requirement of fair and equitable treatment, a tribunal must carefully examine whether the complained of measures were grossly unfair, unjust or idiosyncratic, arbitrary beyond a merely inconsistent or questionable application of administrative or

and *Mobil* v. *Canada*,[37] confirm that a state measure has to pass a high threshold to be found in violation of the FET standard.

The arbitral tribunal in its final award differentiated between the acts of the OPA, as a state enterprise, that are attributable to Canada according to Article 1503 of the NAFTA and those that are not. After analysing the relevant provisions of the Electricity Act, it concluded that all of the challenged OPA actions, except the local content requirements and the FIT Contract itself, were attributable to Canada.[38] With this in mind, the arbitral tribunal turned to the analysis of the FET standard under Article 1105 of the NAFTA, where it largely agreed with the decisions in *Waste Management II* and *Bilcon*.[39] The tribunal considered the following elements to fall within the scope of Article 1105 of the NAFTA: arbitrariness, 'gross' unfairness, discrimination, complete lack of transparency and candour in an administrative process, lack of due process 'leading to an outcome which offends judicial propriety' and 'manifest failure' of natural justice in judicial proceedings.[40] The tribunal also agreed with the previous NAFTA tribunals that 'the failure to respect an investor's legitimate expectations in and of itself does not constitute a breach of Article 1105, but is an element to take into account when assessing whether other components of the standard are breached'.[41] Finally, the tribunal disagreed with Mesa's position on the autonomous FET standard beyond the international minimum standard of treatment and, after assessing all the presented evidence, it concluded that Canada's conduct had not breached Article 1105 of the NAFTA.[42]

legal policy or procedure so as to constitute an unexpected and shocking repudiation of a policy's very purpose and goals or to otherwise grossly subvert a domestic law or policy for an ulterior motive or involved an utter lack of due process so as to offend judicial propriety': Award, 18 September 2009, para. 296.

[37] *Mobil Investments Canada Inc. and Murphy Oil Corporation* v. *Canada*, ICSID Case No. ARB (AF)/07/4.

[38] *Mesa Power Group, LLC* v. *Canada*, PCA Case No. 2012-17, 24 March 2016, para. 371.

[39] *Waste Management, Inc.* v. *United Mexican States* ('Number 2'), ICSID No. ARB(AF)00/3, Award, 30 April 2004; *William Ralph Clayton, William Richard Clayton, Douglas Clayton, Daniel Clayton and Bilcon of Delaware, Inc.* v. *Canada*, PCA Case No. 2009-04, Award on Jurisdiction and Liability, 17 March 2015.

[40] *Mesa Power Group* v. *Canada*, para. 502. [41] Ibid.

[42] *Mesa Power Group* v. *Canada*, paras. 503, 541, 587, and 682. Notably, one of the arbitrators disagreed with this conclusion of the majority with respect to special treatment that the OPA and the Canadian government granted to the Korean consortium: see Concurring and Dissenting Opinion of Judge Charles N. Brower, in the matter of an arbitration under Chapter Eleven of the North American Free Trade Agreement and the UNCITRAL Arbitration Rules between Mesa Power Group LLC and Government of Canada.

2 *PV Investors* v. *Spain*

In *PV Investors* v. *Spain* the investors' claims arose because of a regulatory change to the FIT programme in Spain. During the past decade, Spain has become one of the global leaders in solar energy, ranking second in Europe for solar PV energy. Much of this growth has occurred as a result of Spain's FIT programme introduced in 2004. Favourable financial conditions and long-term agreements attracted foreign investors to the market. However, owing to Spain's difficult economic situation, in 2008 the Spanish government introduced a number of measures targeting the renewable energy sector. Namely, Spain introduced a 30 per cent FIT cut for new PV installations, which applied together with a 400 MW cap.[43] In 2010 an additional cut of 25 per cent was introduced for rooftop PV systems, along with a cut of 45 per cent for ground-mounted PV systems.[44] Between 2011 and 2013, PV energy producers with PV systems installed after 2007 could receive FIT only during limited specified hours, resulting in turn in an additional *de facto* 30 per cent reduction of their income.[45] Finally, in February 2013, the index used to calculate the FIT level of all RES was retrospectively changed from the standard 'Consumer Price Index' (CPI) to a more diffuse one. The 2013 tariffs, resulting in a 2.7 per cent FIT reduction, have been valid since 1 January 2013 and apply to all PV installations connected to the grid between 2009 and 2011. In addition to these renewable energy-specific measures, in January 2012 the Spanish government introduced a 7 per cent tax on income from electricity production. Conventional electricity producers could mitigate the negative impact of the tax through increased market prices, whereas renewable energy producers were limited to conditions under the FIT programme.[46]

[43] Real Decreto 1578/2008, de 26 de septiembre, B.O.E. 2008, 15595.

[44] Real Decreto 1565/2010, de 19 de novembre, B.O.E. 2010, 17976.

[45] Real Decreto-ley 14/2010, de 23 de diciembre, B.O.E. 2010, 19757.

[46] For further details on the Spanish regulatory reforms in the renewable energy sector see R. A. Nathanson, 'The revocation of clean energy investment economic support systems as indirect expropriation post-Nykomb: a Spanish case analysis', *Iowa Law Review*, 98 (2013), 863–904, 894–8; European Photovoltaic Industry Association, *Retrospective Measures at National Level and Their Impact on the Photovoltaic Sector* (10 December 2013), www.photon.info/newsletter/document/83019.pdf. It should be noted that in 2014, Spain passed a new law (Real Decreto 413/2014, de 6 de junio, B.O.E. 2014, 6123) to establish a new remuneration system for the renewable energy sector, composed of 'remuneration to the investment' and 'remuneration to the operation', which should allow renewable electricity producers to achieve a reasonable rate of return. The 'reasonable rate of return' will be re-assessed each six years.

The arbitration proceedings in *PV Investors* v. *Spain* were initiated by a group of sixteen investors on 17 November 2011. They claim a violation of several provisions of the ECT and respective damages caused by the retroactive reductions to the PV FIT rates.[47] The proceedings have not been finalised as of the time of writing.

3 *Charanne B.V. and Construction Investments S.A.R.L.* v. *Spain*

The 2010 regulatory changes in the Spanish FIT regime were also challenged in *Charanne B.V. and Construction Investments S.A.R.L.* v. *Spain.*[48] Charanne B.V. and Construction Investment S.A.R.L. bought shares in T-Solar Global, S.A., which in turn owned thirty-four solar plants. These solar power plants benefited from the FIT as established in Spain in 2007–8. Investors claimed that the Spanish regulatory measures constituted an indirect expropriation, which is prohibited under Article 13 ECT, and a violation of the FET standard as provided for in Article 10 ECT.

After assessing the parties' arguments, the arbitral tribunal found that Spanish regulatory measures did not cause indirect expropriation, since their investment (i.e. T-Solar) was still in operation. A reduction in T-Solar's profitability and in the value of shares did indeed occur; however, it was not sufficient to amount to an indirect expropriation under Article 13 ECT.[49]

As for the FET claim under Article 10(1) ECT, investors argued that the modified regulatory regime frustrated their legitimate expectations, namely their reliance on the conditions of the FIT, which would ensure a certain level of profitability over a set period of time (25–50 years).[50] The tribunal agreed that legitimate expectations can stem both from specific commitments and from the host state's legal system. At the same time, investors cannot expect that the regulatory framework will remain unchanged, unless a host state has made an explicit commitment in this regard. The FIT programme, as a general regulatory framework in this case, did not create specific commitments of the state towards each of

[47] For reasons of confidentiality, the precise claims are not yet known to the public. Updates are available at www.energycharter.org/what-we-do/dispute-settlement/investment-dispute-settlement-cases/31-the-pv-investors-v-spain/.

[48] *Charanne B.V. and Construction Investments S.A.R.L.* v. *Spain*, SCC Arb No. 062/2012.

[49] *Charanne B.V. and Construction Investments S.A.R.L.* v. *Spain*, SCC Arb No. 062/2012, Final Award (in Spanish), 21 January 2016, paras. 455–467.

[50] Ibid., paras. 475–549.

the investors in the renewable energy sector. Any other conclusion would unreasonably limit the host states' policy space. Finally, the tribunal also found that investors could not have reasonably expected that no regulatory changes would occur in the energy sector, which is traditionally a highly regulated sector of the economy. One of the arbitrators dissented on this point, stating that while it is true that legitimate expectations cannot freeze the country's right to regulate, when such regulatory changes occur this should be followed by compensation to the affected investors. The tribunal also did not consider the Spanish measures to be unreasonable, contrary to public interest or disproportionate. Unfortunately, the issue of retroactivity has not been analysed in sufficient depth, and might be a crucial point for tribunals in other cases brought on the same matter.

Notably, these are not the first investment dispute settlement cases in which tribunals have had to apply the FET standard to assess the regulatory changes in electricity markets in general.

D FET Standard as Applied to the Electricity Sector

Two earlier cases relating specifically to the electricity sector illustrate a possible approach to governmental interference with the regulatory framework. These are *AES* v. *Hungary*, where the government changed the pre-liberalisation tariff commitments, and *Total* v. *Argentina*, where Argentina interfered with prices for electricity. In both cases, investors claimed that there was a breach of the FET standard.

In *AES* v. *Hungary*, the claimants argued that Hungary frustrated their 'legitimate and reasonable expectations'. However, the tribunal found that at the moment that investment was made the claimants were not relying on the regulatory framework in place, and therefore Hungary was not prevented from reintroducing administrative electricity prices. The claimants could not show that Hungary had made any explicit commitment to freeze the regulatory conditions.[51]

In *Total* v. *Argentina*, the claimant challenged the FET standard against Argentina's introduction of price caps in the competitive segment of the electricity market. The tribunal supported Argentina's position that '[a] foreign investor must anticipate that circumstances

[51] *AES Summit Generation Ltd.* v. *The Republic of Hungary*, ICSID Case No. ARB/07/22, Award, 23 September 2010.

may change'[52] and that there were no legitimate expectations related to free market pricing. However, in the opinion of the tribunal, the introduction of price caps violated the FET standard.[53] The tribunal assessed the tariffs of the Argentinian electricity market as 'unreasonably low', such that they 'barely allowed [utilities] to recover operating costs'. The tribunal concluded that the price caps violated the 'economic equilibrium principle' which governs electricity markets.[54]

Some tribunals went beyond the approach in *Total* v. *Argentina* and have suggested that stability and consistency of the regulatory framework, as a condition for FET, are absolute, and therefore there can be no case-by-case departure from this rule. However, it has also been suggested that this approach is too narrow, as it does not allow the taking into account of important public policy factors in very specific circumstances (e.g. general economic crises in Argentina, or crucial measures adopted to protect public health).[55] This approach was also expressed in the *ELSI* case, where the ICJ stated that 'the right [to control and manage a business] cannot be interpreted as a sort of warranty that the normal exercise of control and management shall never be distributed. Every system of law must provide, for example, for interference with the normal exercise of rights during public emergencies and the like.'[56] To accommodate such emergency situations, Henckels, for instance, has suggested that a principle of proportionality should apply when assessing case-specific facts, based on the decision in *Glamis* v. *US*.[57]

52 *Total S.A.* v. *Argentine Republic*, ICSID Case No. ARB/04/1, Decision on Liability, 27 December 2010, para. 310. Notably, this case was brought under the Argentina–France BIT.

53 Ibid., para. 327.

54 Ibid., para. 328. For a more detailed analysis of the case see A. Boute, 'Challenging the re-regulation of liberalized electricity prices under investment arbitration', *Energy Law Journal*, 32 (2011), 497–539.

55 The 'absolute stability' approach was adopted in tribunals in a number of cases against Argentina, namely *CMS* v. *Argentina, Enron* v. *Argentina, Sempra* v. *Argentina, BG* v. *Argentina* and *LG&E* v. *Argentina*, in which the investors made claims against the emergency legislation that froze tariffs for utilities. In all five cases, the tribunals found that Argentina had violated the fair and equitable treatment standard by failing to maintain a stable regulatory environment. It should be noted that Argentina also interfered with licences and concession agreements.

56 *Elettronica Sicula S.p.A. (ELSI) (US* v. *Italy)*, [1989] ICJ Report 15 [74].

57 C. Henckels, 'Proportionality and the standard of review in fair and equitable treatment claims: balancing stability and consistency with the public interest', Society of International Economic Law Conference, Singapore (13 July 2012).

E Preliminary Observations from the Case Studies

The FET standard has been interpreted both very narrowly (under NAFTA) and very broadly (e.g. under some BITs), thus resulting in a different level of protection for investors. The possibility that there are different approaches to the FET standard under NAFTA and under other investment protection agreements was also confirmed by the tribunal in *Cargill* v. *Mexico*, where it refers to other awards based on BITs in which the FET standard was found to provide a predictable investment environment not affecting reasonable expectations of investors (but not in NAFTA).[58]

The FET standard under NAFTA, as indicated above, has a high threshold for a violation claim.[59] In addition to several recent awards cited by Canada to support its position in *Mesa* v. *Canada* (e.g. *Glamis Gold* v. *United States*, *Cargill* v. *United Mexican States* and *Mobil* v. *Canada*), the tribunal in *Gami Investments* v. *Mexico* also supported such a high threshold. Namely, the tribunal referred to the famous *Waste Management II* implications:

1. The failure to fulfil the objectives of administrative regulations without more does not necessarily rise to a breach of international law;
2. A failure to satisfy requirements of national law does not necessarily violate international law;
3. Proof of a good faith effort by the Government to achieve the objectives of its laws and regulations may counter-balance instances of disregard of legal or regulatory requirements; and
4. The record as a whole – not isolated events – determines whether there has been a breach of international law. [60]

Thus, under NAFTA, there is a low probability that renewable energy investors will be able to win their FET claims based on changes to the regulatory regime unless these changes are 'grossly unfair, unjust or arbitrary'.[61] Taking into consideration that both the electricity market

[58] *Cargill Inc.* v. *United Mexican States*, ICSID Case No. ARB(AF)/05/2, para. 290.

[59] According to UNCTAD, only 22 per cent of all FET claims in NAFTA disputes had been accepted as of October 2010, whereas in BIT cases about 62 per cent had been accepted. See UNCTAD, *Fair and Equitable Treatment*, UNCTAD Series on Issues in International Investment Agreements II (New York and Geneva: United Nations, 2012), p. 62.

[60] *Gami Investments, Inc.* v. *The Government of the United Mexican States*, Final Award, 15 November 2004, para. 97.

[61] It remains to be seen whether the claims by Mesa, especially those related to the regulatory regime changes linked to a governmental deal with Korean investors, could meet the high FET threshold in NAFTA.

in general and the renewable electricity market in particular are highly regulated, it is probable that, similarly to the *Methanex* case, a tribunal would suggest that the investor entered the market knowing that it is subject to extensive government regulation and the respective changes of the regulatory regime.

Under the ECT, a recent award in *Electrabel* v. *Hungary* seems to suggest that there is no absolute stability requirement, but this does not mean that investors are not protected against arbitrary and inconsistent conduct of the host state. The tribunal found that 'it is well established that the host state is entitled to maintain a reasonable degree of regulatory flexibility to respond to changing circumstances in the public interest'. However, it further explained that 'the requirement of fairness must not be understood as the immutability of the legal framework, but as implying that subsequent changes should be made fairly, consistently and predictably, taking into account the circumstance of the investment. Fairness and consistency must be assessed against the background information that the investor knew and should reasonably have known at the time of the investment and of the conduct of the host State'. 'Article 10(1) ECT not only speaks of fair and equitable treatment and equitable and stable conditions, it also refers to 'favourable and transparent conditions'. The reference to transparency can be read to indicate an obligation to be forthcoming with information about intended changes in policy and regulations that may significantly affect investments, so that the investor can adequately plan its investment and, if needed, engage the host State in dialogue about protecting its legitimate expectations.'[62] Thus, at least under the ECT, FET is more flexible, allowing for case-by-case appreciation and providing investors with additional protection where the state measure is not discriminatory and does not meet the requirements of indirect expropriation.[63]

F Fair and Equitable Treatment and WTO Law

Linkages between international investment law and WTO law have not only been subject to academic discussion but have also been reflected in

[62] *Electrabel* v. *Hungary*, ICSID Case No. ARB/07/19, Award, 25 November 2015, para 7.77–7.79.

[63] K. Hober, 'Investment arbitration and the Energy Charter Treaty', *Journal of International Dispute Settlement*, 1(1) (2010), 158.

dispute settlement. There are at least three aspects to be noted. First, there is the question of an obligation under WTO law similar to the FET standard. Second, there is the question of whether a violation of the core principles of WTO law would automatically lead to a violation of the FET standard (at least under NAFTA). Third, WTO jurisprudence could potentially be used to interpret FET.

With respect to the first aspect, it has been noted that there is a growing tendency for parallel proceedings relating to the same subject matter to be brought by private investors before investment arbitration tribunals and by states before the WTO adjudicating bodies, which in turn can result in 'double jeopardy and double recovery'.[64] Whereas some principles, for example, non-discrimination or prohibition of local content (performance) requirements,[65] are found both in international investment law and in WTO law, [66] the FET standard is primarily reserved to the area of investment law. The WTO Secretariat suggests that the FET principle is rooted in customary international law and encompasses the non-discrimination principles but is more abstract.[67] Moreover, as FET was interpreted in some cases as referring to 'adherence to basic norms of transparency', it could correspond to the transparency obligation under Article X GATT.[68] The question could thus be raised of whether regulatory reforms subject to investment arbitration (with an undesirable outcome) could be also challenged by the home states of investors under WTO law.

Article X GATT contains the transparency requirements for various state measures affecting trade (laws, regulations, judicial decision and administrative rulings of general application). These measures may deal, for example, with taxes or other charges, requirements on imports/exports and conditions for sale, distribution and transportation. Such regulations should be published promptly to allow the actors who are

[64] M. Footer, 'On the laws of attraction: examining the relationship between foreign investment and international trade' in R. Echandi and P. Sauve (eds.), *Prospects in International Investment Law and Policy* (Cambridge: Cambridge University Press, 2013), pp. 127–8.

[65] E.g. under the TRIMS Agreement and under both NAFTA and ECT.

[66] N. DiMascio and J. Pauwelyn, 'Nondiscrimination in trade and investment treaties: worlds apart or two sides of the same coin?', *American Journal of International Law*, 102(1) (2008), 48–89.

[67] WTO, *Working Group on the Relationship between Trade and Investment, Non-Discrimination, Most-Favoured-Nation Treatment and National Treatment*, Note by the Secretariat, WT/WGTI/W/118 (4 June 2002).

[68] WTO, *Working Group on the Relationship between Trade and Investment, Transparency*, Note by the Secretariat, WT/WGTI/W/109 (27 March 2002).

affected (companies and governments) to get acquainted with them.[69] In addition, governments also have to administer such regulations and laws in a uniform, impartial and reasonable manner.[70] In *EC – Bananas III*, the Appellate Body noted that the phrase 'uniform, impartial and reasonable' is interchangeable with 'neutral, fair and equitable'.[71] Thus, it should be noted that changes in the regulatory regime for renewable energy can only be challenged in the WTO under two conditions: first, that they affect trade (types of measures listed in Article X:1 GATT, e.g. the local content requirements); second, that they are not administered in a 'uniform, impartial and reasonable manner'.

As for the second aspect, there have been discussions regarding whether violation of core WTO disciplines would constitute a breach of the FET standard in international investment law.[72] In *SD Myers*, one of the arbitrators noted that when interpreting Article 1105 NAFTA, arbitrators should also look to the 'state of the art' in international trade agreements to render FET objective rather than subjective content.[73] However, the Notes on Interpretation supported a different approach and explicitly stated that '[a] determination that there has been a breach of another provision of the NAFTA, or of a separate international agreement, does not establish that there has been a breach of Article 1105(1)'.[74] Notwithstanding this guidance from the NAFTA FTC, investors often argue that violations of WTO obligations by states constitute treatment not 'in accordance with international law'.[75] One of the most vivid examples of an attempt to export WTO requirements to the FET standard under Article 1105(1) NAFTA is demonstrated in *Methanex* v. *United States,* where the Canadian company Methanex claimed that Californian regulation is incompatible with the Agreement on Technical

[69] Article X:1 GATT. [70] Article X:3 GATT.

[71] Appellate Body Report, *European Communities – Regime for the Importation, Sale and Distribution of Bananas*, WT/DS27/AB/R, adopted 25 September 1997, DSR 1997:II, p. 591, para. 203.

[72] R. Kläger, *Fair and Equitable Treatment in International Investment Law*, pp. 66–7. See also C. O. Verrill Jr, 'Are WTO violations also contrary to the fair and equitable treatment obligations in investor protection agreements?', *ILSA Journal of International and Comparative Law*, 11 (2005), 287.

[73] *SD Myers* v. *Canada*, Final Award on the Merits, 13 November 2000, para. 257–258.

[74] NAFTA Free Trade Commission, 'Notes on interpretation of certain chapter 11 provisions'.

[75] For some examples see G. Verhoosel, 'The use of investor-state arbitration under bilateral investment treaties to seek relief for breaches of WTO law', *Journal of International Economic Law*, 6(2) (2003), 493–506.

Barriers to Trade (TBT) and the Agreement on Sanitary and Phytosanitary Measures (SPS), and therefore also violates Article 1105 NAFTA.[76] In *Mondev*, the tribunal clearly refused to include other international agreements between NAFTA parties, including WTO agreements, for interpretation of the scope of Article 1105(1) NAFTA. The tribunal reasoned, in line with the Notes on Interpretation, that not all treaties between NAFTA parties constitute customary international law.[77] Thus, a violation of transparency or of the non-discrimination obligation under GATT will not automatically mean that there is a breach of FET under an investment agreement, at least not in the case of NAFTA, and most probably also not in the case of other investment agreements, unless they have a similar reference to 'customary international law'.

The third aspect raises the question of whether well-established and more coherent WTO jurisprudence could inform investment arbitration tribunals' interpretations of the FET. Indeed, in *Methanex* v. *United States*, the tribunal agreed that even though WTO adjudicating bodies' reports are not binding precedents, it 'may derive guidance from the way in which a similar phrase in the GATT has been interpreted in the past'.[78]

In particular, investment tribunals could refer to the WTO jurisprudence interpreting the terms 'uniform', 'impartial' and 'reasonable' under Article X:3 GATT. In *China – Raw Materials*, with respect to the term 'uniform', the panel found that 'the lack of any definition, guidelines or standards in how the operation capacity criterion should be applies poses a very real risk to the interests of relevant parties such that this necessarily leads to non-uniform administration inconsistently with Article X:3 (a) of the GATT 1994'.[79] This interpretation might, for instance, be useful when assessing the methodology of the ranking criteria applied by Ontario in *Mesa* v. *Canada*. Further, in *Thailand – Cigarettes (Philippines)*, the panel suggested that 'impartial administration would appear to mean the application of implementation of the relevant laws and

[76] *Methanex Corporation* v. *United States*, Draft Amended Claim, 12 February 2001, para. 64.

[77] *Mondev* v. *United States*, ICSID Case No. ARB(AF)/99/2, Final Award, 11 October 2002, para. 121.

[78] *Methanex Corporation* v. *United States*, Final Award on the Merits, 3 August 2005, paras. 5–6.

[79] WTO Panel Reports, *China – Measures Related to the Exportation of Various Raw Materials*, WT/DS394/R / WT/DS395/R / WT/DS398/R / Add.1 and Corr.1, adopted 22 February 2012, as modified by Appellate Body Reports WT/DS394/AB/R / WT/DS395/AB/R / WT/DS398/AB/R, DSR 2012:VII, p. 3501, paras. 7.751–7.752.

regulations in a fair, unbiased and unprejudiced manner'.[80] The *de facto* impact of the agreement with Korean investors in *Mesa* v. *Canada* might well be unfair, whereas extremely frequent changes of the renewable energy regime between 2010 and 2014 in Spain could potentially meet the standard of 'unreasonable'.

G Lessons Learned from the FET Jurisprudence for the Post-2020 European RES Landscape

With the ambition of making the EU number one in renewable energies worldwide, the European Council has agreed the 2030 Climate and Energy Policy Framework for the EU, targeting a 27 per cent share of renewable energy consumption that is binding for the entire EU. This target will have to be met jointly and does not prevent the individual Member States from setting their own more ambitious national targets that take into account their degree of integration in the internal energy market.[81] The European Commission has announced a new renewable energy package for the period after 2020,[82] to include a new Renewable Energy Directive (RED).[83]

A revision of the RED needs to consider which (new) legal principles or concepts can be introduced in order to strengthen the Directive in the absence of binding national targets, simultaneously enhancing investor certainty and reducing policy risk.[84] The concept of FET might offer some guiding principles for the long-term and stable policy commitments that are crucial to keep the cost of renewable energy low on the one hand, and to encourage Member States to systemically improve their existing mechanisms to adapt to changed market and cost conditions for (certain) renewable energy technologies on the other.

[80] WTO Panel Report, *Thailand – Customs and Fiscal Measures on Cigarettes from the Philippines*, WT/DS371/R, adopted 15 November 2010, as modified by Appellate Body Report WT/DS371/AB/R, para 7.899.

[81] Note of the European Council SN 79/14: Conclusions on 2030 Climate and Energy Policy Framework, 23 and 24 October 2014, p. 5.

[82] Commission Communication, *A Framework Strategy for a Resilient Energy Union with a Forward-Looking Climate Change Policy* (COM/2015/080 final), 25 February 2015.

[83] The European Commission was expected to issue a revised or new RED by the end of 2016 or in early 2017.

[84] T. Wyns, A. Khatchadourian and S. Oberthür, *EU Governance of Renewable Energy Post-2020 – Risks and Options*, report for the Heinrich-Böll-Stiftung European Union, Institute for European Studies – Vrije Universiteit Brussel, December 2014, p. 40.

In its public consultations on the new RED, the Commission reaffirmed the Member States' flexibility with regard to meeting their greenhouse gas reduction targets in the most cost-effective manner in accordance with their specific national circumstances, energy mixes and capacities to produce renewable energy, and emphasised that the new Energy Union Governance will need to provide sufficient transparency and reliability, predictability and stability to spur renewable energy investments.[85]

This confirms the intention to pay due respect to the core elements of the FET. How does the Commission intend to shape the Member States' diverse national support schemes from 2020 onwards? As for the geographical scope of support schemes, the following options were initially under consideration for the new RED:

1. harmonised EU-wide support schemes;
2. regional-level support schemes (group of Member States with joint support scheme);
3. national support schemes fully or partially open to renewable energy producers in other Member States;
4. gradual alignment of national support schemes through common EU rules;
5. national-level support schemes that are only open to national renewable energy producers.[86]

Many political and economic challenges have to be considered in assessing these options (e.g. Member States' subsidiarity with regard to fiscal issues and the choice of generation mix, as well as natural geographical factors, public acceptance and grid interconnections). Each of them might have a different impact on RES investors. While options 1–4 would enhance trade in RES and level the playing field for investors in Europe, option 5 advocates for the current approach taken by the EU Court of Justice (cf. *Sweden* v. *Ålands Vindkraft AB*),[87] which suggests that a territorial limitation may in itself be regarded as necessary in order to attain the legitimate objective – promotion of the RES in the production of electricity. The European Commission's proposal for the revision

[85] See the European Commission Consultation Questionnaire on preparation of a new renewable energy directive for the period after 2020 (held between 18 November 2015 and 10 February 2016), https://ec.europa.eu/energy/en/consultations/preparation-new-renewable-energy-directive-period-after-2020 (accessed 10 February 2016).

[86] Ibid.

[87] Case C-573/12, *Ålands Vindkraft AB* v. *Energimyndigheten*, ECJ judgment of 1 July 2014.

of the RED entails a mixed solution combining options 4 and 5. Namely, it requires EU Member States to open financial support in the amount of 10 per cent of the newly supported capacity annually from 2021 to 2025 to the RES installations in other Member States (and between 2026 and 2030 – 15 per cent annually).[88]

Another question concerns retroactivity – whether to adapt the level of support during the support period or whether market developments or other relevant changes should only be reflected at the time that new RES installations come into operation. In terms of security of investments, the CEER does not recommend changing the level of support in an unexpected manner, as this would lead to additional investment costs and potentially undermine investments already in place. The European Commission also advocates for stable RES support schemes and encourages Member States to prevent revision of the support granted to RES projects where this would have a negative impact on the economic viability of those projects.[89] Moreover, adjustment rates for support levels defined *ex ante* or following a deployment trend should be transparent and communicated in a timely manner to ensure reliable planning for RES investors in terms of expected levels of support.[90]

H Conclusions

The current international investment law framework does not seem to offer significant legal security either for investors or for host states. This being said, in the renewable energy sector there is no clear-cut answer to the question of the policy space available for a government to introduce, amend or replace support measures.

Any type of regulation, including environmental measures, can also be used for discriminatory purposes.[91] Therefore, it is very important to strike a reasonable balance between the interests of an investor and a state's legitimate right to regulate in the renewable energy sector on its territory.

[88] European Commission, 'Proposal for a Directive of the European Parliament and of the Council on the promotion of the use of energy from renewable energy sources (recast)', COM(2016) 767 final, 30 November 2016, Article 5(2).

[89] Ibid., Recital 18 of the Preamble and Article 6.

[90] *Key Support Elements of RES in Europe: Moving Towards Market Integration*, Council of European Energy Regulators, Report C15-SDE-49-03, 26 January 2016, p. 14.

[91] G. Verhoosel, 'Foreign direct investment and legal constraints on environmental domestic policies: striking a reasonable balance between stability and change', *Law and Policy of International Business*, 29 (1998), pp. 451 and 478.

The recent investment arbitration awards under various investment agreements, which are *ad hoc* in nature, have dealt with the minimum standard of behaviour required of host states. However, these awards do not necessarily offer a consistent answer.

Under the current framework, the host state introducing measures in the renewable energy sector should ensure that the new regulations are transparent, objective and flexible; moreover, the regulation may include a transitional period for implementation and the host state should allow for due process. Thus, when host states adopt a regulation they have to stipulate that the amount and duration of support (e.g. FIT) will be subject to revisions, and possibly also indicate in which situations such revisions can be expected.[92]

Independently from the geographical scope of support schemes that will be chosen in Europe for the post-2020 regime (regional or partial opening-up of national support schemes), the reviewed RED could embed the principle that national support mechanisms for renewable energy in place before 2020 cannot be radically altered or abandoned in the short term.[93] Following the cases of *Charanne B.V.* and *Construction Investments S.A.R.L.* v. *Spain* described above, the new RED could explicitly forbid the retroactive abolition of support mechanisms and/or provide for a transition period to ensure transparent and clear legal reform that has the lowest possible negative impact on investors.

Bibliography

Boute, A. (2011a). Challenging the re-regulation of liberalized electricity prices under investment arbitration. *Energy Law Journal*, **32**, 497–539.

(2011b). Combating climate change through investment arbitration changes. *Fordham International Law Journal*, **35**(3), 613–64.

Council of European Energy Regulators (CEER) (2016). *Key Support Elements of RES in Europe: Moving Towards Market Integration*. Report C15-SDE-49-03, 26 January.

DiMascio, N. and Pauwelyn, J. (2008). Nondiscrimination in trade and investment treaties: worlds apart or two sides of the same coin? *American Journal of International Law*, **102**(1), 48–89.

[92] A. Boute, 'Combating climate change through investment arbitration changes', citing Council Directive 2003/96/EC on Restructuring the Community Framework for the Taxation of Energy Products and Electricity, 2003 O.J. L 283/51, Article 16(3), 57.

[93] T. Wyns, A. Khatchadourian and S. Oberthür, *EU Governance of Renewable Energy Post-2020*, p. 41.

Dolzer, R. and Stevens, M. (1995). *Bilateral Investment Treaties*. The Hague: Martinus Nijhoff Publishers.

Dumberry, P. (2013). *The FET Standard: A Guide to NAFTA Case Law on Article 1105*. The Netherlands: Wolters Kluwer.

Footer, M. (2013). On the laws of attraction: examining the relationship between foreign investment and international trade. In R. Echandi and P. Sauve, *Prospects in International Investment Law and Policy*. Cambridge: Cambridge University Press.

Gentry, B. S. and Ronk, J. J. (2003). *International Investment Agreements and Investments in Renewable Energy*. New Haven: Yale Environmental Protection Law Clinic.

Henckels, C. (2012). Proportionality and the standard of review in fair and equitable treatment claims: balancing stability and consistency with the public interest, Society of International Economic Law Conference, Singapore, 13 July.

Hober, K. (2010). Investment arbitration and the Energy Charter Treaty. *Journal of International Dispute Settlement*, **1**(1), 153–90.

IEA. (2010). *World Energy Outlook*. Paris: IEA.

IRENA. (2015). *REthinking Energy: Renewable Energy and Climate Change*, www.irena.org/rethinking/IRENA%20_REthinking_Energy_2nd_report_2015.pdf.

Klaeger, R. (2011). *Fair and Equitable Treatment in International Investment Law*. Cambridge: Cambridge University Press.

Malik, M. (2009). Fair and equitable treatment. *IISD Best Practices Series*, Bulletin No. 3.

Noothout, P., de Jager, D., Tesnière, L., van Rooijen, S., Karypidis, N., Brückmann, R., Jirouš, F., Breitschopf, B., Angelopoulos, D., Doukas, H., Konstantinavičiūtė, I., Resch, G. (2015). Investment risks for renewable energy projects – the case of onshore wind power, *DIA-CORE Policy Brief* (21 May).

Paparinskis, M. (2013). *The International Minimum Standard and Fair and Equitable Treatment*. Oxford: Oxford University Press.

Roe, T. and Happold, M. (2011). *Settlement of Investment Disputes under the ECT*. Cambridge: Cambridge University Press.

UNCTAD. (2007). *Bilateral Investment Treaties, 1995–2006: Trends in Investment Rulemaking* (New York and Geneva: United Nations).

(2010). *World Investment Report 2010: Investing in a Low-Carbon Economy* (New York and Geneva: United Nations).

(2012). *Fair and Equitable Treatment: UNCTAD Series on Issues in International Investment Agreements II* (New York and Geneva: United Nations).

(2014). Recent developments in investor-state dispute settlement (ISDS), *IIA Issue Note no. 1* (April).

Verhoosel, G. (1998). Foreign direct investment and legal constraints on environmental domestic policies: striking a reasonable balance between stability and change. *Law and Policy of International Business*, **29**, 451.

(2003). The use of investor-state arbitration under bilateral investment treaties to seek relief for breaches of WTO law. *Journal of International Economic Law*, **6**(2), 493–506.

Verrill Jr., C. O. (2005). Are WTO violations also contrary to the fair and equitable treatment obligations in investor protection agreements? *ILSA Journal of International and Comparative Law*, **11**, 287.

WTO. (2002a). *Working Group on the Relationship between Trade and Investment, Transparency*, Note by the Secretariat, WT/WGTI/W/109, 27 March.

(2002b). *Working Group on the Relationship between Trade and Investment, Non-Discrimination, Most-Favoured-Nation Treatment and National Treatment*, Note by the Secretariat, WT/WGTI/W/118, 4 June.

Wyns, T., Khatchadourian, A. and Oberthür, S. (2014). *EU Governance of Renewable Energy Post-2020 – Risks and Options*, report for the Heinrich-Böll-Stiftung European Union, Institute for European Studies – Vrije Universiteit Brussel (December).

INDEX